Molecular Biology of Mitochondrial Transport Systems

NATO ASI Series

Advanced Science Institutes Series

A series presenting the results of activities sponsored by the NATO Science Committee, which aims at the dissemination of advanced scientific and technological knowledge, with a view to strengthening links between scientific communities.

The Series is published by an international board of publishers in conjunction with the NATO Scientific Affairs Division

A Life Sciences B Physics	Plenum Publishing Corporation London and New York
C Mathematical and Physical Sciences D Behavioural and Social Sciences E Applied Sciences	Kluwer Academic Publishers Dordrecht, Boston and London
F Computer and Systems Sciences G Ecological Sciences H Cell Biology I Global Environmental Change	Springer-Verlag Berlin Heidelberg New York London Paris Tokyo Hong Kong Barcelona Budapest

NATO-PCO DATABASE

The electronic index to the NATO ASI Series provides full bibliographical references (with keywords and/or abstracts) to more than 30000 contributions from international scientists published in all sections of the NATO ASI Series. Access to the NATO-PCO DATABASE compiled by the NATO Publication Coordination Office is possible in two ways:

- via online FILE 128 (NATO-PCO DATABASE) hosted by ESRIN, Via Galileo Galilei, I-00044 Frascati, Italy.
- via CD-ROM "NATO Science & Technology Disk" with user-friendly retrieval software in English, French and German (© WTV GmbH and DATAWARE Technologies Inc. 1992).

The CD-ROM can be ordered through any member of the Board of Publishers or through NATO-PCO, Overijse, Belgium.

Series H: Cell Biology, Vol. 83

Molecular Biology of Mitochondrial Transport Systems

Edited by

Michael Forte

Vollum Institute for Advanced Biomedical Research
Oregon Health Sciences University
3181 S. W. Sam Jackson Park Rd.
Portland, OR 97201, USA

Marco Colombini

Department of Zoology
University of Maryland
College Park, MD 20742, USA

Springer-Verlag
Berlin Heidelberg New York London Paris Tokyo
Hong Kong Barcelona Budapest
Published in cooperation with NATO Scientific Affairs Division

Proceedings of the NATO Advanced Research Workshop on Molecular Biology of Mitochondrial Transport Systems, held at Il Ciocco, Italy, September 17–21, 1992

ISBN 3-540-57908-7 Springer-Verlag Berlin Heidelberg New York
ISBN 0-387-57908-7 Springer-Verlag New York Berlin Heidelberg

CIP data applied for

Printed in Germany

Typesetting: Camera ready by authors
31/3130 - 5 4 3 2 1 0 - Printed on acid-free paper

TABLE OF CONTENTS

List of Participants

Directors

COLOMBINI, Marco
University of Maryland
Department of Zoology
College Park, MD 20742, U.S.A.

FORTE, Michael
Vollum Institute for Advanced Biomedical Research
Oregon Health Sciences University
3181 S. W. Sam Jackson Park Rd.
Portland, OR 97201, U.S.A.

Speakers

AZZI, Angelo
Universität Bern
Institut für Biochemie und Molekularbiologie
Bühlstrasse 28, Postfach 98
CH-3000 Bern 9, Switzerland

BEAVIS, Andrew D.
Department of Pharmacology
Medical College of Ohio
P. O. Box 10008
Toledo, OH 43699-0008, U.S.A.

BRANDOLIN, Gerard
DBMS/Biochimie
Centre d'Etudes Nucléaires, 85X
38041 Grenoble cedex, France

BRDICZKA, Dieter
Faculty of Biology
University of Konstanz
D-7750 Konstanz, Germany

COLOMBINI, Marco, see Directors

DE PINTO, Vito
Department of Pharmaco-Biology
Laboratory of Biochemistry and Molecular Biology
University of Bari
Bari, Italy

DIWAN, Joyce J.
Biology Department and Center for Biophysics
Rensselaer Polytechnic Institute
Troy, NY 12180-3590, U.S.A.

FORTE, Michael, see Directors

GELLERICH, F.N.
Institut für Biochemie
Medizinische Akademie Magdeburg
Leipziger Str. 44
0-3090 Magdeburg, Germany

HENRY, Jean-Pierre
Institute de Biologie Physico-Chimique
Fondation Edmond de Rothschild
13, Rue Pierre et Marie Curie
75005, Paris, France

KINNALLY, Kathleen W.
Department of Biological Sciences
State University of New York at Albany
1400 Washington Ave.
Albany, NY 12222, U.S.A.

KLINGENBERG, Martin
Institute for Physical Biochemistry
University of Munich
Goethestrasse 33
8000 München 2, Germany

KRÄMER, Reinhard
Institute für Biotechnologie
Forschungszentrum Jülich
5170 Jülich, Germany

MANNELLA, Carmen A.
Wadsworth Center for Laboratories and Research
New York State Department of Biomedical Sciences
School of Public Health
State University of New York at Albany
Empire State Plaza, Box 509
Albany, NY 12201-0509, U.S.A.

MCCABE, Edward R.B.
Department of Pediatrics
and Institute for Molecular Genetics
Baylor College of Medicine
One Baylor Plaza
Houston, TX 77030, U.S.A.

McEnery, Maureen W.
Department of Physiology & Biophysics
Case Western Reserve University
School of Medicine
10900 Euclid Street
Cleveland, OH 44106, U.S.A.

Michejda, Jan
Poznań University
Department of Bioenergetics
61-701 Poznań
ul. Fredry 10, Poland

Nałęcz, Katarzyna A.
Nencki Institute of Experimental Biology
Pasteur str.3
02-093 Warsaw, Poland

Papa, Sergio
Institute of Medical Biochemistry and Chemistry
University of Bari
Piazza G. Cesare
70124 Bari, Italy

Ricquier, Daniel
Centre de Recherche sur l'Endocrinologie
Moléculaire et le Dévelopment
9, rue Jules Hetzel
92190 Meudon, France

Sorgato, Catia
Università Degli Studi di Padova
Dipartimento di Chimica Biologica
Via Trieste, 75
35121 Padova, Italy

Nałęcz, Maciej
Department of Muscle Biochemistry
Nencki Institute of Experimental Biology
Polish Academy of Sciences
Pasteur str., 3
02-093 Warsaw, Poland

Thinnes, Friedrich P.
Max-Planck-Institut für Experimentelle Medizin
Abteilung Immunochemie
D-3400 Goettingen, Germany

THOMAS, Lorie
Department of Biochemistry
University of Alberta
474 Medical Sciences Bldg.
Edmonton, Alberta
T6H 2H7, Canada

WALKER, John E.
M.R.C. Laboratory of Molecular Biology
Hills Road
Cambridge CB2 2QH, U.K.

WILSON, John E.
Department of Biochemistry
Michigan State University
East Lansing, MI 48823, U.S.A.

ZORATTI, Mario
CNR Unit for the Physiology of Mitochrondria
Department of Biomedical Sciences
Via Trieste 75
35121 Padova, Italy

MOLECULAR STUDIES OF THE MITOCHONDRIAL UNCOUPLING PROTEIN

B. Miroux, A.M. Doulcier-Cassard, L. Casteilla, S. Raimbault, C. Levi-Meyrueis, C. Gelly, S. Klaus, S. Prieto*, E. Rial*, F. Bouillaud, D. Ricquier

Centre de Recherche sur l'Endocrinologie Moléculaire et le Développement
Centre National de la Recherche Scientifique
9, rue Jules Hetzel 92190 Meudon - FRANCE
and
* Centro de Investigaciones Biologias
Consejo Superior de Investigaciones Cientificas
Velasquez, 144
28006 Madrid - SPAIN

INTRODUCTION

The brown adipocytes play an important role in the regulation of body temperature in hibernating as well as in small and newborn mammals. These thermogenic adipocytes express an Uncoupling Protein - UCP. This mitochondrial carrier, unique to these cells, uncouples mitochondrial ATP synthesis from the respiratory chain activity and is responsible for heat production by brown adipocytes. Besides its important physiological role, UCP is a proton membranous transporter proving the validity of Mitchell's chemiosmotic theory (Nicholls and Locke 1984, Klingenberg 1990). UCP is also able to transport Cl^- ions electrophoretically (Garlid 1990). Both H^+ and Cl^- transports are inhibited by purine nucleotides and free fatty acids activate H^+ translocation (Rial and Nicholls 1989). Another characteristic property of UCP, is that it is a member of an expanding family of mitochondrial ion transporters sharing sequence and structural homologies. In addition to UCP, the main members forming this family are the Adenine Nucleotide Translocator, the Phosphate Carrier and the Oxoglutarate Carrier (Walker 1992).

NATO ASI Series, Vol. H 83
Molecular Biology of Mitochondrial Transport Systems
Edited by M. Forte and M. Colombini

We are interested in the mechanisms controlling the unique expression of UCP in brown adipocytes and in the functional organization of this transporter.

MOLECULAR MECHANISMS RESTRICTING UCP GENE EXPRESSION TO BROWN ADIPOCYTES

One main characteristic of UCP is its unique expression in brown adipocytes. UCP synthesis is positively controlled by norepinephrine released at the surface of cells and interacting with ß-adrenoceptors. UCP gene is known to be strongly regulated at the level of transcription by cAMP and thyroid hormones (Ricquier et al. 1986, Bianco et al. 1988, Rehnmark et al. 1990). Mouse, rat and human UCP gene have been isolated and DNAse I hypersensitive sites have been proposed (Bouillaud et al. 1988, Boyer and Kozak 1991, Cassard-Doulcier et al. 1990, Kozak et al. 1988). We are interested in understanding the activation of UCP gene transcription and especially in mechanims that strictly limit UCP gene transcription to brown adipose cells. To identify cis-regulatory elements regulating rat UCP gene expression, 4.5 kb upstream of the transcription start site were analysed using DNA constructs fused to the chloramphenicol-acetyl-transferase gene (CAT). The 4.5 kb CAT plasmid was analysed by transfecting either primary cultures of brown adipocytes expressing UCP or several other types of cells that do not express UCP. A strong CAT activity was assayed only in cells able to transcribe the UCP gene. The 4.5 kb CAT DNA was also used to construct transgenic mouse. Interestingly, CAT activity was restricted to brown adipose tissue of transgenic animals. Taken together, the data from transfected cells and transgenic animals indicate that the genomic elements that control the specific expression of UCP gene are located in the 4.5 kb of the 5' -flanking region.

To identify short regions controlling UCP gene, the original 4.5 kb plasmid was deleted and 12 new constructs were used to transfect cultured cells. We observed that a 200 bp element located at -2.3 kb is a powerful activator of UCP gene transcription. In addition, deletions revealed that an inhibitory element is placed downstream of the activatory element. The functionality and the specificity of these two regulatory elements will be checked by gel-retard experiments and new transgenic animals.

TOPOGRAPHY OF UCP

UCP and the other members of the family of mitochondrial transporters have a polypeptide chain consisting of three tandemly repeated related sequences of approximately 100 amino acids (Klingenberg 1990, Walker 1992). It has been predicted that each third of protein can be formed by two α-helices linked by a polar loop (Fig.1). Interestingly, the distribution of introns in mouse, rat and human UCP genes confirms that UCP gene derives by triplication from an ancestral gene (Ricquier at al. 1991). The structure of UCP has been approached using chemical modifications, digestion by trypsin or chymotrypsin (see ref. in Ferdandez et al. 1987 and Klingenberg 1990), infrared spectroscopy (Rial et al. 1989).
An other approach is to express wild or mutated UCP in heterologous systems. In comparison with the other mitochondrial carriers, the strict tissue specificity of UCP makes much simpler the design of recombinant expression systems for it, because there is no interference with activity of endogenous gene.

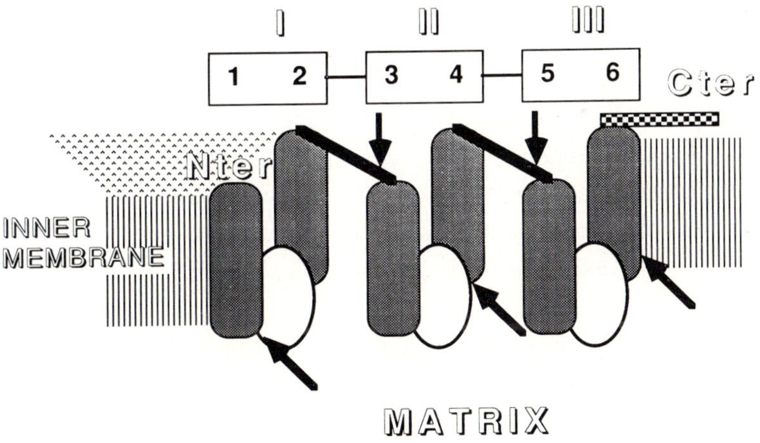

Figure 1 : Model of the transmembrane arrangement of UCP as predicted from the primary sequence and the hydrophobic profile. According to this model, UCP is made of three repeats and six α-helices linked by long (open oval area) and short (black segment) hydrophilic domains. The arrows indicate position of introns. The checkered segment represents the C-terminal hydrophilic extremity of UCP.

Expression of UCP in CHO cells

To study the functional organization of UCP, this protein was initially expressed in mammalian cells and CHO cells expressing permanently UCP were cloned. Mitochondria isolated from these cells exhibited a partially uncoupled respiration that could be recoupled by addition of purine nucleotides (Casteilla et al. 1990). Then UCP mutated in an expected nucleotide-binding region was transiently expressed in CHO cells but no cell line expressing such mutants could be isolated. An explanation for the failure of isolating these mutants could be that the mutated protein was lethal.

Expression of UCP in yeasts

The obtention of a large number of mutants analysable by biochemical and biophysical methods prompted us to express UCP in yeasts (Prieto et al. 1990) as also reported by others (Murdza-Inglis et al. 1991, Bathgate et al. 1992). Several expression vectors were used and more recently a high expression level was obtained using the pYEDP 1/8 10 plasmid. This high expression level of UCP was expected to induce a modification of the proton permeability of the inner membrane, observable in osmotic swelling experiments. The setting of control experiments led to the unexpected conclusion that yeast mitochondria which do not express UCP have an endogenous uncoupling pathway activated by ATP (Prieto et al. 1992). This discovery makes the UCP activity itself very difficult to observe in recombinant yeast mitochondria. This new uncoupling pathway is presently under study.

Expression of MalE-UCP fusion proteins in Escherichia coli

Expression of hydrophobic proteins in bacteria generally results in protein aggregates. We tried to circumvent this problem by constructing chimaeric proteins made up of a fragment of UCP fused to the MalE protein of E. coli (Miroux et al. 1992). Production of restricted part of UCP should be considered as a valuable tool to study the structure of this protein respectively to the triplicated structure of carriers. Moreover, the observation of surprising homologies with DNA binding proteins suggests smaller structural domains (Bouillaud et al.1992). DNA corresponding to different regions of rat UCP was linked to DNA encoding MalE protein. After transformation of bacteria, the presence of fusion proteins was checked by antibodies against MalE or UCP.

A fusion protein encoding residues 253-290 was purified by amylose chromatography. This purified fusion protein was attached to Affigel 15 activated biobeads. These biobeads were used as affinity chromatography columns to screen polyclonal antibodies against rat UCP. Using this procedure, antibodies specific for this part of UCP were isolated. Comparison of the reactivity of different fusion proteins with these antibodies led to the conclusion that the sequence recognized lies between residues 253 and 279. According to prediction structure, this sequence corresponds to the end of the third hydrophilic domain and to the first half of the last (sixth) transmembranous alpha helix. The reactivity of these purified antibodies against freeze-thawed mitochondria or sonicated mitochondria (submitochondrial particles) was then analysed using back titration. They were unable to recognize UCP in freeze-thawed mitochondria but were reactive towards sonicated mitochondria.

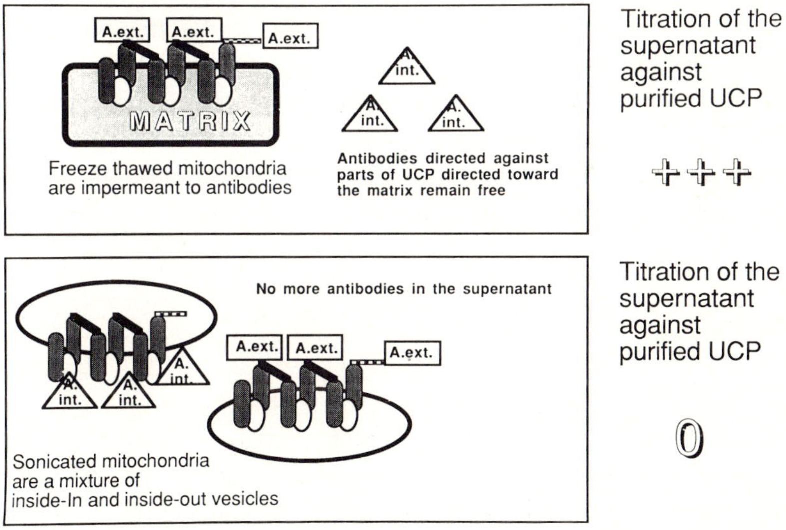

Figure 2 : Shema illustrating the back-titration of antibodies present in whole serum or in purified serum fraction. Freeze-thawed mitochondria or submitochondrial particles were exposed to antibodies. After incubation and centrifugation, unreacted antibodies present in the supernatant were assayed by ELISA against UCP or fusion proteins.

This result led us to conclude that the N-terminal extremity of the sixth α-helix of UCP is facing the matrix (Miroux et al. 1992). This conclusion agrees with the prediction designed in Figure 1 and made by Aquila et al. (1985) and Klingenberg (1990).

The UCP moiety in a fusion protein can be of variable size and a fusion protein made of the entire UCP fused to MalE has been recently obtained.

CONCLUSION AND PERSPECTIVES

Antibodies against synthetic peptides and proteases have been used to study the topology of the ADP/ATP carrier and the mitochondrial phosphate carrier (Brandolin et al. 1989, Capobianco et al. 1991, Ferreira et al. 1990, Marty et al. 1992). Presently, most data suggest that the different mitochondrial transporters have a similar arrangement in the membrane with six α-helices. However, such a conclusion does not agree with data obtained from limited proteolysis and photolabeling of the ADP/ATP translocator (Marty et al. 1992). Use of fusion proteins and selection of specific antibodies is an alternate strategy to study the topology of mitochondrial transporters. Recently, we constructed a DNA library corresponding to a large number of fusion proteins. The screening of this library will provide fusion proteins identifying the different epitopes in UCP and will allow to purify specific antibodies. We expect that these antibodies will be useful to study the organization of UCP.

REFERENCES

Aquila H, Link T, Klingenberg M (1985) The uncoupling protein from brown fat mitochondria is related to the mitochondrial ADP/ATP carrier. Analysis of sequence homologies and of folding of the protein in the membrane. EMBO J 4:2369-2376

Bathgate B, Freebairn EM, Greenland AJ, Reid GA (1992) Functional expression of the rat brown adipose tissue uncoupling protein in Saccharomyces-cerevisiae. Mol Microbiol 6:363-370

Bianco AC, Sheng X, Silva JE (1988) Triiodothyronine amplifies norepinephrine stimulation of uncoupling protein gene transcription by a mechanism not requiring protein synthesis. J Biol Chem 263:18168-18175

Bouillaud F, Casteilla L, Ricquier D (1992) A conserved domain in mitochondrial transporters is homologous to a Zinc-finger knuckle of nuclear hormone receptors. Mol Biol Evol 9:970-975

Bouillaud F, Raimbault S, Ricquier D (1988) The gene for rat uncoupling protein : complete sequence, structure of primary transcript and evolutionary relationship between exons. Biochem Biophys Res Commun 157:783-792

Boyer BB, Kozak LP (1991) The mitochondrial uncoupling protein gene in brown fat - Correlation between DNase-I hypersensitivity and expression in transgenic mice. Mol Cell Biol 11:4147-4156

Brandolin G, Boulay F, Dalbon P, Vignais PV (1989) Orientation of the N-terminal region of the membrane-bound ADP/ATP carrier protein explored by antipeptide antibodies and an arginine-specific endoprotease. Evidence that the accessibility of the N-terminal residues depends on the conformational state of the carrier. Biochemistry 28:1093-1100

Capobianco L, Brandolin G, Palmieri F (1991) Transmembrane topography of the mitochondrial phosphate carrier explored by peptide-specific antibodies and enzymatic digestion. Biochemistry 30:4963-4969

Cassard AM, Bouillaud F, Mattei MG, Hentz E, Raimbault S, Thomas M, Ricquier D (1990) Human uncoupling protein gene : structure, comparison with rat gene and assignment to the long arm of the chromosome 4. J Cell Biochem 43:255-264

Casteilla L, Blondel O, Klaus S, Raimbault S, Moreau F, Bouillaud F, Ricquier D (1990) Stable expression of functional mitochondrial uncoupling protein in chinese hamster ovary cells. Proc Natl Acad Sci USA 87:5124-5128

Fernandez M, Nicholls DG, Rial E (1987) The uncoupling protein from brown-adipose-tissue mitochondria - Chymotrypsin-induced structural and functional modifications. Eur J Biochem 164:675-680

Ferreira GC, Pratt RD, Pedersen PL (1990) Mitochondrial proton/phosphate transporter - An antibody directed against the COOH terminus and proteolytic cleavage experiments provides new insights about its membrane topology. J Biol Chem 265:21202-21206

Garlid KD (1990) New insights into mechanism of anion uniport through the uncoupling protein of brown adipose tissue mitochondria. Biochim Biophys Acta 1018:151-154

Klingenberg M (1990) Mechanism and evolution of the uncoupling protein of brown adipose tissue. TIBS 15:108-112

Kozak LP, Britton JH, Kozak UC, Wells JM (1988) The mitochondrial uncoupling protein gene. Correlation of exon structure to transmembrane domains. J Biol Chem 263:12274-12277

Marty I, Brandolin G, Gagnon J, Brasseur R, Vignais PV (1992) Topography of the membrane-bound ADP/ATP carrier assessed by enzymatic proteolysis. Biochemistry 31:4058-4065

Miroux B, Casteilla L, Klaus S, Raimbault S, Grandin S, Clement JM, Ricquier D, Bouillaud F (1992) Antibodies selected from whole antiserum by fusion proteins as tools for the study of the topology of mitochondrial membrane proteins - Evidence that the N-terminal extremity of the 6th alpha-helix of the uncoupling protein is facing the matrix. J Biol Chem 267:13603-13609

Murdza-Inglis D, Patel HV, Freeman KB, Jezek P, Orosz DE, Garlid KD (1991) Functional reconstitution of rat uncoupling protein following its high level expression in yeast. J Biol Chem 260:11871-11875

Nicholls DG, Locke RM (1984) Thermogenic mechanisms in brown fat. Physiol Rev 64:1-64

Prieto S, Bouillaud F, Raimbault S, Rial E, Gimenez G, Ricquier D (1990) Expression of the brown fat mitochondrial uncoupling protein in S. cerevisiae. EBEC Rep 6:32

Prieto S, Bouillaud F, Ricquier D, Rial E (1992) Activation by ATP of a proton conducting pathway in yeast mitochondria. Eur J Biochem in press

Rehnmark S, Néchad M, Herron D, Cannon B, Nedergaard J (1990) α- and ß-adrenergic induction of the expression of the uncoupling protein thermogenin in brown adipocytes differentiated in culture. J Biol Chem 265:16464-16471

Rial E, Muga A, Valpuesta JM, Arrondo JLR, Goni FM (1990) Infrared spectroscopic studies of detergent-solubilized uncoupling protein from brown adipose tissue mitochondria. Eur J Biochem 188:83-89

Rial E, Nicholls DG On the mechanisms of transport by the uncoupling protein from brown-adipose-tissue mitochondria. Azzi A et al. (eds) (1989) Anion Carriers of Mitochondrial Membranes. Springer-Verlag Berlin-Edelberg-New York

Ricquier D, Bouillaud F, Toumelin P, Mory G, Bazin R, Arch J, Pénicaud L (1986) Expression of uncoupling protein mRNA in thermogenic or weakly thermogenic brown adipose tissue. Evidence for a rapid ß-adrenoceptor-mediated and transcriptionally regulated step during activation of thermogenesis. J Biol Chem 261:13905-13910

Ricquier D, Casteilla L, Bouillaud F (1991) Molecular studies of the uncoupling protein. FASEB J 5:2237-2242

Walker JE (1992) The mitochondrial transporter family. Curr Opin Struct Biol 2:519-526

STRUCTURE-FUNCTION RELATIONSHIP IN THE UNCOUPLING PROTEIN OF BROWN ADIPOSE TISSUE

M. Klingenberg, E. Winkler
Institute for Physical Biochemistry
University of Munich
Goethestrasse 33
8000 München 2
Fed. Rep. of Germany

UCP is the simplest H^+ carrier known

The uncoupling protein (UCP) is the key factor in the thermogenesis of brown adipose tissue. UCP recycles the H^+ generated by the oxidation of fatty acids in the respiratory chain through the inner mitochondrial membrane and generates neutralization heat. Thus, a large proportion of the oxidative energy is funneled into heat instead of ATP. By transporting H^+ or OH^-, UCP is functionally the most simple carrier known in the biological kingdom. Surprisingly, the primary structure of UCP is similar to that of the ADP/ATP carrier and the other members of the mitochondrial carrier family (Aquila et al., 1985). Since UCP occurs only in mammalians it can be regarded to be the latest member in the evolution of this protein family (Klingenberg, 1990).

The H^+-transport activity of UCP is inhibited by purine nucleotides. The binding of nucleotides was instrumental in identifying and isolating the protein from mitochondria (Heaton et al., 1978). Besides being inhibited and therefore regulated by ATP or ADP, H^+-transport activity in mitochondria was shown to be activated by fatty acids (Locke et al., 1982). In addition, a pH increase diminishes the affinity and the inhibition of H^+-transport activity by nucleotides (Klingenberg, 1988). As a result there is an interplay of various parameters which maintain the H^+-conducting activity of the UCP under tight control in the brown adipose tissue. In the following we shall attempt to elucidate the principle problem of H^+ transport in UCP, the role of fatty acids therein, and present a tentative model of the H^+ transport by UCP.

The problem of low H^+ concentrations

H^+ transport has a particular position among solute transports. At the most common physiological pH the concentration of H^+ or OH^- is below 10^{-6} M. This is about 10^{-6} lower than the concentration of Na^+. Whereas with these concentrations Na^+ can access a translocation channel at a sufficient rate, the virtual absence of H^+ severely limits the diffusion and entry to a translocation path (Läuger, 1976; Akeson and Deamer, 1991). This fundamental problem can be overcome in H^+ translocating systems by providing H^+ buffering with appropriate groups, in particular carboxyl

NATO ASI Series, Vol. H 83
Molecular Biology of Mitochondrial Transport Systems
Edited by M. Forte and M. Colombini

groups. The best researched example is bacteriorhodopsin where at least two aspartyl groups on both sides of the H^+-switching centre provide this role (Butt et al., 1989; Cao et al., 1991). It will be our concern in the following how this problem has been solved in the simplest H^+/OH^- carrier known, in UCP.

The transport modes of UCP

UCP permits a "facilitated" diffusion of H^+/OH^- by what we believe a carrier type of transport. There are two arguments which favour the carrier type instead of channel or pore type of mechanism: first, the relatively slow turnover observed under optimum conditions in the reconstituted system of UCP and second the similarity of the UCP protein to the other mitochondrial carriers. A third argument is the very high abundance of UCP molecules in the mitochondrial membranes which contrasts to the low occurrence of Na^+ channels in plasma membranes.

UCP is known to transport also anions like Cl^- (Nicholls, 1974). This would classify UCP as an anion transporter as most of the other members of the mitochondrial carrier family, and it might suggest that UCP transports OH^- instead of H^+. However, both H^+ and OH^- are equivalent for transport. A differentiation between OH^- or H^+ transport may become useful when seeing the role of UCP in context with other cation or anion carriers, in particular the net charge which is moved across the membrane driven by the membrane potential. The problem is illustrated in Fig. 1. Two possible modes of charge transfer during the H^+/OH^--transport cycle of UCP are compared in an elementary translocation cycle. In the first case a positive charge is generated at the binding centre which can be driven by the membrane potential. In the second case a H^+ neutralizes a negative charge, or vice-versa OH^- generates a negative charge so that the membrane potential will drive a net negative charge (Klingenberg, 1988).

Probably the H^+/OH^- translocating group is a carboxyl group in the protein. As further shown in Fig. 1, the positive charge in the binding centre can be generated by assuming that the CO_2^--group forms a salt bridge with a lysine or arginine group. When the carboxyl group becomes neutralized, a net positive charge is unmasked. In the second case the carboxyl group is independent and either neutral or negatively charged. The first model would have the advantage to explain the transport of anions by a positively charged binding centre. However it might be questioned in view of evidence for other Cl^- channels whether Cl^- actually is bound by positive charge or only by induced dipoles, for example by OH groups in the channel. The second case stipulates that Cl^- is transported only when the carboxyl group is in the neutral form, since a negative charge in the translocation site would prevent Cl^- uptake. In this case one would expect a pH dependency of Cl^- transport. However, no evidence for this pH

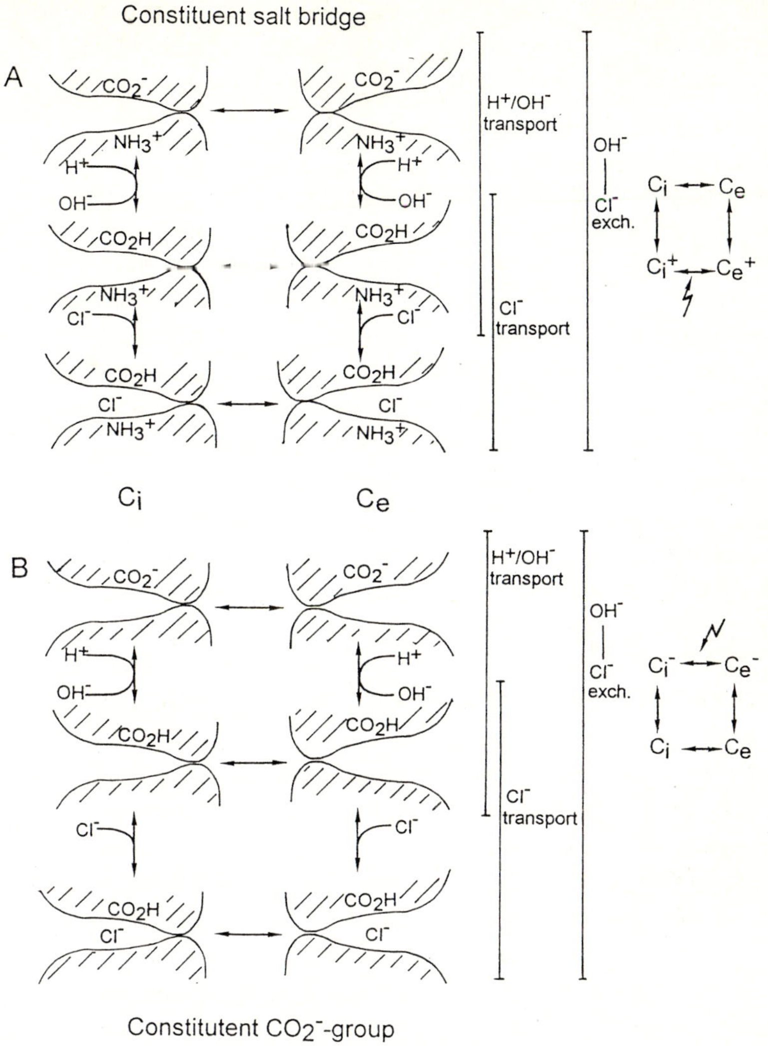

Fig. 1. The modes of H^+/OH^- and of Cl^- translocation in the uncoupling protein.
A. The binding centre is visualized to contain an ion pair in which an acidic-(carboxyl) and basic group are either neutralized or, after uptake of H^+, contain a net positive charge. Cl^- binds to the net positive charge. The membrane potential drives the positively charged branch of the translocation cycle. B. The binding centre contains a carboxylate group which is neutralized on H^+ uptake. Cl^- binds to the neutral centre. In both the H^+ and Cl^- translocation cycles the membrane potential drives the negatively charged branch of the translocation cycle.

dependency has been found in detailed experimentation (Winkler and Klingenberg, unpublished).

Free fatty acids are essential activators of H^+ transport in UCP

The most striking finding for the H^+/OH^- transport of UCP is the dependence on free fatty acids. H^+ transport in mitochondria from brown fatty adipose tissue was first shown to be dependent on free fatty acids by Nicholls and his group (Locke et al., 1982; Rial et al., 1983). We have concentrated on the elucidation of the influence of fatty acids in the reconstituted system of proteoliposomes prepared with purified phospholipids and UCP (Winkler and Klingenberg, in preparation). A variety of parameters of the H^+ transport are investigated, such as pH, fatty acid concentration, the fatty acid specificity, sidedness of fatty acid effect, and the interplay of fatty acid activation and inhibition by GTP or ATP. Both the H^+ influx and the H^+ efflux were used to examine the influence of these parameters.

The specificity for fatty acids is relatively broad. The principle requirement is a chain length of above C-10 carbon atoms. The highest and best activities are obtained with C-12 (lauric acid) and C-14 (myristic acid). Chain lengths above C-16 are limited by their solubility. Saturated and unsaturated fatty acids, including linoleic acid, activate equally well. Fatty acids which are -substituted with hydrophilic groups, such as OH, Br, carboxyl or spinlabel, also activate. The amounts of fatty acids required for activation are surprisingly high. At maximum activation the fatty acid concentration may reach 10 mol % of the phospholipids in the proteoliposomes. At first sight this indicates that the activation is due to an interaction between the lipid and the protein phase. Although this interpretation cannot be ruled out, a specific binding of fatty acids to the UCP in the translocation channel is a more attractive model. The high amounts of fatty acids produce only a minimum concentration of free fatty acids. Thus, despite the high amounts of fatty acids required in the reconstituted system, the putative specific binding centre in the translocation channel can have a high affinity. A further argument for a specific effect of fatty acids in H^+ translocation is the requirement of fatty acids only for the H^+/OH^- transport, but not for Cl^- transport. Transport of Cl^- or of other anions is insensitive to fatty acid additions, although it is fully inhibited by nucleotides in the same manner as the H^+ transport. Again, this argues for a specific and intimate participation of free fatty acids in the H^+-translocation process in UCP.

The pH dependence of H^+ translocation under fatty acid activation is a particularly important but experimentally intricate problem. In the reconstituted system not only the external and internal pH can be varied separately, but also the pH. Exact control of the internal pH and the pH is difficult in view of the unspecific H^+ transfer by fatty acids. The dependence of H^+ transport on external pH is critically linked to the chain-length of the activating fatty acid. With short-chain fatty acids as C-8, the maximum is around pH 5 and with long-chain fatty acids as C-16, the maximum is above pH 8. This is interpreted to be caused by a shift of the pK with increasing chain-lenghts of fatty acids due to binding in a hydrophobic environment. Thus, the pK of palmitic acid can reach 7.5 when bound to a phospholipid bilayer (Ptak et al., 1980). Obviously, for optimum activation, the pH of H^+ transport should be around the pK of the fatty acids. Another interesting finding is a sidedness of the pH influence in UCP. Careful evaluation of the internal versus external pH influence lead to the conclusion that in the proteoliposomes primarily the internal pH controls the UCP activity. Since most of the incorporated UCP molecules are rightside-outside it seems that fatty acids are bound to the matrix side of UCP which are under control of pH. This may indicate that the activating fatty acid, when bound to the translocation channel, faces more the matrix side than the cytosolic side.

How do fatty acids function in H^+ transport of UCP?

For the fatty acid activation of H^+ transport in UCP several mechanisms have been proposed. The group of Nicholls considered the possibility that the carboxyl group is a cofactor and therefore a direct H^+ carrier, but rejected this possibility because the nucleotides do inhibit the transport (Rial et al., 1983). An allosteric activation by fatty acids was considered by Rial (1984). All the measurements were based on H^+ efflux by swelling or by measuring respiration in mitochondria. In the reconstituted system a much more detailed analysis of the fatty acid specificity and of other parameters, such as concentration, pH etc., could be determined. Although the very high concentration of fatty acids seemed to indicate a more general effect originating from the lipid-protein interaction on the membrane interface, a specific cofactor type effect of fatty acids in the H^+ transport is also here our preferred interpretation.

In the simplest model (Fig. 2) fatty acid with its carboxyl group is the H^+ translocating "prosthetic" group of the UCP. In this function the carboxyl group accepts H^+ from one side and delivers it to the other side according to the single binding centre-gated pore mechanism, first demonstrated to exist for the ADP/ATP carrier. In this model the H^+ will diffuse from one side through the open gate to the carboxyl group and after opposite opening and closing of the gates it is delivered to the other side. A difficulty in this model is the acceptance and delivery of H^+ from the aqueous bulk phase and in the half channels to the carboxylic group. H^+ wires, similar as in gramicidin, are not possible in a carrier type of gated channel with one or more H^+ accepting and delivering groups. In order to overcome these restrictions in H^+ acceptance and intrachannel delivery, we prefer the concept that specifically bound fatty acids may act in the delivery of H^+ to a central constituent carboxyl group in the UCP

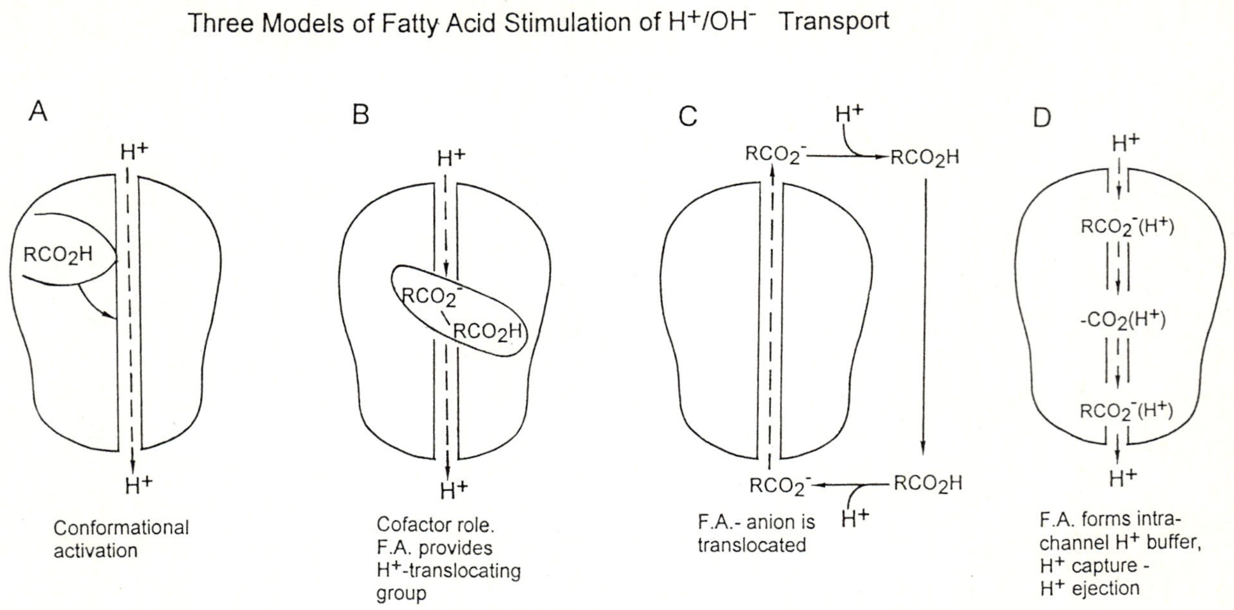

Fig. 2. The mechanism of fatty acids in the H^+/OH^- transport by the uncoupling protein. A. Specifically bound fatty acids or fatty acids from the lipid bilayer activate the H^+ transport by UCP via a conformational change. B. Specifically bound fatty acid is a prosthetic group of H^+ transport by providing the central H^+-translocating group. C. Fatty acids are carrying H^+ by being transported as anions through UCP and by diffusion as undissociated acids through the lipid bilayer (Andreyev et al., 1989). D. Specifically bound fatty acids are involved in H^+ capture and H^+ injection from and to aqueous phase. This buffering role is necessary in view of the 10^{-7} M H^+ in aqueous phase. The fatty acid carboxylates are mediating between the very H^+ poor aqueous phases and the central constituent translocating group in the H^+ transfer channel.

(Fig. 2). Thus, fatty acid carboxylates located either on one or both sides of the translocation centre, preferably on the matrix side, accept and deliver H^+ at a sufficient rate to the central H^+-translocating group which is constituent in UCP. We may also classify this role as a H^+-buffering effect of fatty acids. The observed pH profile of the fatty acid activation supports this concept. A maximum buffering H^+ capacity exists at a pH around the pK of the carboxyl group. Fatty acids, when bound to a membrane or fixed in a hydrophobic environment, have an increased pK and thus binding is enhanced with the increasing chain length. Thus the strong chain length dependency of the pH maximum of the activation agrees with the pK shift of the bound fatty acids and supports the concept of a buffering role of the fatty acids. Although in this role the fatty acid is not at the translocating centre, it participates in H^+ transport and is therefore an essential cofactor.

This mechanism is obviously analogous to that of bacteriorhodopsin. Here the central switch formed by the Schiff-base band of retinal to lysine is flanked by two aspartyl groups. These are located in the opposite half channels with a high pK at the H^+-donor site and a low pK at the H^+-acceptor site. UCP can be visualized to be constructed such that these donor-acceptor carboxyl groups are not constituent, but only inserted by the addition of free fatty acids. In this way the requirement of a strong regulation in UCP of H^+-transport activity is met by the particular necessity in H^+ transport for H^+-delivering or buffering groups. The inhibition of H^+ transport by nucleotides is superimposed to the fatty acid regulation. This is understandable in view of the evidence that the nucleotide-binding site is different from the translocation channel and that the inhibitory nucleotide binding induces a major conformational change in the UCP. In the inhibited state UCP is considered to be much less flexible and unable to undergo the conformational changes involved in the gating process independent of the presence of fatty acids.

References

Akeson M, Deamer DW (1991) Proton conductance by the gramicidin water wire. Biophys J 60: 101-109

Andreyev AY, Bondareva TO, Dedukhova V, Mokhova EN, Skulachev VP, Tsofina LM, Volkov NI, Vygodina TV (1989) The ATP/ADP-antiporter is involved in the uncoupling effect of fatty acids on mitochondria. Eur J Biochem 182: 585-592

Aquila H, Link TA, Klingenberg M (1985) The uncoupling protein from brown fat mitochondria is related to the mitochondrial ADP/ATP carrier. Analysis of sequence homologies and of folding of the protein in the membrane. EMBO J 4: 2369-2376

Butt HJ, Fendler K, Bamberg E, Tittor J, Oesterhelt D (1989) Aspartic acids 96 and 85 play a central role in the function of bacteriorhodopsin as a proton pump. EMBO J 8: 1657-1663

Cao Y, Vàrò G, Chang M, Baofu N, Needleman R, Lanyi JK (1991) Water is required for proton transfer from aspartate-96 to the bacteriorhodopsin schiff base. Biochemistry 30: 10972-10979

Heaton GM, Wagenvoord, RJ, Kemp A, Nicholls DG (1978) Brown-adipose-tissue mitochondria: Photoaffinity labelling of the regulatory site of energy dissipation. Eur J Biochem 82: 515-521

Klingenberg M (1988) Nucleotide binding to uncoupling protein. Mechanism of control by protonation. Biochemistry 27: 781-791

Klingenberg M (1990) Mechanism and evolution of the uncoupling protein of brown adipose tissue. TIBS 15: 108-112

Läuger P (1976) Diffusion-limited ion flow through pores. Biochim Biophys Acta 455: 493-509

Locke RM, Rial E, Scott ID, Nicholls DG (1982) Fatty acids as acute regulators of the proton conductance of hamster brown-fat mitochondria. Eur J Biochem 129: 373-380

Nicholls DG (1974) Hamster brown-adipose-tissue mitochondria. Eur J Biochem 49: 585-593

Ptak M, Egret-Charlier M, Sanson A, Bouloussa O (1980) A NMR study of the ionization of fatty acids, fatty amines and n-acylamino acids incorporated in phosphatidylcholine vesicles. Biochim Biophys Acta 600: 387-397

Rial E, (1984) The regulation of thermogenesis in brown adipose tissue mitochondria.Ph.Thesis, Basque Country University, Bilbao/Dundee

Rial E, Poustie A, Nicholls DG (1983) Brown-adipose-tissue mitochondria: the regulation of the 32000-Mr uncoupling protein by fatty acids and purine nucleotides. Eur J Biochem 137: 197-203

THE GATING AND COUPLING FUNCTION OF MITOCHONDRIAL H^+ ATP SYNTHASE. ROLE OF F_o AND F_1 SUBUNITS.

S. Papa
Institute of Medical Biochemistry and Chemistry
University of Bari
Piazza G. Cesare
70124 Bari,
Italy

INTRODUCTION

The F_oF_1 H^+-ATP synthase (E.C. 3.6.1.34) is the key energy transfer enzyme, which mediates ATP synthesis driven by trans membrane proton motive force in coupling membranes (inner mitochondrial membrane, plasma membrane of bacteria, thylakoid membrane of chloroplasts) (Senior, 1988; Papa, 1989).

In prokaryotes the enzyme functions both as ATP synthase and hydrolase, in mitochondria it functions essentially as ATP synthase, thus sparing ATP synthesized in mitochondria for export to other cell compartments where it is utilized (Pedersen and Carafoli, 1987; Papa, 1989).

The eukaryotic enzyme has, in addition to the 5 subunits of the catalytic F_1 sector and the 3 subunits of the H^+ translocating membrane F_o sector of the prokaryotic enzyme (Senior, 1988), 5 or 6 supernumerary subunits (Walker et al, 1991)(Fig. 1). One of these is the F_1 inhibitor protein (Harris, 1984), which inhibits specifically ATP hydrolysis, the others belong to the F_o sector (Papa, 1989).

A critical region in the H^+-ATP synthase is the stalk connecting the F_1 head-piece to F_o (Cogol et al., 1987; Papa, 1989). This structure is likely to contribute the ion filter,

NATO ASI Series, Vol. H 83
Molecular Biology of Mitochondrial Transport Systems
Edited by M. Forte and M. Colombini

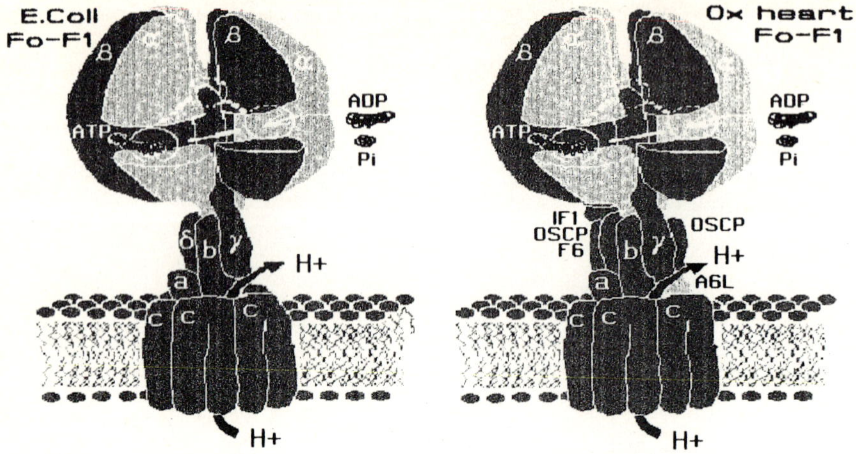

Fig 1 - Architecture of F_oF_1 ATP synthase in E. coli and bovine heart mitochondria. F_oI-PVP protein is indicated as b in both complexes.

the gate and the coupling device between H^+ transport and hydro-anhydro catalysis. In E. coli the stalk is apparently formed by segments of F_1 gamma, delta and epsilon and F_o b subunit (Senior, 1988; Takeyoma et al., 1988; Jounouchi et al., 1992)(subunit c seems also to contribute to the binding of F_o to F_1)(Fraga et al., 1991)(Fig. 1). In mitochondria a larger number of subunits are involved in the connection of F_o to F_1 (Papa, 1989)(Fig. 1), indicating that this region has undergone evolutionary adaptation to environmental conditions.

In this paper experimental observations from our laboratory on the role of F_o and F_1 subunits in the gate and coupling function of the mitochondrial ATP synthase are presented.

RESULTS

<u>The role of mitochondrial F_o and F_1 subunits</u>

The carboxyl terminal region of the F_oI-PVP subunit, from C-

Met 214 to Lys 202, extends out of the mitochondrial membrane at the matrix side where it is covered by F_1 (Fig. 1)(Houstek et al., 1988; Papa et al., 1989). In inside-out vesicles of the membrane, after F_1 removal (USMP), this segment is cleaved off by trypsin digestion; the truncated protein remains in the membrane (Zanotti et al., 1988). The cleavage results in loss of the correct binding of F_1 to F_o, depression of transmembrane H^+ conduction and loss of inhibition of ATP hydrolysis and H^+ conduction by oligomycin (Zanotti et al., 1988; Guerrieri et al., 1991). These activities can be restored by giving back the isolated native F_oI-PVP protein to the digested enzyme (Zanotti et al., 1988).

In inside-out F_1 depleted membrane, trypsin digestes also OSCP and F_6. The role of these subunits in the binding of F_1 to F_o was tested by reconstituting isolated soluble F_1 with F_o in membrane vesicles. The addition of F_1 to undigested USMP reconstituted a membrane associated ATPase activity which was inhibited, like the native F_oF_1 complex, by oligomycin (Fig. 2). Oligomycin binds to the F_o sector and inhibits H^+ conduction; extension of the inhibitory effect to ATP hydrolysis depends on a functionally correct binding of F_1 to F_o. Binding of F_1 to F_o was altered after digestion of USMP by trypsin, as shown by loss of oligomycin sensitivity of the ATPase activity of reconstituted F_1 (Fig. 2). Oligomycin sensitivity of ATP hydrolase could, however, be restored by adding, to digested USMP reconstituted with soluble F_1, F_oI-PVP and OSCP or F_oI-PVP and F_6 (see Guerrieri et al., 1991).

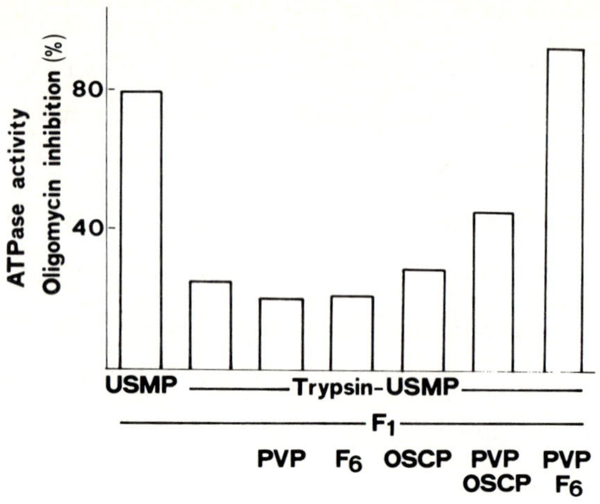

Fig.2- Oligomycin sensitivity of ATPase activity in trypsinized USMP reconstituted with purified F_1.
For preparation of USMP, F_1, trypsin digestion, reconstitution of ATPase activity, see Houstek et al., 1988, preparation of F_oI-PVP, OSCP and F_6, see Zanotti et al., 1992. Where indicated F_oI-PVP (4 µg/mg particle proteins), OSCP (2 µg/mg particle proteins), F_6 (4 µg/mg particle proteins) or combination of the same amounts of these proteins were preincubated 10 min with USMP before addition of F_1.

Treatment of "inside-out" submitochondrial particles, containing both F_o and F_1, with diamide, which oxidizes vicinal dithiol to disulfide bridges, causes oligomycin sensitive enhancement of H^+ conduction (Zanotti et al., 1992). This enhancement of H^+ conduction produced by diamide appeared to be symptomatic of the opening of the gate of the H^+ channel (Zanotti et al., 1992). The stimulatory effect exerted by diamide in F_oF_1 submitochondrial particles (ESMP) decreased upon partial removal of F_1 subunits from the particles (USMP), and changed into inhibition in liposomes reconstituted with a preparation of F_o devoid of F_1 subunits (Zanotti et al., 1992).

The addition of soluble gamma subunit to F_o liposomes

markedly depressed H^+ conduction (Table I). The addition of soluble alpha and beta subunits had, on the other hand, no significant effect on H^+ conduction (Table I). Immunoblot analysis showed that both alpha and beta subunits become bound to F_o liposomes (Table I). Treatment with diamide caused inhibition of H^+ conduction in F_o liposomes both in the absence and presence of added alpha and beta subunits, but produced a dramatic enhancement of H^+ conduction in the presence of added gamma subunit. Diamide, which caused disappearance of the band of gamma subunit, did not reduce those of alpha and beta subunits (Table I).

TABLE I. EFFECT OF ADDED F_1 SUBUNITS ON H^+ CONDUCTION IN F_o LIPOSOMES IN THE ABSENCE AND PRESENCE OF DIAMIDE. Where indicated isolated polypeptide, at the concentration of 4 µg/mg protein F_o were added to F_o liposomes (incubation time 10 min) before treatment with 80 µM diamide (incubation time 20 min). For details see Zanotti et al., 1992. Reproduced with permission from Zanotti et al., 1992.

Added polypeptides	H^+ release $\frac{\text{ng ion } H^+}{\text{min·mg protein}}$		Immunoblot (arbitrary units)			
			alpha/beta		gamma	
	-	+Diamide	-	+Diamide	-	+Diamide
none	720	579	-	-	-	-
alpha/beta	669	594	1.05	1.10	-	-
gamma	309	1414	-	-	2.14	1.20
gamma+alpha/beta	321	771	1.51	1.34	1.46	1.10

In Fig. 3 the time course of the effect of 80 µM diamide in ESMP is shown. Diamide caused a progressive enhancement of H^+ conduction (Fig. 3A). The enhancement of H^+ conduction was directly correlated with decrease of the original bands of F_oI-

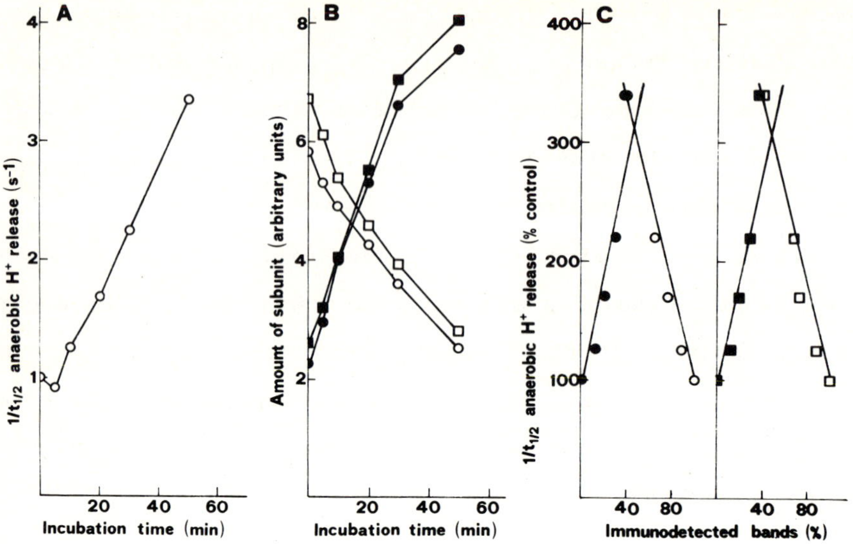

Fig. 3 - Relationship between stimulation of anaerobic H^+ release by diamide and cross-linking of F_oI-PVP and gamma subunits in ESMP.

ESMP were treated with 80 µM diamide for the intervals given in the figure. Areas of F_oI-PVP and gamma subunit immunoreactive bands were determined by densitometric analysis of immunoblots. For determination of F_oI-PVP protein and gamma subunit in cross-linking product, ESMP, treated with diamide, were subjected to SDS/PAGE. The section of the gels above the gamma subunit was cut and electroeluted. The electroeluted proteins were treated with 20 mM dithiothreitol and subjected to a second electrophoretic run. (B)(□) Immunodetected gamma subunit; (○) immunodetected F_oI-PVP subunit in SDS/PAGE of ESMP; (●) immunodetected F_oI-PVP and (■) gamma subunit released by dithiothreitol from high-molecular mass cross-linking products in diamide-treated ESMP. (C) Apparent rate of anaerobic H^+ release in diamide-treated ESMP as a function of the residual areas of original bands of gamma (□) and F_oI-PVP (○) subunits and of the appearance of gamma (■) and F_oI-PVP (●)in cross-linking products. For details see Zanotti et al., 1992. Reproduced with permission from Zanotti et al., 1992.

PVP and gamma subunit (Fig. 3B and 3C). The correspondence of disappearance of the original band of the F_oI-PVP protein with that of the gamma subunit, suggests their cross-linking through disulfide bridging of the single thiol residues present in each of these two proteins (Walker et al., 1985; Walker et al., 1987). The proteins with apparent mass higher than that of the gamma subunit (33 kDa) were electroeluted from gel slabs of diamide treated preparations. This material, after treatment with dithiothreitol, to cleave disulfide-bridges, was electrophoresed again. Immunoblotting with anti F_oI-PVP protein and anti F_1 sera revealed release from the material of diamide treated ESMP, electroeluted from gels, of both the F_oI-PVP and gamma subunits (Fig. 3B). The progressive enhancement of H^+ conduction, caused in ESMP by diamide, correlated equally well with the disappearance of the original bands, of both F_oI-PVP and gamma subunits and their appearance in the cross-linking product (Fig. 3C).

F_oF_1 zybrids

Soluble F_1 isolated from E. coli (EcF_1) could be reconstituted with F_1-depleted E. coli membrane (UP_{Ec})(Steffens et al., 1984) and also with F_1-depleted submitochondrial vesicles (USMP) (zybrid). The binding of EcF_1 to UP_{Ec} (followed by measuring the ATPase activity appearing in the sedimented membranes) followed saturation kinetics (Fig. 4). USMP also bound EcF_1 with saturation kinetics. The reconstituted activity in this case was higher, but the affinity of binding lower. The same reconstitution pattern was obtained when soluble F_1 from bovine heart (BHF_1) was reconstituted with UP_{Ec} or USMP.

Addition to UP_{Ec}-EcF_1 or UP_{Ec}-BHF_1 of soluble F_6, isolated from the mitochondrial ATP synthase, induced a significant

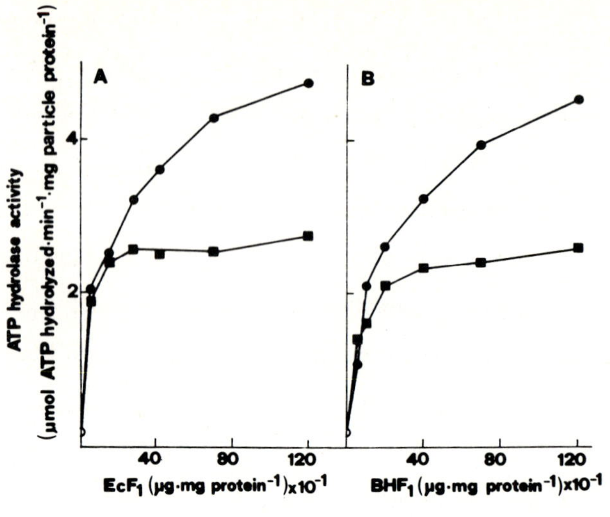

Fig. 4 - Reconstitution of ATPase activity in UP_{Ec} and USMP by addition of purified EcF_1 (panel A) or BHF_1 (panel B). Preparation of everted F_1-stripped vesicles and of EcF_1 from E. coli ML 308-225 were carried out as reported in Futai et al., 1974. For preparation of USMP, BHF_1 and reconstitution of ATPase activity see legend to Fig. 2. Symbols: (■) UP_{Ec}; (●) USMP.

enhancement of the ATPase activity (Table II). On the contrary OSCP and F_oI-PVP and their combination induced depression of the ATPase activity. These effects were not due to a change in the amount of F_1 bound to the membrane as shown by immunoblot analysis and were specific. Addition of the isolated 31 kDa band or albumin, had, in fact, no effect at all on the ATPase activity. When the ATPase activity of reconstituted UP_{Ec}-EcF_1 or of the zybrid UP_{Ec}-BHF_1 was tested for its sensitivity to oligomycin, the activity resulted insensitive to the inhibitor (Fig. 5) differently from USMP reconstituted with BHF_1 or EcF_1, whose ATPase activity was inhibited by oligomycin. The addition to reconstituted UP_{Ec}-EcF_1 or to UP_{Ec}-BHF_1 zybrid, of OSCP or F_oI-PVP alone, did not induce significant oligomycin sensitivity of the reconstituted ATP hydrolase activity.

TABLE II. EFFECT OF MITOCHONDRIAL F_o SUBUNITS ON ATP HYDROLASE ACTIVITY OF RECONSTITUTED UP_{Ec}-EcF_1 OR UP_{Ec}-BHF_1.
For purification of F_oI-PVP, OSCP and F_6 and for reconstitution of ATPase activity, see legends to Fig.s 2 and 4. 31 kDa protein was isolated by electroelution from SDS-PAGE of USMP. Isolated proteins (4 µg/mg UP_{Ec}) were incubated with UP_{Ec} for 10 min before reconstitution with purified EcF_1 or BHF_1 (0.8 mg/mg UP_{Ec}).

Additions	ATP hydrolase activity µmol ATP hydrolyzed·min^{-1}·mg $protein^{-1}$	
	Reconstitution with	
	BHF_1	EcF_1
UP_{Ec}	2.67	2.54
$+F_6$	3.06	3.28
+OSCP	2.00	1.78
$+F_oI$-PVP	1.40	1.88
+OSCP$+F_oI$-PVP	0.96	0.43
+OSCP$+F_6$	1.72	1.92
$+F_oI$-PVP$+F_6$	1.90	2.18
+OSCP$+F_6+F_oI$-PVP	0.96	0.63
+31 kDa protein	-	2.49
+ Albumin	-	2.50

However, the combined addition of these two subunits together with F_6, which per se had no effect (not shown), largely restored oligomycin sensitivity (Fig. 5).

Fig. 6 shows that the ATP hydrolase activity of reconstituted UP_{Ec}-EcF_1 and of the zybrid UP_{Ec}-BHF_1 exhibited a lower sensitivity as compared to USMP-BHF_1 or USMP-EcF_1 to inhibition by DCCD, a lipophilic chemical modifier of acidic aminoacid residues which is a potent inhibitor of H^+ translocation by F_o both in E. coli and mitochondria. However, the inhibition by DCCD of ATPase activity was enhanced by the combined addition of OSCP and F_oI-PVP. F_6 had no effect in this respect.

Fig. 7 shows that UP_{Ec} and UP_{Ec}-subunit c had a lower

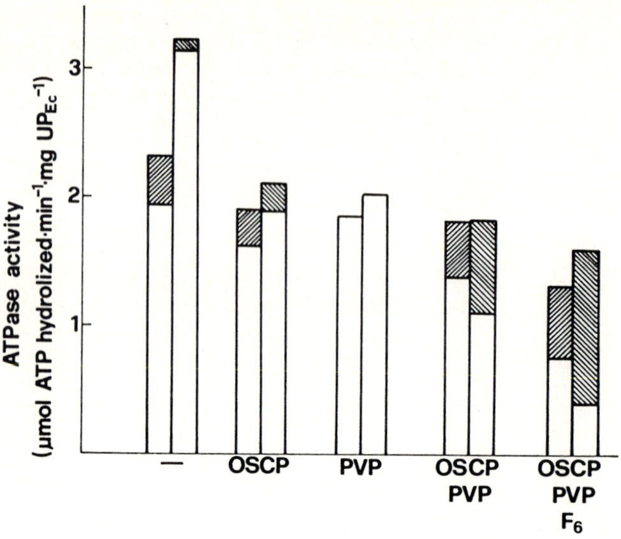

Fig. 5 - Effect of isolated bovine heart OSCP, F_oI-PVP and F_6 on oligomycin sensitivity of ATPase activity in UP_{Ec} reconstituted with EcF_1 or BHF_1.
UP_{Ec}, BHF_1 and EcF_1 were prepared as described in the legend to Fig. 4. Reconstitution and determination of ATPase activity were carried out as described in the legend to Fig. 4. In each condition the left columns refer to reconstituted UP_{Ec}-EcF_1 and the right columns to the reconstituted UP_{Ec}-BHF_1. The dashed portions of the columns represent the inhibition induced by oligomycin (1 μg/mg UP_{Ec}).

affinity for binding of ^{14}C-DCCD than USMP and USMP-subunit c. The binding of ^{14}C-DCCD to subunit c in UP_{Ec} was enhanced by addition of F_oI-PVP together with OSCP.

H^+ conduction in UP_{Ec} was markedly depressed by reconstitution with soluble EcF_1 or BHF_1 (Fig. 8). A plot of H^+ conduction in UP_{Ec} against DCCD bound to subunit c shows that the addition of isolated F_oI-PVP and F_oI-PVP together with OSCP enhanced the inhibitory action of DCCD bound to subunit c on H^+ conduction.

These observations, thus, show that OSCP and F_oI-PVP of

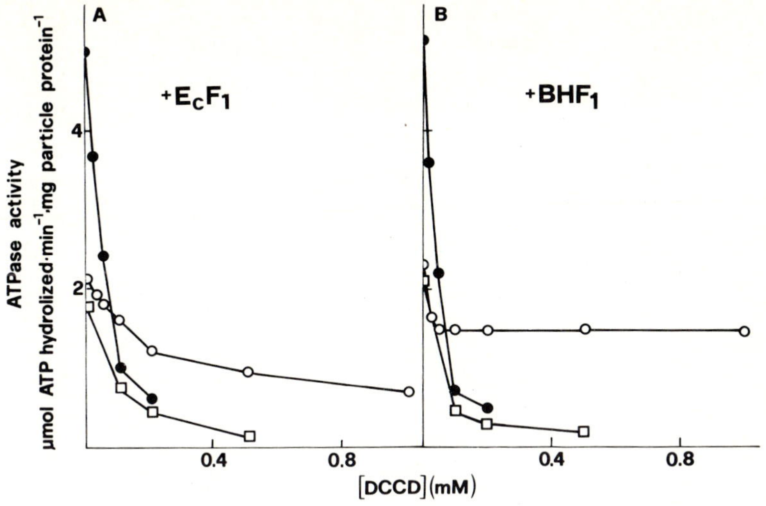

Fig. 6 - Titration of the inhibitory effect of DCCD on ATP hydrolase activity in UP_{EC} and USMP reconstituted with purified EcF_1 (panel A) or BHF_1 (panel B). Effect of addition of isolated OSCP and F_oI-PVP.
For experimental procedure see legend to Fig. 4 and Table II. Symbols: (○-○) UP_{EC}-EcF_1 or BHF_1 (0.8 mg/mg UP_{EC}); (●-●) USMP-EcF_1 or BHF_1 (0.8 mg/mg USMP); (□-□) UP_{EC} preincubated 10 min with OSCP (4 µg/mg protein of UP_{EC}) and F_oI-PVP (4 µg/mg protein of UP_{EC}) before reconstitution with EcF_1 or BHF_1.

mitochondrial F_o increase the reactivity with DCCD of subunit c in the E. coli F_o sector and its inhibitory action.

Regulation of ATPase activity and H^+ conduction by IF_1

The ATPase inhibitor protein (IF_1) has an amphiphatic nature and binds to the beta subunit, thus inhibiting the catalytic activity (Harris, 1984). The binding of IF_1 to the ATPase complex is suppressed by respiratory $\Delta\mu H^+$, thus allowing ATP synthesis to proceed in the respiring steady-state (Lippe et al., 1983).

In the absence of $\Delta\mu H^+$ IF_1 binds to the complex and

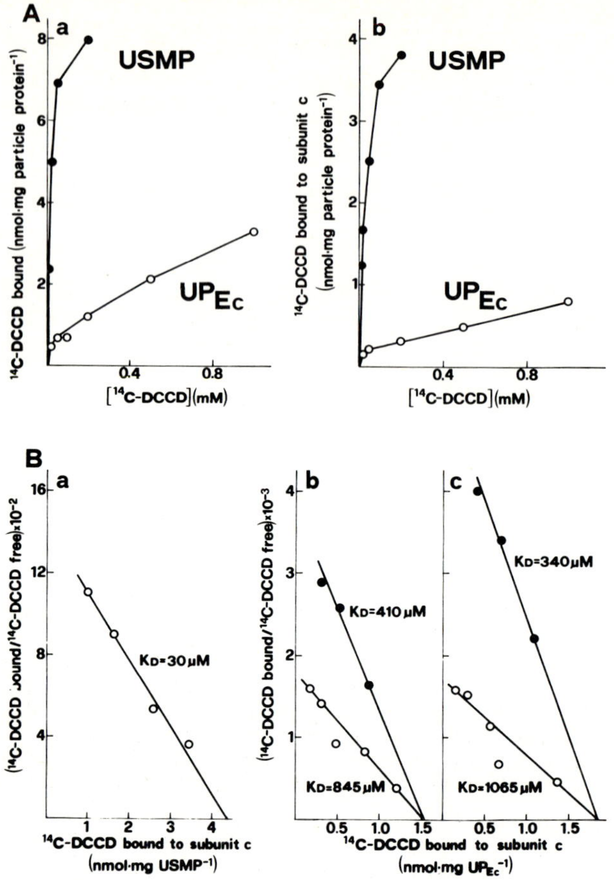

Fig. 7 - Binding of ^{14}C-DCCD to membrane particles and to subunit c.

UP_{Ec} or USMP were incubated with ^{14}C-DCCD at the concentrations reported at 25°C for 30 min. The total binding of ^{14}C-DCCD to vesicles or binding of ^{14}C-DCCD to subunit c was determined as in Kopecky et al. 1983.
Panel A. a) Binding of ^{14}C-DCCD on USMP (●-●) or UP_{Ec} (○-○) or b) on subunit c of USMP (●-●) or of UP_{Ec} (○-○).
Panel B. Scatchard plots for binding of ^{14}C-DCCD to subunit c. a) USMP; b)(○-○) UP_{Ec}, (●-●) UP_{Ec}+OSCP+F_oI-PVP (4 µg/mg UP_{Ec}); c) (○-○) UP_{Ec}-EcF_1, (●-●) UP_{Ec}-EcF_1+OSCP+F_oI-PVP (4 µg/mg UP_{Ec}).

prevents ATP hydrolysis. It has been found in our laboratory that IF_1 also exerts an inhibitory action on passive H^+

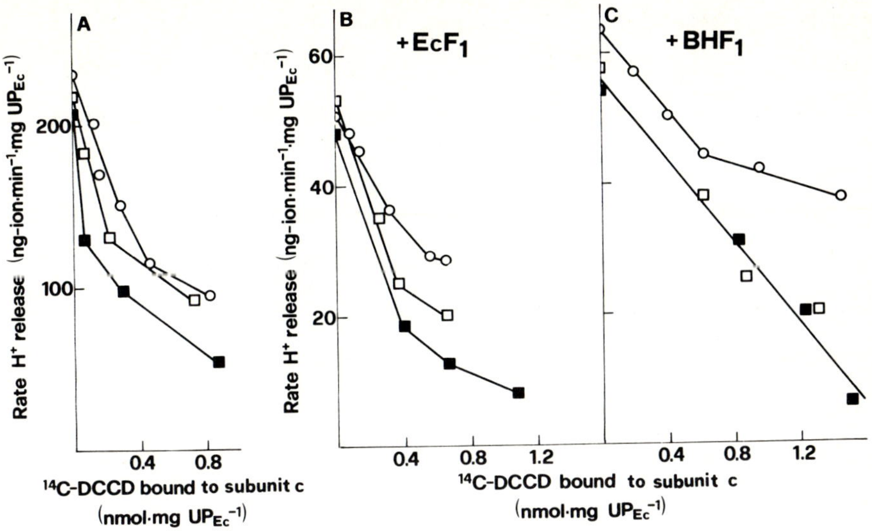

Fig. 8 - Relationship between inhibition of H^+ conduction by ^{14}C-DCCD and its binding to subunit c in UP_{Ec} (panel A) and in UP_{Ec} reconstituted with EcF_1 (panel B) or BHF_1 (panel C).

1 mg UP_{Ec}/ml (panel A); 1 mg UP_{Ec}/ml reconstituted with EcF_1 (panel B); 1 mg UP_{Ec}/ml reconstituted with BHF_1 (panel C) were incubated at 25°C for 30 min in absence or in presence of ^{14}C-DCCD in a range of concentrations between 50 µM to 1 mM. For details see legend to Fig. 7. Symbols: (○ - ○) UP_{Ec} or UP_{Ec}-EcF_1 or UP_{Ec}-BHF_1; (□ - □) F_oI-PVP (4 µg/mg UP_{Ec}) or (■ - ■) F_oI-PVP+OSCP (4 µg/mg UP_{Ec}) added to UP_{Ec} or UP_{Ec}-EcF_1 or UP_{Ec}-BHF_1 respectively 10 min before addition of ^{14}C-DCCD.

conduction by F_o (Guerrieri et al., 1987). This dual inhibitory effect exerted by IF_1 on the ATP synthase is clearly shown by its activity in submitochondrial particles from which the protein is removed by Sephadex chromatography (Guerrieri et al., 1986). Both activities are inhibited by giving back the isolated soluble IF_1 to depleted particles. The inhibitory effects decrease with pH, with an apparent pK around 7, suggesting a critical role of hystidine residues in IF_1 (Guerrieri et al., 1987).

The bovine heart IF_1 has, in fact, 5 hystidine residues (Fig. 9) one of which is conserved in IF_1 from yeast mitochondria (Harris, 1984). Modification of ESMP with ethoxyformic anhydride, a reagent for hystidine residues, was found to cause a dramatic enhancement of passive H^+ conduction. This effect was lost in the particles from which IF_1 was removed by Sephadex (Guerrieri et al., 1987). Treatment of the isolated IF_1 with ethoxyformic anhydride under conditions leading to modification of hystidine residues, prevented the inhibitory activity of isolated IF_1 on H^+ translocation in Sephadex particles (Table III).

TABLE III - EFFECT OF NATIVE ISOLATED IF_1 AND ITS SYNTHETIC SEGMENT (RESIDUE 42-62) ON ANAEROBIC H^+ CONDUCTION IN SEPHADEX-TREATED ESMP. EFFECT OF DIETHYLPYROCARBONATE (EFA).

The synthetic polypeptide was produced by f-MOC chemical synthesis using a 431A Peptide Synthesizer (Applied Biosystems). For purification of synthetic IF_1 peptide, its treatment with EFA and experimental details see legend to Fig. 10. 50μM of IF_1 or synthetic peptide were incubated at pH 6.2 with Sephadex submitochondrial particles (incubation time 10 min).

	Anaerobic release of respiratory proton gradient $1/\frac{1}{2}$ (s^{-1})
Control particles	2.00
+ATPase inhibitor protein	0.75
+ATPase inhibitor protein treated with diethylpyrocarbonate	2.60
+Synthetic peptide	0.75
+Synthetic peptide treated with diethylpyrocarbonate	1.75

The five hystidine residues of IF_1 are located in two putative elices in the carboxyl-terminal half of the protein (Fig. 9). We have synthesized a segment of IF_1 extending from arginine 62

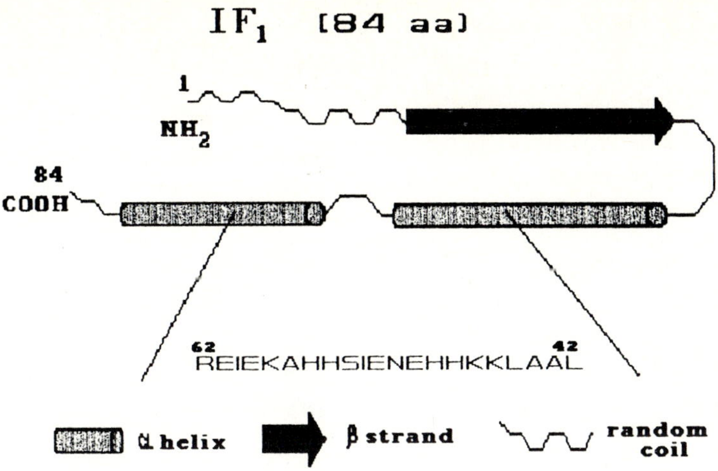

Fig. 9 - Secondary structure of IF_1.

The structure was obtained by application of the Chou-Fasman rules (Chou et al., 1978) through the "nodo EMBnet" (C.N.R. - Bari) program to the aminoacid sequence obtained from Frangione et al., 1981.

to leucine 42 which contains four hystidines at positions 48, 49, 55, 56. The polypeptide was purified by HPLC, the purity checked by capillary chromatography (Fig. 10) and the sequence verified (Fig. 9). The results presented in Table III show that the synthetic peptide was effective in inhibiting H^+ conduction when reconstituted with depleted submitochondrial particles. Treatment of the synthetic polypeptide with ethoxyformic anhydride, under conditions leading to modification of hystidine residues (Fig. 10), resulted, as in the case of the native IF_1, in loss of its inhibitory activity. These results seem therefore to identify the 42-62 segment of F_1, which contains four hystidine residues, as critical for functional activity of the protein.

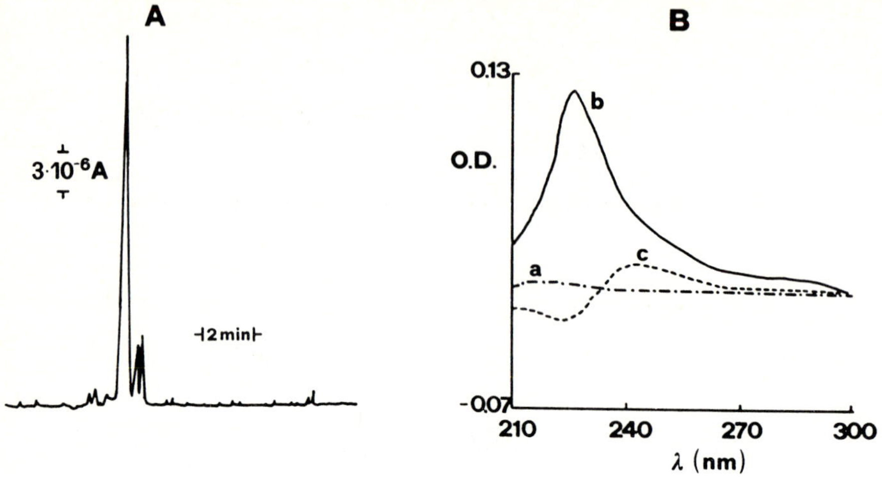

Fig. 10 - Capillary electrophoresis of IF_1 synthetic peptide (residues 42-62),A. Effect of modification by diethyl pyrocarbonate on the absorption spectrum of the IF_1 peptide in the ultraviolet region, B.

A. Synthesis was carried out with Model 270A Applied Biosystem synthetizer with 20 mM sodium citrate buffer, pH 8.5 at 35°C. The eluate was monitored at 200 nm for 30 min run time.
B. Incubation of IF_1 peptide (4 mg/ml) at room temperature in the absence and in presence of 0.5 mM diethylpyrocarbonate. After 10 min the peptide was diluted and its absorption spectrum recorder. a) baseline; b) spectral difference between the absorbance of diethylpyrocarbonate-treated peptide and control; c) spectral difference between the absorbance of the peptide treated with diethylpyrocarbonate followed by 3mM NH_2OH and control. The reversal by NH_2OH of the absorption peak at 232 nm induced by treatment of the peptide with diethylpyrocarbonate documents modification of hystidine residues.

The dual inhibitory effects exerted by IF_1 on both catalytic activity of H^+ conduction qualify this protein as a component of the gate of ATP synthase.

CONCLUSIONS

The membrane peripheral, carboxyl-terminal region of the

F_oI-PVP protein from C-terminal Met 214 to Cys 197 plays an important role as a component of the gate in the functional binding of F_1 to F_o, modulation of H^+ conduction by F_o and its sensitivity to oligomycin and DCCD (Papa et al., 1988; Guerrieri et al., 1991). This protein thus appears, after subunit c, as the second component of mitochondrial F_o for which definite functional activities are identified (Zanotti et al., 1988).

The carboxyl-terminal segment of F_oI-PVP, 214-197, a segment around Cys 91 of the gamma subunit, OSCP and F_6 contribute to the stalk of the mitochondrial F_oF_1 H^+-ATP synthase (Fig. 1). F_6, OSCP and the carboxyl-terminal region of F_oI-PVP are essential for the correct functional binding of F_1 to F_o, being all of them necessary for oligomycin sensitivity of the hydrolase activity of the F_oF_1 complex (Fig. 3 and ref. Guerrieri et al., 1991). The gamma subunit specifically depresses H^+ conduction by F_o (alpha and beta have per se no direct effect on this process). Disulfide cross-linking of the F_oI-PVP protein with the gamma subunit results in dramatic enhancement of proton conduction by F_o. This effect, which is accompanied by stimulation of ATPase activity and depression of ^{32}Pi-ATP exchange (Zanotti et al., 1992), seems to reflect the opening of the mouth of the H^+ channel at the junction with F_1 with consequent decoupling of H^+ transport from hydro-anhydro catalysis (Fig. 10).

On the basis of the resemblance of the respective hydropathy plots, Walker et al., 1987 have proposed that the mitochondrial

F_oI-PVP protein is analogous to the b subunit of E. coli. It should, however, be noted that no direct homology exists in the residue sequences of the bacterial b subunit and mitochondrial F_oI-PVP subunit. In addition, structural predictions would indicate the existence in the F_oI-PVP of a beta strand domain which is absent in the E. coli b subunit (Senior et al., 1983). It can also be noted that the single Cys 197 of F_oI-PVP protein is located in a membrane embedded part close to the matrix surface of the protein only seventeen residues apart from the carboxyl-terminal region. In the b subunit of E. coli the single Cys 21 is located in the N-terminal hydrophobic helix, which apparently anchors the protein to the membrane (Senior et al., 1983). The conserved Cys 197 in the mitochondrial F_oI-PVP protein appears to be in vicinal position with the conserved Cys 91 in the F_1 gamma-subunit as shown by diamide-induced cross linking of the two proteins and enhancement of passive H^+ conduction by F_o. The location and reactivity of Cys 21 in the b subunit of E. coli appears to be different from that in F_oI-PVP as indicated by the fact that diamide instead of causing enhancement of H^+ conduction induces depression of proton conduction in E. coli vesicles both in the absence and presence of added EcF_1 or BHF_1 (Fig. 11).

Oligomycin is a bulky lipophilic substance which is likely to interact with the F_o sector at the boundary with the hydrophobic core of the phospholipid bilayer (Hoppe et al., 1986). Also DCCD which is lipophilic is likely to react with the side chain of the acidic residue (Asp in E. coli; Glu in mitochondria) at the protein-lipid interphase.

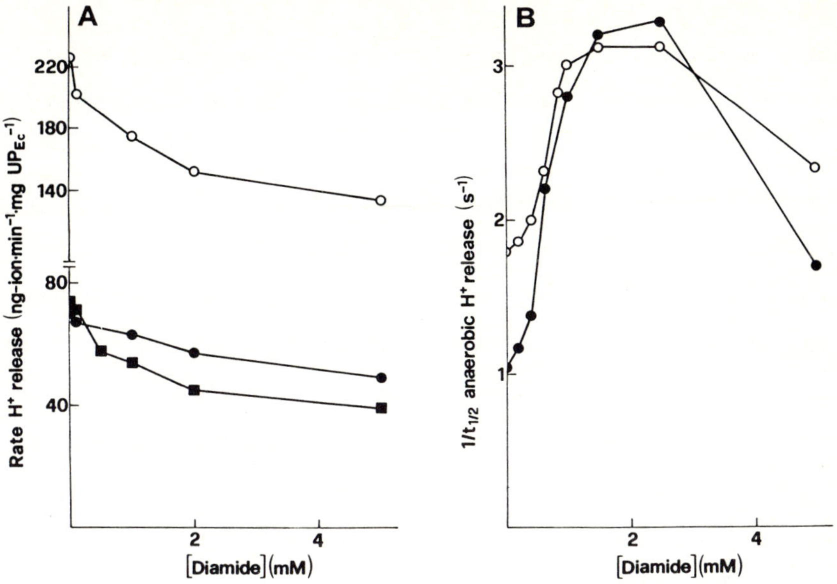

Fig. 11 - Effect of diamide on proton conduction in UP_{Ec} and USMP reconstituted with EcF_1 or BHF_1.

For measurement of H^+ conduction see Zanotti et al., 1987. For other experimental procedure see legend to Fig. 4. Symbols: Panel A. (○-○) UP_{Ec} (1 mg protein/ml) or (●-●) UP_{Ec} reconstituted with EcF_1 (1 mg protein/ml) or (■-■) UP_{Ec} reconstituted with BHF_1 (1 mg protein/ml) were incubated for 2 min in KCl (150mM) in absence or in presence of diamide at the concentrations reported in figure before the addition of valinomycin.
Panel B. (○-○) USMP (3 mg protein/ml) or (●-●) USMP reconstituted with BHF_1 (3 mg protein/ml) were incubated for 2 min in absence or in presence of diamide at the concentrations reported in figure before measurement of H^+ conduction.

The DCCD-binding acidic residue of subunit c appears to be directly involved in transmembrane H^+ translocation. It has been postulated that H^+ translocation takes place in a pore in the middle of a boundle of multiple copies of rotating subunits c, with the acidic residue at 65 interacting with protolytic residues of a membrane spanning helix of subunit a of E. coli (ATPase 6 of mitochondria)(Cox et al., 1986).

Rotation of subunit c, on its long axis perpendicular to the plane of the membrane, will expose during revolution the acidic residue subsequently to the interior of the pore and to the sursounding lipids, thus allowing its participation to H^+ conduction in the pore and reaction with lipophilic reagents at the protein-lipid interphase. The apparent affinity for the binding of oligomycin and DCCD to subunit c might depend on the rate of rotation of this subunit. The different affinity for oligomycin and DCCD exhibited by UP_{Ec} and USMP could, in fact, result from difference in the rotation rate of subunit c in the two membranes. Mitochondrial F_oI-PVP and OSCP could affect the rotation rate of subunit c in the E. coli membrane and consequently enhance its apparent binding affinity for ^{14}C-DCCD. Effect on the rotation rate of subunit c (enhancement or depression?) could also be responsible for a tigher coupling of F_1 with F_o as indicated by the inhibition that these two subunits exert on the ATP hydrolase activity when added to UP_{Ec}-EcF_1 or UP_{Ec}-BHF_1.

ACKNOWLEDGEMENTS

This work was supported, in part, by C.N.R. Grant 91/04203ST75.

REFERENCES

Chou PY, Fasman GD (1978) Ann Rev Biochem 7:251-276
Cogol EP, Lucken U and Capaldi R (1987) FEBS Lett 219:274-278
Cox GB, Fimmel AL, Gibson F and Hatch L (1986) Biochim Biophys Acta 849:62-69
Fraga D, Fillingame RH (1991) J Bacteriol 173:2639-2643
Frangione B, Rosenwasser E, Penefsky HS and Pullman ME (1981) Proc Natl Acad Sci 78:7403-7407
Futai M, Sternweis P and Heppel L (1974) Proc Natl Acad Sci 71:2725-2729
Guerrieri F, Zanotti F, Capozza G, Colaianni G, Ronchi S and Papa S (1991) Biochim Biophys Acta 1059:348-354

Guerrieri F, Zanotti F, Che YW, Scarfò R and Papa S (1987) Biochim Biophys Acta 892:284-293

Harris DA (1984) in "H^+-ATPase (ATP synthase): Structure, Function and Biogenesis. The F_oF_1 Complex of Coupling Membranes" (Papa S, Altendorf K, Ernster L and Packer L eds), ICSU Miami/Adriatica Editrice Bari, pp 387-394

Hoppe J, Gatti D, Weber H and Sebald W (1986) Eur J Biochem 155:259-264

Houstek J, Kopecky J, Zanotti F, Guerrieri F, Jirillo E, Capozza G and Papa S (1988) Eur J Biochem 173:1-8

Klein G, Satre M, Dianoux AC and Vignais PV (1980) Biochemistry 19:2919-2927

Kopecky J, Guerrieri F and Papa S (1983) Eur J Biochem 131:17-24

Jounouchi M, Takeyoma M, Chairprasert P, Noumi T, Moryama Y, Maeda M and Futai M (1992) Arch Biochem Biophys 292: 376-381

Lippe G, Sorgato MC and Harris DA (1988) Biochim Biophys Acta 973:1-11

Papa S (1989) in "Organelles in Eukaryotic Cells: Molecular Structure and Interactions" (Tager JM, Azzi A, Papa S and Guerrieri F eds), Plenum Publ Co, New York and London, pp 9-26

Papa S, Guerrieri F, Zanotti F, Altendorf K and Deckers-Hebestreit G (1992) in "Adenine Nucleotides in Cellular Energy Transfer and Signal Transduction" (Papa S, Azzi A and Tager JM eds), Birkhauser Verlag, Basel, pp 13-32

Papa S, Guerrieri F, Zanotti F, Houstek J, Capozza G and Ronchi S (1989) FEBS Lett 249:62-66

Pedersen PL and Carafoli E (1987) TIBS 12:146-150

Senior AE (1988) Physiol Rev 68:177-231

Senior AE, Wise JG (1983) J Membr Biol 73:105-124

Steffens K, Schneider E, Herkenoff B, Schmid R and Altendorf K (1984) Eur J Biochem 138:617-622

Takeyoma M, Moumi M, Maeda M and Futai M (1988) J Biol Chem 263:16103-16112

Walker J, Fearnley IM, Gay NJ, Gibson BW, Northrop FD, Powell SJ, Runswick MJ, Saraste M and Tybulewicz VLJ (1985) J Mol Biol 184:677-701

Walker J, Lutter R, Dupuis A and Runswick JM (1991) Biochemistry 30:5369-5378

Walker J, Runswick MJ and Poulter L (1987) J Mol Biol 197:89-109

Zanotti F, Guerrieri F, Capozza G, Fiermonte M, Berden J and Papa S (1992) Eur J Biochem 208:9-16

Zanotti F, Guerrieri F, Capozza G, Houstek J, Ronchi S and Papa S (1988) FEBS Lett 237:9-14

Zanotti F, Guerrieri F, Che YW, Scarfò R and Papa S (1987) Eur J Biochem 164:517-523

EXTENSION OF THE MITOCHONDRIAL TRANSPORTER FAMILY

J. E. Walker and M. J. Runswick
M. R. C. Laboratory of Molecular Biology
Hills Road
Cambridge CB2 2QH
U. K.

The sequences of the ADP/ATP, phosphate and oxoglutarate/malate carriers from the inner membranes of mitochondria, and of the uncoupling protein (UCP) from brown fat mitochondria, demonstrate that they are a family of related proteins [reviewed by Walker (1992)]. Their sequences are all about 300 amino acids long and they are characterized by the presence of a three-fold sequence repeat of about 100 amino acids, first noted in the sequence of the ADP/ATP carrier (Saraste & Walker, 1982), and subsequently in the UCP (Aquila et al., 1985), in the phosphate carrier (Runswick et al., 1987) and in the oxoglutarate/ malate carrier (Runswick et al., 1990; see Fig. 1). The various sequence repeats in the family are related to each other, albeit rather weakly in some cases, and it is a reasonable assumption that each of the related sequences will be folded into the same structural element. The secondary structure of that element is not known, as no member of this family of proteins has been induced to form either 2-dimensional or 3-dimensional crystals, but clues are present in their hydrophobic profiles, which are also conserved (Fig. 2). These profiles suggest that the second and third repeats in the sequence could consist of two hydrophobic transmembrane α-helices joined by an extensive hydrophilic region, with each repeated element joined by a shorter hydrophilic region. This secondary structure is less evident in the first repeat, as the region that corresponds to the second helix of the first repeat tends to be less hydrophobic than in the subsequent repeats, but the requirement that the sequence repeats should have the same fold makes it likely that this region will also form a transmembrane α-helix, and that the same structural motif will be present in the first third of the polypeptide chain as in the second and third repeats.

These considerations have led to a common structural model for the mitochondrial transport proteins in which the predominant features are six transmembrane α-helices and three extensive hydrophilic regions exposed to the aqueous environment (see Fig. 3). A prediction of this model is that the N- and C-termini lie on the same side of the inner mitochondrial membrane, and, in confirmation, it has been demonstrated that both extremities of the phosphate carrier are exposed on the cytoplasmic surface (Capobianco et al., 1991; Ferreira et al., 1990). Other evidence concerning the topology of the carriers is also consistent with the general model,

NATO ASI Series, Vol. H 83
Molecular Biology of Mitochondrial Transport Systems
Edited by M. Forte and M. Colombini

although it is becoming clear that the conformations of the large hydrophilic regions change during transmembrane transport [see Walker (1992)].

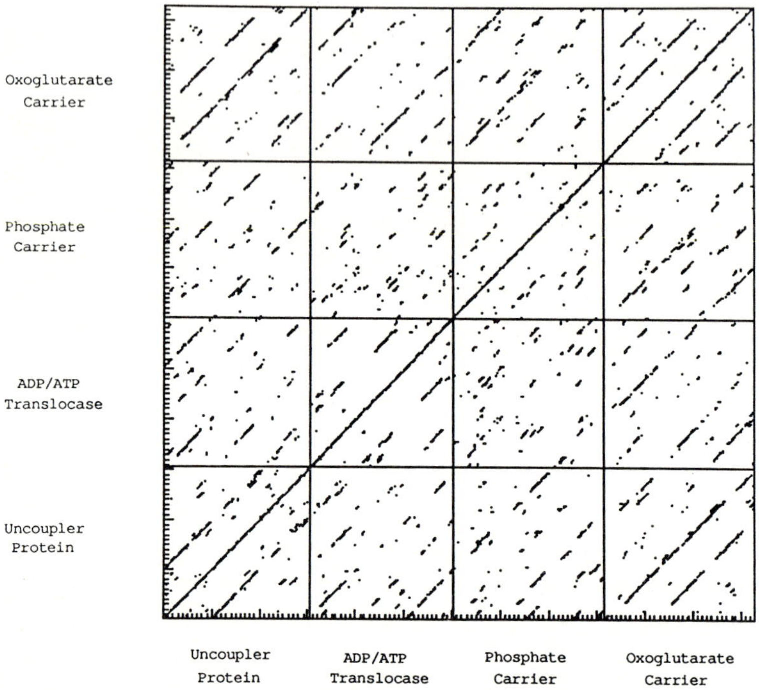

Figure 1. Pairwise comparison of sequences of mitochondrial transport proteins. The comparisons were made with the computer program DIAGON. Dots are recorded where the score exceeds a threshold of 280 in a window 25 amino acids long, using the MDM78 scoring matrix.

Alignment of the repeated sequences in the family shows that a small number of amino acids are absolutely conserved. Most notable amongst these are conserved proline residues at the end of proposed helices I, III and V, followed after an interval of one amino acid by an acidic residue (Walker, 1992). It is likely that these amino acids are important in the overall structure of the family of transporters. This simple sequence motif present in each of the threefold 100 amino acid sequence repeats combined with the repetitive feature and the hydrophobic profile provide characteristic features for the recognition of new family members of unknown biochemical function. They have been noted in two closely related proteins MRS3 and MRS4 from *Saccharomyces cerevisiae* (Weisenberger *et al.*, 1991), in a protein detected by screening a human cDNA expression library with a circulating autoantigen associated with Graves'

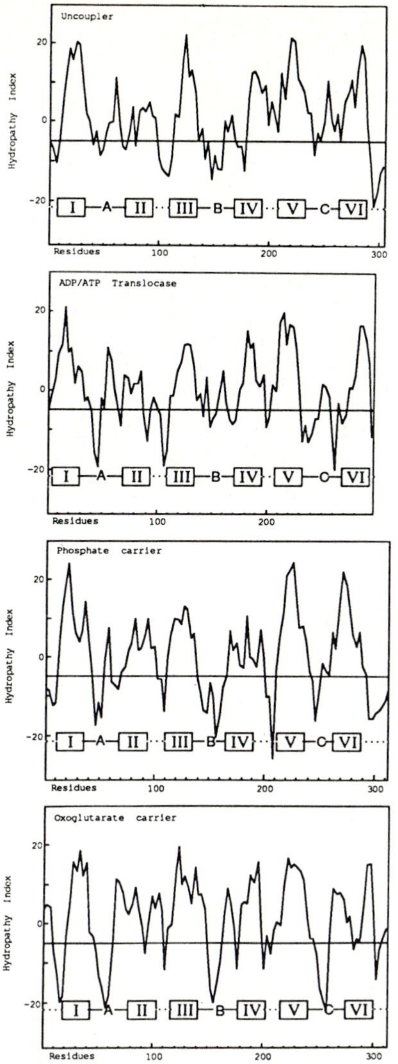

Figure 2. Hydrophobic profiles of four mitochondrial transport proteins of known sequence. Beneath each profile, boxes I-IV indicate the positions of possible α-helical membrane spanning segments. Segments A, B and C represent more hydrophilic regions. The three sequence repeats present in each sequence consist of I-A-II, III-B-IV and V-C-VI.

disease (Zarrilli et al., 1989), and in a protein encoded in a hypotrychous ciliated protozoan, *Oxytricha fallax* (Williams & Herrick, 1991). Yeast MRS3 and MRS4 proteins are targetted to the inner membrane of the organelle, where their exact function is unknown, but they

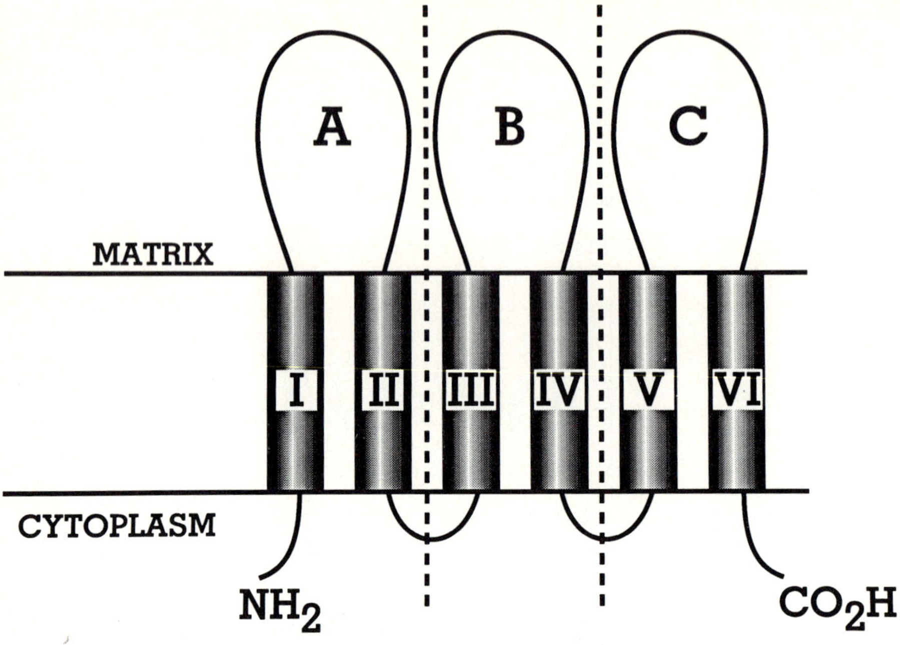

Figure 3. A six-helix model of mitochondrial transport proteins. The dotted lines indicate that the proteins have a threefold structural repeat which ids proposed to consist of pairs of transmembrane α-helices (I and II, III and IV, V and VI) linked by extensive hydrophilic regions (A, B and C). The cytoplasmic locations of the N- and C-terminal extremities in the phosphate carrier, of the N-terminal of the ADP/ATP carrier and of the C-terminal of the uncoupling protein have been demonstrated.

influence RNA splicing in the organelle, perhaps by translocating cationic polyamines. The Graves' disease protein is expressed primarily in the thyroid gland. Transcripts of the bovine homologue are also present in other tissues, but at lower levels than are found in the thyroid (Fiermonte et al., 1992). Its subcellular location and function are not known, nor are they for the protozoan protein. In the absence of further information, it is a reasonable assumption that these proteins will also prove to be transport proteins in the inner membranes of mitochondria.

The sequences of other new members of the mitochondrial carrier family are beginning to emerge from genomic sequencing projects. For example, the sequences of two yeast carriers of unknown function have been determined (Colleaux *et al.*, 1992; B. G. Barrell personal communication), and in the course of the determination of partial sequences of cDNAs from the nematode worm *Caenorhabditis elegans* (Waterston *et al.*, 1992) not only have the ADP/ATP and phosphate carriers been discovered, but in addition four other new carriers have been found. The sequences of these nematode carriers, including the ADP/ATP and phosphate carriers (see Figs. 4 and 5), have been completed. One of the unknown carrier sequences (that of carrier O) is rather similar to that of the bovine oxoglutarate/ malate carrier (see Fig. 6), but it is less conserved than are either the ADP/ATP and phosphate carriers from the cow and the nematode. Therefore, it is less certain that carrier O is the nematode oxoglutarate/malate carrier. The other three carriers (C1, C2 and C4) are not closely related to any of the carriers of known sequence (M. J. Runswick and J. E. Walker, unpublished work), and the question arises of what their transport specificities might be. It is possible that these sequences are those of carriers of unknown sequence that have been biochemically characterized from mammalian mitochondria.

There are at least seven such mitochondrial carriers (see Fig. 7). For example, pyruvate and acyl-carnitine are exchanged for OH^- and carnitine respectively by their carriers. The aspartate/glutamate carrier, together with the oxoglutarate/malate carrier, participates in the aspartate-malate shuttle, a device for reducing NAD^+ in the matrix, and additionally there are carriers for citrate, glutamate, ornithine and dicarboxylate. Some of them, such as those involved in oxidative phosphorylation, are found in mitochondria in all tissues, whereas others have a more limited tissue distribution; for example, the dicarboxylate and citrate carriers have little if any activity in heart mitochondria, but are particularly active in liver (LaNoue & Schoolwerth, 1984). It is likely that these carriers also belong to the carrier family since they have features in common with the established family members. For example, their molecular weights fall within the same range of values (28-32 kDa) as known family members, and they also tend to co-chromatograph with them (Krämer & Palmieri, 1989, 1992). However, until their sequences are known, it will remain uncertain that they are members of the family.

A number of practical problems make the sequence determination of some carriers a somewhat difficult undertaking; many of the carriers are can only be isolated in a homogeneous state with difficulty (Krämer & Palmieri, 1989), and often in insufficient amount for direct protein sequencing, particularly as their N-terminals may be modified, making it necessary to isolate peptides from digests. Therefore, another strategy of cloning and sequencing these proteins is being investigated and is described as follows. The nematode DNA sequences are being used to try and determine the sequences of their bovine homologues by using the polymerase chain reaction to amplify the homologues from bovine cDNA and genomic DNA. It is assumed from the high degeree of conservation of the ADP/ATP and phosphate carriers that the other carriers will be sufficiently conserved. Once these sequences have been determined, the cDNA sequences will be cloned into an appropriate expression vector, and the proteins will be

```
                                       1
H. sapiens 1                         M GDHAWSFLKD FLAGGVAAAV SKTAVAPIER VKLLLQVQ-H ASKQISAEKQ
B. taurus 1                          M SDQALSFLKD FLAGGVAAAI SKTAVAPIER VKLLLQVQ-H ASKQISAEKQ
H. sapiens 2                         M TEQAISFAKD FLAGGIAAAI SKTAVAPIER VKLLLQVQ-H ASKQIAADKQ
B. taurus 2                          M TEQAISFAKD FLAGGIAAAI SKTAVAPIER VKLLLQVQ-H ASKQIAADKQ
H. sapiens 3                         M TDAALSFAKD FLAGGVAAAI SKTAVAPIER VKLLLQVQ-H ASKQITADKQ
N. crassa                      M AEQQK VLGMPPFVAD FLMGGVSAAV SKTAAAPIER IKLLVQNQDE MIRAGRLDRR
S. cerevisiae 1                M SHTET QTQQSHFGVD FLMGGVSAAI AKTGAAPIER VKLLMQNQEE MLKQGSLDTR
S. cerevisiae 2                 M SSDA KQQETNFAIN FLMGGVSAAI AKTAASPIER VKILIQNQDE MIKQGTLDKK
S. cerevisiae 3     M SSNAQ VKTPLPPAPA PKKESNFLID FLMGGVSAAV AKTAASPIER VKLLIQNQDE MIKQGTLDKK
Z. mays                   M QTPLCANAPA EKGGKNFMID FMMGGVSAAV SKTAAAPIER VKLLIQNQDE MIKSGRLSEP
C. elegans                      M SKEK SFDTKKFLID LASGGTAAAV SKTAVAPIER VKLLLQVQDA SKAIAVDKR-
C. rheinhardtii                      M AKEEKNFMVD FLAGGLSAAV SKTAAAPIER VKLLIQNQDE MIKQGRLASP
                                             *       **  **   **   ****  *** * *
                                         -----------------------------
                                                       I

                    50                                                    100
H. sapiens 1        YKGIIDCVVR IPKEQGFLSF WRGNLANVIR YFPTQALNFA FKDKYKQLFL GGVDRHKQFW RYFAGNLASG
B. taurus 1         YKGIIDCVVR IPKEQGFLSF WRGNLANVIR YFPTQALNFA FKDKYKQIFL GGVDRHKQFW RYFAGNLASG
H. sapiens 2        YKGIVDCIVR IPKEQGVLSF WRGNLANVIR YFPTQALNFA FKDKYKQIFL GGVDKHTQFW RYFAGNLASG
B. taurus 2         YKGIVDCIVR IPKEQGVLSF WRGNLANVIR YFPTQALNFA FKDKYKQIFL GGVDKRTQFW RYFAGNLASG
H. sapiens 3        YKGIIDCVVR IPKEQEVLSF WRGNLANVIR YFPTQALNFA FKDKYKQIFL GGVDKRTQFW RYFAGNLASG
N. crassa           YNGIIDCFKR TTADEGVMAL WRGNTANVIR YFPTQALNFA FRDKFKKMFG YKKDV-DGYW KWMAGNLASG
S. cerevisiae 1     YKGILDCFKR TATHEGIVSF WRGNTANVLR YFPTQALNFA FKDKIKSLLS YDRER-DGYA KWFAGNLFSG
S. cerevisiae 2     YSGIVDCFKR TAKQEGLISF WRGNTANVIR YFPTQALNFA FKDKIKLMFG FKKE--EGYG KWFAGNLASG
S. cerevisiae 3     YSGIVDCFKR TAKQEGLISF WRGNTANVIR YFPTQALNFA FKDKIKLMFG FKKE--EGYG KWFAGNLASG
Z. mays             YKGIVDCFKR TIKDEGFSSL WRGYTANVIR YFPTQALNFA FKDYFKRLFN FKKDR-DGYW KWFAGNLASG
C. elegans          YKGIMDVLIR VPKEQGVAAL WRGNLANVIR YFPTQAMNFA FKDTYKAIFL EGLDKKKDFW KFFAGNLASG
C. rheinhardtii     YKGIGECFVR TVREEGFGSL WRGNTANVIR YFPTQALNFA FKDKFKRMFG FNKD--KEYW KWFAGNMASG
                    ****     *            ***  *** * ****** *** * *  *                   *** ***
                                     -----------------------------                 -------------
                                                   II                                        III

                                                      150
H. sapiens 1        GAAGATSLCF VYPLDFARTR LAAD--VGKG AA-QREFHGL GDCIIKIFKS DGLRGLYQGF NVSVQGIIIY
B. taurus 1         GAAGATSLCF VYPLDFARTR LAAD--VGKG AA-QREFTGL GNCITKIFKS DGLRGLYQGF NVSVQGIIIY
H. sapiens 2        GAAGATSLCF VYPLDFARTR LAAD--VGKS GT-EREFRGL GDCLVKITKS DGIRGLYQGF SVSVQGIIIY
B. taurus 2         GAAGATSLCF VYPLDFARTR LAAD--VGKS GS-EREFRGL GDCLVKITKS DGIRGLYQGF NVSVQGIIIY
H. sapiens 3        GAAGATSLCF VYPLDFARTR LAAD--VGKA GA-EREFRGL GDCLVKIYKS DGIKGLYQGF NVSVQGIIIY
N. crassa           GAAGATSLLF VYSLDYARTR LANDAKSAKK GG-ERQFNGL VDVYRKTIAS DGIAGLYRGF GPSVAGIVVY
S. cerevisiae 1     GAAGGLSLLF VYSLDYARTR LAADARGSKS TS-QRQFNGL LDVYKKTLKT DGLLGLYRGF VPSVLGIIVY
S. cerevisiae 2     GAAGALSLLF VYSLDFARTR LAADAKSSKK GGA-RQFNGL TDVYKKTLKS DGIAGLYRGF MPSVVGIVVY
S. cerevisiae 3     GAAGALSLLF VYSLDYARTR LAADSKSSKK GGA-RQFNGL IDVYKKTLKS DGVAGLYRGF LPSVVGIVVY
Z. mays             GAAGASSLFF VYSLDYARTR LANDAKAAKG GG-ERQFNGL VDVYRKTLKS DGIAGLYRGF NISCVGIIVY
C. elegans          GAAGATSLCF VYPLDFARTR LAAD--IGKA --NDREFKGL ADCLIKIVKS DGPIGLYRGF FVSVQGIIIY
C. rheinhardtii     GAAGAVSLSF VYSLDYARTR LANDAKSAKK GGGDRQFNGL VDVYRKTIAS DGIAGLYRGF NISCVGIVVY
                    ****  ** * ** ** **** ** *    *        * **  *   *     **  *** **   *  **  *
                    ---------------
                                                                         -------------------
                                                                                    IV

                                 200                                                       250
H. sapiens 1        RAAYFGVYDT AK-GMLPDPK NVHIFVSWMI AQSVTAVAGL VSYPFDTVRR RMMMQSGRKG ADIMYTGTVD
B. taurus 1         RAAYFGVYDT AK-GMLPDPK NVHIIVSWMI AQTVTAVAGL VSYPFDTVRR RMMMQSGRKG ADIMYTGTVD
H. sapiens 2        RAAYFGVYDT AK-GMLPDPK NTHIVVSWMI AQTVTAVAGV VSYPFDTVRR RMMMQSGRKG ADIMYTGTVD
B. taurus 2         RAAYFGIYDT AK-GMLPDPK NTHIVVSWMI AQTVTAVAGV VSYPFDTVRR RMMMQSGRKG ADIMYKGTVD
H. sapiens 3        RAAYFGIYDT AK-GMLPDPK NTHIVISWMI AQTVTAVAGL TSYPFDTVRR RMMMQSGRKG TDIMYTGTLD
N. crassa           RGLYFGLYDS IKPVLLVGDL KNNFLASFAL GWCVTTAAGI ASYPLDTIRR RMMMTSG--- EAVKYKSSFD
S. cerevisiae 1     RGLYFGLYDS FKPVLLTGAL EGSFVASFLL GWVITMGAST ASYPLDTVRR RMMMTSG--- QTIKYDGALD
S. cerevisiae 2     RGLYFGMFDS LKPLVLTGSL DGSFLASFLL GWVVTTGAST CSYPLDTVRR RMMMTSG--- QAVKYNGAID
S. cerevisiae 3     RGLYFGMYDS LKPLLLTGSL EGSFLASFLL GWVVTTGAST CSYPLDTVRR RMMMTSG--- QAVKYDGAFD
Z. mays             RGLYFGLYDS IKPVVLTGNL QDNFFASFAL GWLITNGAGL ASYPIDTVRR RMMMTSG--- EAVKYKSSLD
C. elegans          RAAYFGMFDT AKMVFASDGQ KLNFFAAWGI AQVVTVGSGI LSYPWDTVRR RMMMQSGRK- -DILYKKHPR
C. rheinhardtii     RGLYFGMYDS LKPVVLVGPL ANNFLAAFLL GWGITIGAGL ASYPIDTIRR RMMMTSG--- SAVKYNSSFH
                    *  ***  *                            *       *** ** ** **** **        *
                    --------            -----------------------------
                                                      V

H. sapiens 1        CWRKIAKDEG AKAFFKGAWS NVLRGMGGAF VLVLYDEI-- ----KKYV
B. taurus 1         CWRKIAKDEG PKAFFKGAWS NVLRGMGGAF VLVLYDEI-- ----KKFV
H. sapiens 2        CWRKIFRDEG GKAFFKGAWS NVLRGMGGAF VLVLYDEL-- ----KKVI
B. taurus 2         CWRKILKDEG GKAFFKGAWS NVLRGMGGAF VLVLYDEL-- ----KKVI
H. sapiens 3        CWRKIARDEG GKAFFKGAWS NVLRGMGGAF VLVLYDEI-- ----KKYT
N. crassa           AASQIVAKEG VKSLFKGAGA NILRGVAGAG VLSIYDQLQV LLFGKAFKGG SG
S. cerevisiae 1     CLRKIVQKEG AYSLFKGCGA NIFRGVAAAG VISLYDQLQL IMFGKKFK
S. cerevisiae 2     CLKKIVASEG VGSLFKGCGA NILRSVAGAG VISMYDQLQM ILFGKKFK
S. cerevisiae 3     CLRKIVAAEG VGSLFKGCGA NILRGVAGAG VISMYDQLQM ILFGKKFK
Z. mays             AFQQILKKEG PKSLFKGAGA NILRAIAGAG VLSGYDQLQI LFFGKKYGSG GA
C. elegans          LRKEDHPNEG MSAMFKGALS NVFRGTGGAL VLAIYDEI-- ----QKFL
C. rheinhardtii     CFQEIVKNEG MKSLFKGAGA NILRAVAGAG VLAGYDQLQV ILLGKKYGSG EA
                            **     ***    *  *    *  *   **
                               -----------------------------
                                             VI
```

Figure 4. Sequences mitochondrial ADP/ATP carrier proteins. Asterisks denote conserved residues, and I-VI are the positions of proposed α-helical segments.

expressed in *Escherichia coli*. The expressed proteins will be isolated and used to prepare antibodies, and these antibodies will be employed in Western blot analyses to follow carrier activities during their isolation.

```
                    1                                                     50
H. sapiens          AVEE-YSCEF GSAKYYALCG FGGVLSCGLT HTAVVPLDLV KCRMQVDPQK YKGIFNGFSV TLKEDGVRGL
B. taurus           AVEEQYSCDY GSGRFFILCG LGGIISCGTT HTALVPLDLV KCRMQVDPQK YKSIFNGFSV TLKEDGFRGL
R. norvegicus       AVEQGYSCEF GSMKYYALCG FGGVLSCGLT HTAVVPLDLV KCRMQVDPQK YKGIFNGFSI TLKEDGVRGL
S. cerevisiae     M SVSAAPAIPQ YSVSDYMKFA LAGAIGCGST HSSMVPIDVV KTRIQLEPTV YNKGMVGSFG IIAGEGAGAL
C. elegans          AVSAPGQVEF GSGKYYAYCA LGGVLSCGIT HTAIVPLDLV KCRIQVNPEK YTGIATGFRT TIAEEGARAL
                                *           *   ** * *   ** * * * * *  *   *     *         *   *
                                -----------------------------                               ----
                                             I

                                               100
H. sapiens          AKGWAPTFLG YSMQGLCKFG FYEVFKVLYS NMLGEENTYL WRTSLYLAAS ASAEFFADIA -LAPMEAAKV
B. taurus           AKGWAPTFIG YSLQGLCKFG FYEVFKVLYS NMLGEENAYL WRTSLYLAAS ASAEFFADIA -LAPMEAAKV
R. norvegicus       AKGWAPTLIG YSMQGLCKFG FYEVFKALYS NILGEENTYL WRTSLYLAAS ASAEFFADIA -LAPMEAAKV
S. cerevisiae       LTGFGPTLLG YSIQGAFKFG GYEVFKKFFI DNLGYDTASR YKNSVYMGSA AMAEFIADIA -LCPLEATRI
C. elegans          VKGWAPTLLG YSAQGLGKFG FYEIFKNVYA DMLGEENAYL YRTSLYLAAS ASAEFFADIL LLAPMEATKV
                      *  **  * ** **  ***  ** **       **          * *     * *** ***   * * **
                    -------------------------                 ------------------------------
                           II                                                 III

                              150                                                        200
H. sapiens          RIQTQPGYAN TLRDAAPKMY KEEGL--KAF YKGVAPLWMR QIPYTMMKFA CFERTVEALY KFVVPKPRSE
B. taurus           RIQTQPGYAN TLRDAAPKMY KEEGL--KAF YKGVAPLWMR QIPYTMMKFA CFERTVEALY KFVVPKPRSE
R. norvegicus       RIQTQPGYAN TLREAVPKMY KEEGLNALAF YKGVAPVWMR QIPYTMMKFA CFERTVEALY KFVVPKPRSE
S. cerevisiae       RLVSQPQFAN GLVGGFSRIL KEEGI--GSF YSGFTPILFK QIPYNIAKFL VFERASEFYY GFAGPK--EK
C. elegans          RIQTSPGAPP TLRGCAPMIY KAEGL--TGF YKGLPPLWMR QIPYTMMKFA CFEKTVEALY QYVVPKPRAE
                    *    *                *  **     * * *  *     ****   **   **   *  *     **
                                              --------------------------------
                                                              IV

                                                                         250
H. sapiens          CSKPEQLVVT FVAGYIAGVF CAIVSHPADS VVSVLNKEK- ------GSSA SLVLKRLGFK GVWKGLFARI
B. taurus           CSKPEQLVVT FVAGYIAGVF CAIVSHPADS VVSVLNKEK- ------GSSA SEVLKRLGFR GVWKGLFARI
R. norvegicus       CTKAEQLVVT FVAGYIAGVF CAIVSHPADS VVSVLNKEK- ------GSTA SQVLQRLGFR GVWKGLFARI
S. cerevisiae       LSSTSTTLLN LLSGLTAGLA AAIVSQPADT LLSKVNKTKK APGQSTVGLL AQLAKQLGFF GSFAGLPTRL
C. elegans          CSKAEQLVVT FVAGYIAGVF CAIVSHPADT VVSKLNQDS- ------QATA GGILKKLGFA GVWKGLVPRI
                                  *  **    **** ***    *  *                      ***  *   **  *
                      -----------------------------                                ------------
                                    V

                                               300
H. sapiens          IMIGTLTALQ WFIYDSVKVY FRLPRPPPPE MPESLKKKLG LTQ
B. taurus           IMIGTLTALQ WFIYDSVKVY FRLPRPPPPE MPESLKKKLG YTQ
R. norvegicus       IMIGTLTALQ WFIYDSVKVY FRLPRPPPPE PMESLKKKLG LTE
S. cerevisiae       VMVGTLTSLQ FGIYGSLKST LGCPPTIEIG GGGH
C. elegans          IMIGTLTALQ WFIYDSVKVA LNLPRPPPPE MPASLKAKL
                     * **** **   ** * *      *
                    ---------------
                    VI
```

Figure 5. Sequences of mitochondrial phosphate carrier proteins.

In trial experiments, it has been possible to express both the bovine ADP/ATP and oxoglutarate/malate carriers to reasonable levels (about 10 mg/litre) in *E. coli* (in collaboration with G. Fiermonte and F. Palmieri, University of Bari, Italy). In both instances the protein makes insoluble inclusion bodies, which is at the same time both an advantage and a disadvantage; the inclusion bodies are easy to isolate by centrifugation, and so the expressed protein is readily purified, but in order to demonstrate directly its activity (which would be desirable), the inclusion body proteins must be refolded correctly and inserted into a membrane. This could be a difficult practical problem to overcome, but if this hurdle can be surmounted it opens the possibility of a wide range of other experiments. For example, the carriers associated with Graves' disease and the carriers of unknown specificity in yeast and in the nematode, and others that doubtless will be discovered in the near future in the course of genomic sequencing of various species could then be expressed and their transport properties studied (in the case of the yeast proteins it is perhaps better to employ yeast expression systems). Bacterial expression of active or reconstituted carriers also opens the possibility of study of functional residues by site directed mutagenesis, and of the provision of sufficient material of rare carriers for crystallization trials to be undertaken as a prelude to structure determination by X-ray crystallography.

```
                  1                                                  50
H. sapiens      M AATASAGAGG MDGKPRTSPK SVKFLFGGLA GMGATVFVQP LDLVKNRMQL SGEGAKTREY KTSFHALTSI
B. taurus       M AATASPGASG MDGKPRTSPK SVKFLFGGLA GMGATVFVQP LDLVKNRMQL SGEGAKTREY KTSFHALISI
C. elegans                      AEDKTKR LGRWYFGGVA GAMAACCTHP LDLLKVQLQT QQQGKLTIGQ LS-----LKI
                                             *** * *  *     * *** *   *     *  *             *
                                  -----------------------------
                                               I

                                               100
H. sapiens        LKAEGLRGIY TGLSAGLLRQ ATYTTTRLGI YTVLFERLTG ADGTPPGFLL KAVIGMTAGA TGAFVGTPAE
B. taurus         LRAEGLRGIY TGLSAGLLRQ ATYTTTRLGI YTVLFERLTG ADGTPPGFLL KAVIGMTAGA TGAFVGTPAE
C. elegans        YKNDGILAFY NGVSASVLRQ LTYSTTRFGI YETVKKQLPQ D--QPLPFYQ KALLAGFAGA CGGMVGTPGD
                              * *   ***  ** *** ** *      *       *  *   **     ***  *  ****
                       -----------------------------             ----------------------------
                                      II                                        III

                         150                                                    200
H. sapiens        VALIRMTADG RLPADQRRGY KNVFNALIRI TREEGVLTLW RGCIPTMARA VVVNAAQLAS YSQSKQFLLD
B. taurus         VALIRMTADG RLPVDQRRGY KNVFNALFRI VQEEGVPTLW RGCIPTMARA VVVNAAQLAS YSQSKQFLLD
C. elegans        LVNVRMQNDS KLPLEQRRNY KHALDGLVWI TREEGFMKMF NGATMATSRA ILMTIGQLSF YDQIKQTLIS
                      **  *   **  *** * *     *  *   ***              **       **   *   ** *
                                                        -----------------------------
                                                                       IV

                                                          250
H. sapiens        SGYFSDNILC HFCASMISGL VTTAASMPVD IAKTRIQNMR MIDGKPEYKN GLDVLFKVVR YEGFFSLWKG
B. taurus         SGYFSDNILC HFCASMISGL VTTAASMPVD IVKTRIQNMR MIDGKPEYKN GLDVLVKVVR YEGFFSLWKG
C. elegans        SGVAEDNLQT HFASSISAAS VATVMTQPLD VMKTR----- MMNAGPGEFK GILDCFMFTA KLGPMGFFKG
                  **   **    **  *      * *    * *   ***      *    *     *            *     **
                     ----------------------------                                      -------
                                   V

                                    300
H. sapiens        FTPYYARLGP HTVLTFIFLE QMNKAYKRLF LSG
B. taurus         FTPYYARLGP HTVLTFIFLE QMNKAYKRLF LSG
C. elegans        FIPAWARLAP HTVLTFIFFE QLRLKFGYAP PVKAS
                  * *  *** * ******** *
                  --------------------
                        VI
```

Figure 6. Sequences of the human and bovine mitochondrial oxoglutarate/malate carrier proteins aligned with the sequence of a related protein, referred to in the text as carrier O, from the nematode worm *C. elegans*.

The mitochondrial transporter family has an additional complexity since isoforms differing in sequence have been detected for the ADP/ATP translocase. Two such isoforms, named T1 and T2, were first characterized in cows (Walker et al., 1987; Powell et al., 1989), and subsequently in humans (Neckleman et al., 1987; Houldsworth et al.,1988; Cozens et al., 1989), where a third expressed gene T3 and at least seven pseudo-genes were also found (Battini et al., 1987; Ku et al., 1990). In addition, three expressed genes (AAC1, AAC2 and AAC3) for the translocase are present in *Saccharomyces cerevisiae* (Lawson & Douglas, 1988; Kolarov et al., 1990) and two genes have been characterized in *Zea mays* (Bathgate et al., 1989). In contrast, it appears from Southern blots and from genomic DNA sequencing experiments that the UCP (Ricquier et al., 1991) and the phosphate and oxoglutarate/malate

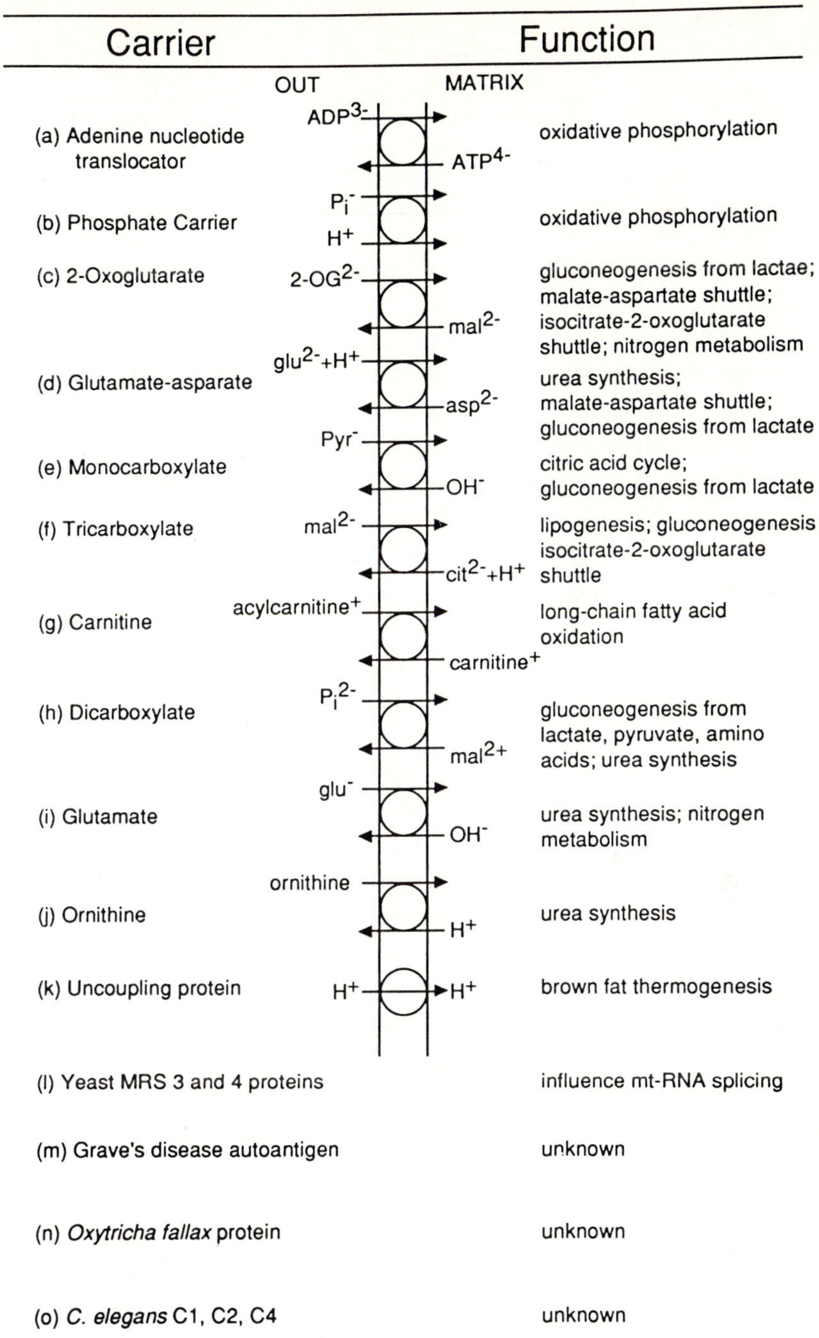

Figure 7. Carrrier proteins transporting metabolites across the inner membranes of mitochondria.

carriers (Runswick et al., 1990; Iacobazzi et al., 1992) are each encoded by a single gene in mammals.

The mammalian isoforms of the ADP/ATP translocase are expressed differently in various tissues. The T1 isoform appears to be generally expressed in all bovine tissues that have been examined, but predominantly in heart tissue (Powell et al., 1989). In human cell lines grown in tissue culture, T2 and T3 were expressed in equal amounts, but T1 expression was not observed (Lunardi & Attardi, 1991). During human muscle development, T2 was found at high levels in myoblasts and myotubes, but was decreased markedly in mature adult muscle; T3 was present in high amounts in myoblasts, was decreased in myotubes and was barely detectable in adult muscle; T1 expression was found in adult muscle but not in myoblasts or myotubes. These differences in the expression of the three human genes may reflect the adaptation of muscle cells to changing energetic requirements during their development (Lunardi et al, 1992). In the case of the yeast genes it was found by gene disruption that AAC1 was non-essential for growth on a respiratory carbon source. If AAC2 was disrupted the mutant could not grow on glycerol as a carbon source, but the deletion was compensated by over-expression of AAC1. Expression of AAC3 occurs under anaerobic conditons (Kolarov et al., 1990). The biochemical properties of the yeast AAC1 and AAC2 isoforms have been compared. The transport activity of AAC1 was 40% that of AAC2 (Gawaz et al., 1990).

So far, all members of the mitochondrial carrier family of characterised specificity are confined to mitochondria, but chloroplasts and bacteria also have a very active transport of metabolites across their membranes. The outer membrane of chloropasts is permeable to small molecules, but important metabolites are transported through the inner membrane by proteins. The major transport protein is the triose phosphate-3-phosphoglycerate / phosphate translocator, which provides the cell with fixed carbon in the form of triose phosphate. In its role of moving the main product of the organelle to the cytosol, the chloroplast phosphate transporter is analogous to the mitochondrial ADP/ATP translocase, and it also has a similar apparent molecular mass (29 kDa in spinach chloroplasts). However, its protein sequence does not contain the three-fold repeat that is characteristic of the mitochondrial carriers, and appears to be unrelated to the mitochondrial family (Flügge et al., 1989). Nor are any of the three types of bacterial transport proteins (Lengeler et al., 1990; Higgins et al, 1990; Henderson, 1991) related in sequence to the mitochondrial carriers

It has been pointed out by Maloney that many membrane transport proteins have two common characteristics: they operate as a dimer, and the dimer has usually 12 transmembrane α-helices. The monomers may be separate chains as in the mitochondrial carriers, or within the same chain, as in the bacterial sugar transporters, for example. Therefore, it is suggested that common structural and functional principles unify this wide range of transport proteins

(Maloney, 1990). The validity of this proposal will be assessed when more structural evidence becomes available.

Acknowledgements

We thank Drs. J. Sulston, A. R. Coulson and M. Craxton for supplying the nematode cDNA clones.

References

Aquila, H., Link, T. A. & Klingenberg, M. (1985). The uncoupling protein from brown fat mitochondria is related to the mitochondrial ADP/ATP carrier. Analysis of sequence homologies and of folding of the protein in the membrane. EMBO J. 4: 2369-2376.

Bathgate, B., Baker, A. & Leaver, C. J. (1989). Two genes encode the adenine nucleotide translocator of maize mitochondria. Eur. J. Biochem. 183: 303-310.

Battini, R., Ferrari, S., Kaczmarek, L., Calabretta, B., Chen, S. T. & Baserga, R. (1987). Molecular cloning of a cDNA for a human ADP/ATP carrier which is growth regulated. J. Biol. Chem. 262: 4355-4359.

Capobianco, L., Brandolin, G. & Palmieri, F. (1991). Transmembrane topography of the mitochondrial phosphate carrier explored by peptide specific antibodies and enzymic digestion. Biochemistry 30: 4963-4969.

Colleaux, L., Richard, G. F., Thierry, A. & Dujon, B. (1992). Sequence of a segment of yeast chromosome XI identifies a new mitochondrial carrier, a new member of the G protein family, and a protein with the PAAKK motif of the H1 histones. Yeast 8: 325-326.

Cozens, A. L., Runswick, M. J. & Walker, J. E. (1989). DNA sequences of two differently expressed nuclear genes for human mitochondrial ADP/ATP translocase. J. Mol. Biol. 206: 261-280.

Ferreira, G. C., Pratt, R. D. & Pedersen, P. L. (1990). Mitochondrial proton/phosphate transporter. An antibody directed against the COOH terminus and proteolytic cleavage experiments provides new insights about its membrane topology. J. Biol. Chem. 265: 21202-21206.

Fiermonte, G., Runswick, M. J., Walker, J. E. & Palmieri, F. (1992). Sequence and pattern of expression of a bovine homologue of a human mitochondrial transport protein associated with Graves' disease. DNA Sequence 3: 71-78.

Flügge, U. I., Fisher, K., Gross, A., Sebald, W., Lottspeich, F. & Eckershorn, C. (1989). The triose phosphate-3-phosphoglycerate-phosphate translocator from spinach chloroplasts: nucleotide sequence of a full length cDNA and import of the in vitro synthesized precursor protein into chloroplasts. EMBO J. 8: 39-46.

Gawaz, M., Douglas, M. G. & Klingenberg, M. (1990). Structure-function studies of adenine nucleotide transport in mitochondria. II. Biochemical analysis of distinct AAC1 and AAC2 proteins in yeast. J. Biol. Chem. 265: 14202-14208.

Henderson, P. J. F. (1991). Sugar transport proteins. Current Opinion in Struct. Biol. 1: 590-601.

Higgins, C. F., Hyde, S. C., Mimmack, M. M., Gileadi, U., Gill, D. R. & Gallagher, M. P. (1990). Binding protein dependent transport systems. J. Bioenerget. Biomembr. 22: 571-592.

Houldsworth, J. & Attardi, G. (1988). Two distinct genes for ADP/ATP translocase are expressed at the mRNA level in adult human liver. Proc. Natl. Acad. Sci. U.S.A. 85: 377-381.

Iacobazzi, V., Palmieri, F., Runswick, M. J. & Walker, J. E. Sequences of the human and bovine genes for the mitochondrial 2-oxoglutarate carrier. DNA Sequence 3: 79-88.

Kolarov, J., Kolarova, N. & Nelson, N. (1990). A third ADP/ATP translocator in yeast. J. Biol. Chem. 265: 12711-12716.

Krämer, R. & Palmieri, F. (1989). Molecular aspects of isolated and reconstituted carrier proteins from animal mitochondria. Biochim. Biophys. Acta 974: 1-23.

Krämer, R. & Palmieri, F. (1992). Metabolite carriers in mitochondria. In Molecular mechanisms in bioenergetics (Ernster, L., Ed.) In the press. Elsevier Science Publishers, Amsterdam.

Ku, D.-H., Kagan, J., Chen, S.-T., Chang, C.-D., Baserga, R. & Wurzell, J. (1990). The human fibroblast adenine nucleotide translocator gene. J. Biol. Chem. 265: 16060-16063.

LaNoue, K. F. & Schoolwerth, A. C. (1984). Metabolite transport in mammalian mitochondria. In Bioenergetics (Ernster, L., Ed.), pp. 221-268. Elsevier Science Publishers, Amsterdam.

Lawson, J. E. & Douglas, M. G. (1988). Separate genes encode functionally equivalent ADP/ATP carrier proteins in Saccharomyces cerevisiae. J. Biol. Chem. 263: 14812-14818.

Lengeler, J. W., Titgemayer, F., Vogler, A. P. & Wöhrl, B. M. (1990). Structures and homologies of the carbohydrate phosphotransferase system (PTS) proteins. Phil. Trans. Roy. Soc. B 326: 489-504.

Lunardi, J. & Attardi, G. (1991). Differential regulation of expression of the multiple ADP/ATP translocase genes in human cells. J. Biol. Chem. 266: 16534-16540.

Lunardi, J., Hurko, O., Engel, W. K. & Attardi, G. (1992). The multiple ADP/ATP translocase genes are differentially expressed during human muscle development. J. Biol. Chem. 267: 15267-15270.

Maloney, P. C. (1990). A consensus structure for membrane transport. Res. in Microbiol 141: 374-383.

Neckelmann, N., Li, K., Wade, R. P., Shuster, R. & Wallace, D. C. (1987). cDNA sequence of a human skeletal muscle ADP/ATP translocator: lack of a leader peptide, divergence from a fibroblast translocator cDNA, and co-evolution with mitochondrial DNA genes. Proc. Natl. Acad. Sci. U.S.A. 84: 7580-7584.

Powell, S. J., Medd, S. M., Runswick, M. J. & Walker, J. E. (1989). Two bovine genes for mitochondrial ADP/ATP translocase expressed differently in various tissues. Biochemistry 28: 866-873.

Ricquier, D., Casteilla, L. & Bouillard, F. (1991). Molecular studies of the uncoupling protein. FASEB J. 5, 2237-2242.

Runswick, M. J., Powell, S. J., Nyren, P. & Walker, J. E. (1987). Sequence of the bovine mitochondrial phosphate carrier protein: structural relationship to ADP/ATP translocase and the brown fat mitochondrial uncoupling protein. EMBO J. 6: 1367-1373.

Runswick, M. J., Walker, J. E., Bisaccia, F., Iacobazzi, V. & Palmieri, F. (1990). Sequence of the bovine 2-oxoglutarate / malate carrier protein: structural relationship to other mitochondrial transport proteins. Biochemistry 29: 11033-11040.

Saraste, M. & Walker, J. E. (1982). Internal sequence repeats and the path of polypeptide in mitochondrial ADP/ATP translocase. FEBS Lett. 144: 250-254.

Walker, J. E. (1992). The mitochondrial transporter family. Curr. Opinion Struct. Biol. 2: 519-526.

Walker, J. E., Cozens, A. L., Dyer, M. R., Fearnley, I. M., Powell, S. J. & Runswick, M. J. (1987). Studies of the genes for ATP synthases in eubacteria, chloroplasts and mitochondria: implications for structure and function of the enzyme. Chem. Scr. 27B: 97-105.

Waterston, R., Martin, C., Craxton, M., Huynh, C., Coulson, A., Hillier, L., Durbin, R., Green, P., Showkeen, R., Halloran, N., Metzstein, M., Hawkins, T., Wilson, R., Berks, M., Du, Z., Thomas, K., Thierry-Mieg, J. & Sulston, J. (1992). A survey of the expressed genes in *Caenorhabditis elegans*. Nature Genetics 1: 114-122.

Weisenberger, G., Link, T. A., von Ahsen, U., Waldherr, M. & Schweyen, R. J. (1991). MRS3 and MRS4, two suppressors of mtRNA splicing defects in yeast, are members of the mitochondrial carrier family. J. Mol. Biol. 217: 23-37.

Williams, K. R. & Herrick, G. (1991). Expression of the gene encoded by a family of macronuclear chromosomes generated by alternative DNA processing in *Oxytricha fallax*. Nucl. Acids Res. 19, 4717-4724.

Zarrilli, R., Oates, E. L., McBride, O. W., Lerman, M. I., Chan, J. Y., Santisteban, P., Ursini, M. V., Notkins, A. L. & Kohn, L. D. (1989). Sequence and chromosomal assignment of a novel cDNA identified by immunoscreening of a thyroid expression library: similarity to a family of mitochondrial solute carrier proteins. Molec. Endocrinol. 3: 1498-1508.

THE TRICARBOXYLATE CARRIER

A. Azzi, M. Glerum, R.Koller and W. Mertens
Institut für Biochemie und Molekularbiologie
Universität Bern
Bühlstrasse 28
3012 Bern
Switzerland

1. Summary

The citrate carrier has recently been purified from rat liver mitochondria by several groups using different methods. A 37-38 kDa protein has been prepared by Silica Gel 60 chromatography by our group (Claeys and Azzi, 1989; Glerum et al., 1990). The specific citrate transport activity of this preparation is not significantly different from that measured in mitochondria and it is inhibitable by 1,2,3- benzenetricarboxylic acid. Bisaccia et al. (1990) have reported the isolation of a 30 kDa protein by Celite 535 chromatography, and Kaplan et al. (1990) have isolated a 32.5 kDa protein by Matrex Orange, Matrex Blue and Affi-Gel chromatography. The clarification of the structural diversities of the above-described preparations has been one of the objectives of the present work. Peptide mapping, after cleavage of the different protein preparations with Endoproteinase Glu-C (*Staphylococcus aureus* V8 protease), cyanogen bromide and hydroxylamine, has failed to support any structural homologies. The 38 kD protein is N-terminally blocked. The peptides obtained from enodproteinase Glu-C or hydroxylamine lysis, after separation on SDS-PAGE, and those obtained after CNBr cleavage and HPLC purification have been partially sequenced. The protein sequence information is being used to obtain a full-length cDNA, which will allow further characterisation of the carrier at the molecular level.

2. Introduction

The discovery of the tricarboxylate carrier of liver mitochondria was made by Chappell and Haarhoff (1967), who also analysed the features required for transport of various substrates. Citrate, cis-aconitate, threo-D-isocitrate, and D- and L-tartrate (but not trans-aconitate) can cross the inner mitochondrial membrane. Malate, phosphoenolpyruvate and succinate, but not α-ketoglutarate and malonate, are also substrates for this carrier (Robinson et al., 1971). Inhibition of citrate transport was brought about by

NATO ASI Series, Vol. H 83
Molecular Biology of Mitochondrial Transport Systems
Edited by M. Forte and M. Colombini

1,2,3-benzenetricarboxylic acid (BTA) but not by its 1,2,4- and 1,3,5-isomers (Robinson et al., 1972). Butylmalonate and N-ethylmaleimide are much less potent inhibitors of the transport of dicarboxylic acids by the citrate carrier, than by the dicarboxylate carrier, suggesting that the di- and tricarboxylate carriers are different molecular entities (Palmieri et al., 1972).

The export of citrate from mitochondria, *via* the tricarboxylate carrier, is fundamental for the cytosolic biosynthesis of fatty acids and cholesterol. Inhibition of citrate transport by long-chain acyl-CoA esters may be significant in the control of fatty acid and cholesterol biosynthesis (Halperin et al., 1972). The role of the tricarboxylate carrier in the biosynthesis of cholesterol, especially in tumours, has been studied by Coleman and co-workers (1986) who have observed an activating effect of cholesterol on the carrier.

Gluconeogenesis and the shuttle of reducing equivalents also require an effective transport of citrate and phosphoenolpyruvate, via the tricarboxylate carrier, across the mitochondrial membrane (for a review, see Robinson, 1972).

The tricarboxylate carrier has been most widely studied in liver mitochondria, although it is also found in brain and kidney mitochondria, albeit with an apparently lower activity. There appears to be no carrier activity in heart mitochondria (Sluse et al., 1971).

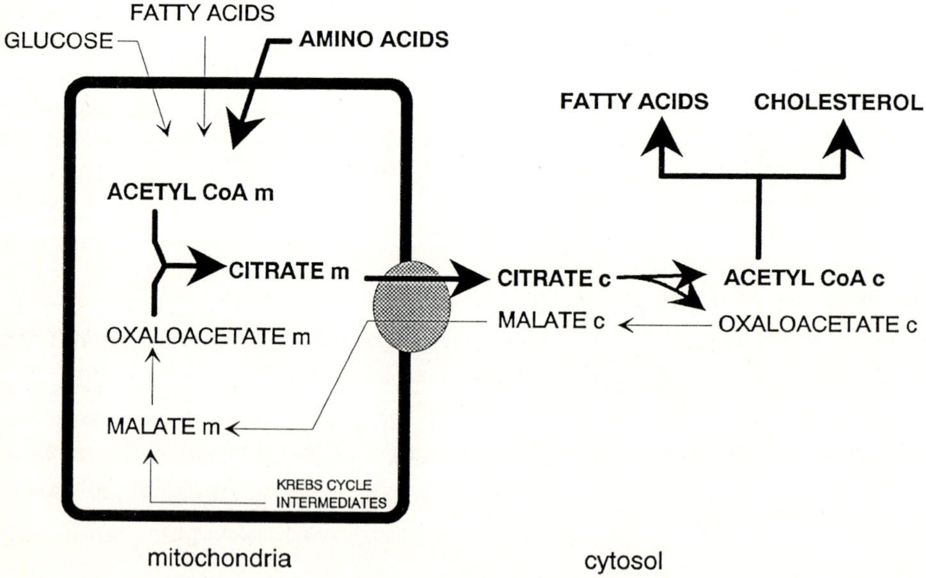

Fig. 1. The role of citrate in cell metabolism.

3. Methods

Solubilisation of Mitochondria: Rat liver mitochondria were prepared according to Fleischer et al. (1979), using leupeptin (5 mM) and phenylmethylsulfonyl fluoride (200 mM) as protease inhibitors in the initial steps of the isolation procedure. Solubilisation of rat liver mitochondria and hydroxylapatite chromatography were carried out as described previously (Claeys and Azzi, 1989). Mitochondria were pre-extracted in a 0.2% (mass/vol.) Triton X-100 buffer and, following centrifugation were subsequently solubilized in a 3% Triton X-114 buffer, containing 2 mg/ml cardiolipin (diphosphatidylglycerol). Mitoplasts were prepared as described by Ragan et al. (1987).
Hydroxylapatite and Silica Gel Chromatography: The mitochondrial extract was loaded onto 400 mg of cold, dry hydroxylapatite and eluted with a 2% Triton X-114 buffer. The hydroxylapatite eluate was adjusted to pH 8.0, and then loaded onto an equilibrated Silica gel 60 column. Prior to equilibration, the column was extensively washed with distilled water. Before eluting the bound carrier, the column was washed with 1 ml of equilibration buffer, 2 ml of a buffer containing 25 mM Mops pH 6,7, 5 mM citrate, 2.2 % Triton X-114, 2.5% ethylene glycol and then 1 ml of the same buffer with 6 mM Mops pH 6.7. The active tricarboxylate carrier was eluted with 1.1 ml of the 6 mM Mops buffer containing 0.5-0.7 mg/ml cardiolipin. Isocitrate (5 mM), instead of citrate, was employed when activity determinations were carried out using the coupled enzymatic assay. Reconstitution of the carrier was carried out as previously described (Glerum et al., 1990).
Measurements of citrate transport: The uptake of [14C]citrate was monitored as described previously (Claeys and Azzi, 1989), measuring the uptake of 0.2 mM [14C]citrate at 30° C for 10 min. The measurement of tricarboxylate carrier activity using the enzymatic assay was developed by Lüthy and Azzi (1989). In order to remove the isocitrate that was not trapped inside the liposomes, 700 ml proteoliposomes were loaded on a Dowex 1-X8 column. The eluted proteoliposomes (1 ml) were mixed with 170 mM sucrose, in a cuvette which also contained 100 mM $MnCl_2$, 50 mM NaCl, 6 mM Tris, 150 mM $NADP^+$ and 1.8 units of isocitrate dehydrogenase. Transport was initiated by the addition of 0.2 mM citrate (unless otherwise noted), and the formation of NADPH was measured spectrophotometrically at 340-400 nm, using an Aminco DW-2a UV/VIS spectrophotometer, as described previously (Glerum et al., 1990).
Hydroxylamine cleavage: Chemical cleavage with hydroxylamine was carried out as described by Saris et al. (Saris et al., 1983).
Endoproteinase Glu-C digestion: After Silica gel chromatography, the purified protein (50 µg) was precipitated with acetone, dried under nitrogen, redissolved in Laemmli buffer, and then applied to SDS-PAGE. The gel was stained with Coomassie Brilliant Blue B and a gel slice, containing the major protein band of apparent molecular mass of 38 kD, was excised and equilibrated for one hour with a buffer consisting of 125 mM Tris/HCl, 3 mM EDTA, 1.1% SDS, 1% glycerol, 0.3% 2-mercaptoethanol, pH 6.8.

Subsequently, the gel slice was placed in a well of a second polyacrylamide gel and overlaid with Laemmli buffer containing 100 ng of Endoproteinase Glu-C (Boehringer). Electrophoresis through the stacking gel was performed slowly to allow complete digestion. The separated peptides were electroblotted onto an Immobilon-P membrane and stained with Coomassie. The peptide bands were cut out and applied directly to an Applied Biosystems 477A sequencer for protein sequencing.

CNBr-cleavage: 200 mg of acetone-precipitated protein was redissolved in 100 µl 50% formic acid. The protein was further purified by reversed phase HPLC on a Nucleosil 300-5C_4 column (2 x 110 mm), using an isopropanol gradient from 0 to 30%, in 60% formic acid. The fractions containing the 38kD protein (as analysed by SDS-PAGE) were collected, pooled, and concentrated to a volume of 10 µl by vacuum centrifugation. Then, 100 µl of 150 mM CNBr in 70% formic acid were added and cleavage was carried out at 25°C in the dark. After 6 hours, the reaction mixture was concentrated to 10 µl and applied to RP HPLC on the same column, but with a 5 to 80% acetonitrile gradient in 0.1 % trifluoroacetic acid. The peptide-containing fractions were collected, concentrated, and directly applied to the protein sequencer.

Miscellaneous methods: The Silica eluate (3 ml) was precipitated with acetone (12-15 ml) at -20°C for 2h, and the pellet redissolved in 1 ml 1% SDS. Sodium dodecyl sulphate/polyacrylamide gel electrophoresis (SDS-PAGE) of acetone-precipitated samples was carried out as described by Bisaccia et al. [3]. Staining was performed with silver nitrate according to the BioRad method. Protein was determined using a modified Lowry method, with the inclusion of 1% SDS (Markwell et al., 1981).

4. Results and Discussion

Isolation and characterisation of the tricarboxylic acid carrier.

The purification method we reported (Claeys and Azzi, 1989; Glerum et al., 1990) results in a reconstitutively active tricarboxylate carrier, with either bovine or rat liver as starting tissue. As shown in Fig 2, the protein appears as a single band in SDS-PAGE.

The two preparations from rat and bovine liver differ by about 1 kDa in molecular mass. The higher apparent molecular mass of the protein from rat liver could represent a species-specific isoform of the tricarboxylate carrier. The localisation of the protein in the inner mitochondrial membrane was established by preparing it from mitoplasts, obtained by the digitonin procedure. As Fig. 2B shows, the protein, which migrates at 38 kDa, can be equally well purified from mitoplasts, and may not be ascribed to the outer membrane or to other organelles known to contaminate mitochondrial preparations.

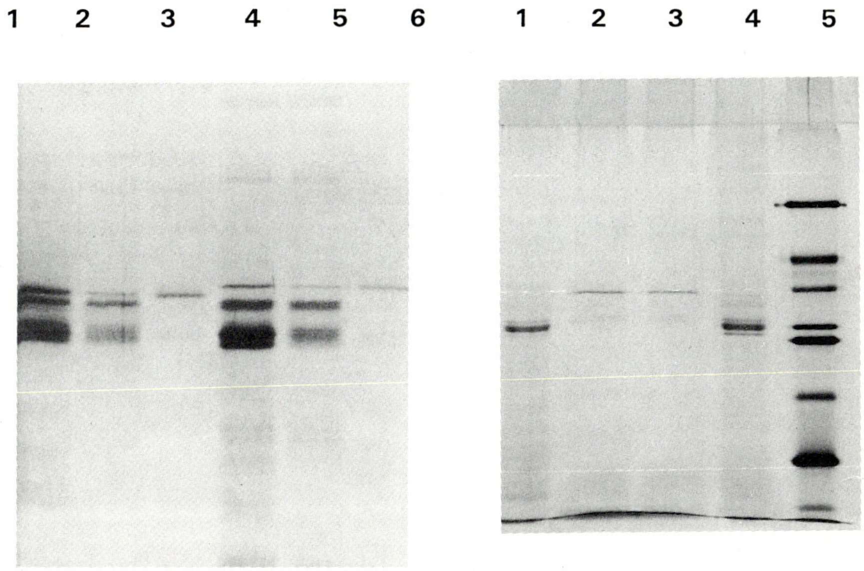

Fig. 2A- Polypeptide pattern of the different purification steps for the citrate carrier. SDS-PAGE and silver staining were carried out as described in Materials and Methods. Lanes 1 and 4, hydroxylapatite eluate; lanes 2 and 5 fraction not bound to Silica gel; lanes 3 and 6, Silica gel eluate. Lanes 1-3 represent the purification from bovine liver mitochondria, and lanes 4-6 from rat liver mitochondria.
Fig.2B The preparation of the tricarboxylic acid carrier from mitoplasts.Lane 1 and 2 correspond to mitoplasts HTP and Silica gel preparations, lane 3 and 4 correspond to mitochondria HTP and Silica gel preparations, lane 5 corresponds to Mw standards (from top 66, 45, 38, 29, 24, 20.1, 14.2 kDa, respectively).

In Fig. 3, the substrate transport specificity of the isolated bovine liver carrier is shown. The data indicate close similarity between the substrate specificity measured with the isolated carrier and with intact mitochondria. Fig. 4 shows the inhibition of the carrier at different 1,2,3-benzenetricarboxylic acid concentrations. Also in this case, the half-inhibitory concentration is similar to that obtained with intact liver mitochondria.

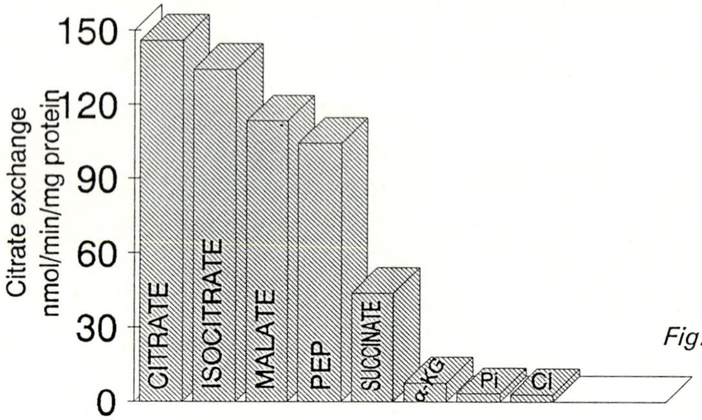

Fig. 3.

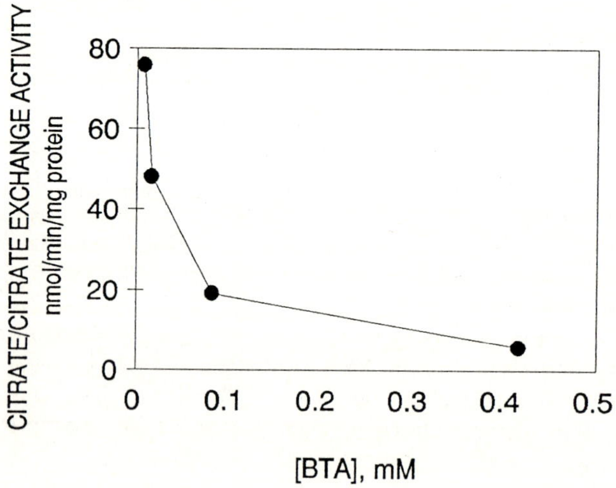

Fig. 4.

Fig. 3. The substrate transport specificity of the isolated tricarboxylate carrier. Activity measurements were performed as described in Methods.
Fig. 4. Benzenetricarboxylic acid inhibition of citrate transport by the isolated tricarboxylate carrier. Activity measurements were performed as described in Methods.

Comparison of three different methods of citrate carrier preparation

Two alternative methods to isolate the tricarboxylic acid carrier have been described. Bisaccia et al. (1990) reported the isolation of a 30 kDa protein by Celite 535 chromatography, and Kaplan et al. (1990) have isolated a 32.5 kDa protein by Matrex Orange, Matrex Blue and Affi-Gel chromatography. The differences in the three preparations are shown in Table 1. They have similar specific activities, and all of them are equally sensitive to 1,2,3 benzenetricarboxylic acid. The difference in molecular mass appears, however, to be more significant than the differences in inhibition, which may be caused by dissimilar amounts of detergent or phospholipid remaining associated with the protein after purification. In order to ascertain whether the lower molecular weight polypeptides from the Kaplan (Kaplan et al., 1990) and Palmieri (Bisaccia et al., 1990) laboratories originated from proteolytic digestion of a higher molecular weight precursor, such as the protein obtained in our laboratory, the sensitivity of the preparations to Endoproteinase Glu-C protease and to chemical splitting were analysed.

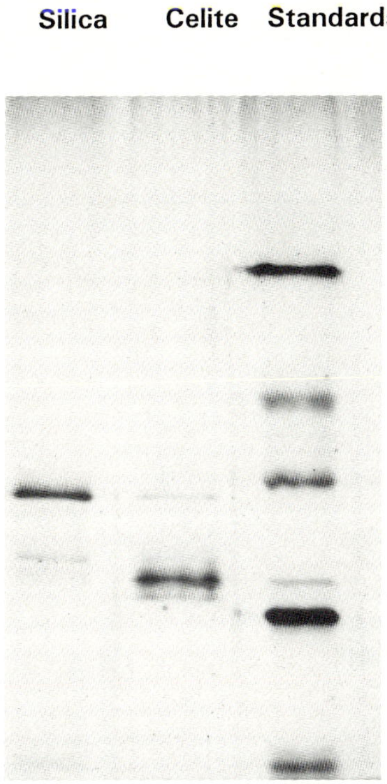

Fig. 5. Comparison between the Palmieri and the Silica gel preparations, as obtained in our laboratory, upon analysis by SDS-PAGE.

Fig. 5 shows the comparison between the Palmieri and the Silica gel preparations, as prepared in our laboratory, upon analysis by SDS-PAGE. The results suggest that the difference in molecular mass between the two preparations are not due to differences in detection methods, but are rather intrinsic properties of the polypeptides isolated.

Table I. Differences between three different preparations of the tricarboxylic acid carrier.

Laboratory	Azzi	Kaplan	Palmieri
Procedure after HTP	Silica gel	2 dye columns	celite
Detergent	Triton X 114	Triton X 114	Triton X 100
Mr	37,000	32,500	30,000
Activity (nmol/min/mg)	1 ,625	810	1 ,350
Inhibition by:			
Phenylsuccinate (2 mM)	29%	-5%	48%
Phenylglyoxal (2 mM)	59	100%	
NEM ** (0.2 mM)	26%	79%	1 6%
Cyano OH cinnamate (1 mM)	72%	-2%	

**low inhibition in mitochondria

1 2 3 4 5

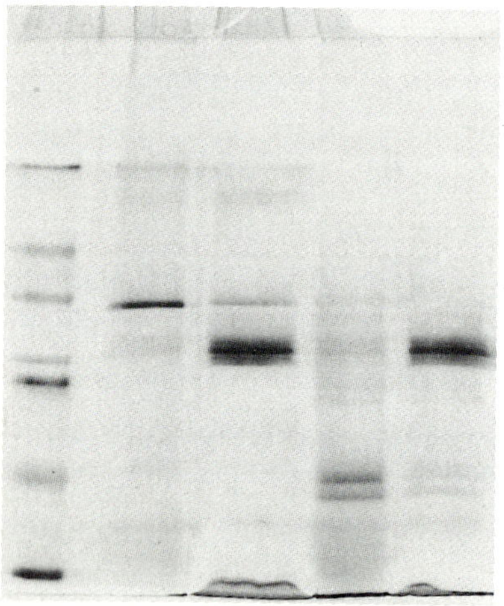

Fig. 6

1 2 3

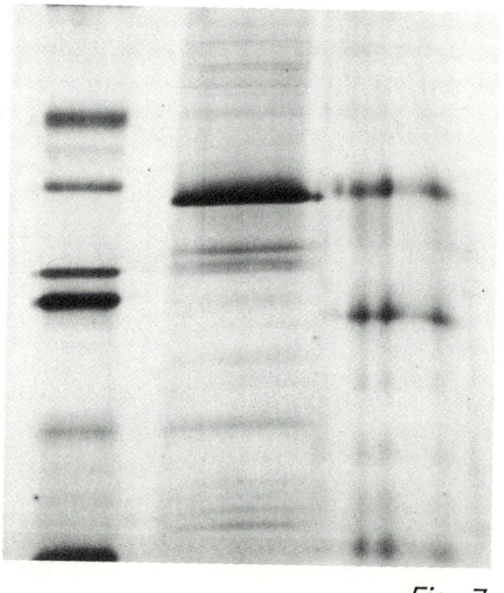

Fig. 7

Fig. 6 Splitting of the Silica gel preparation into two fragments by the Endoproteinase Glu-C. Lane 1 Mw standards, Lane 2 and 4 Silica gel preparation before and after Endoproteinase Glu-C splitting, respectively, Lane 3 and 5 Celite preparation before and after Endoproteinase Glu-C splitting, respectively.

Fig. 7.Cleavage of the Silica gel protein (but not of the celite preparation) with hydroxylamine produces two fragments. Lane 1 Mw standards, Lane 2 Silica gel preparation, Lane 3 Hydroxylamine fragments

Fig. 6 shows that the Silica gel preparation was split into two fragments by the Endoproteinase Glu-C, the peptides obtained were 18 and 20 kDa in molecular mass, respectively. The latter could be partially sequenced (see below) while the former was blocked at the amino terminus. The preparation of Palmieri's group (Bisaccia et al., 1990), in contrast to ours (Glerum et al., 1990), was not split by Endoproteinase. Cleavage of the Silica gel protein at asparagine-glycine bonds with hydroxylamine produces two fragments, of 14 and 24 kDa masses, respectively. The former fragment could again be partially sequenced (see below), while the latter was blocked at the amino terminus. Also in this case, no splitting was obtained with the celite preparation (Fig. 7). Finally, CNBr fragmentation produced a large number of peptides of small molecular weight from the Silica preparation, and a major fragment of molecular mass 24 kDa with the celite preparation.
In conclusion, the two preparations from Celite and Silica gel are structurally different, and the former cannot be considered a proteolytic degradation product of the latter. On the basis of the similarities pointed out by Kaplan et al. (1990), between their preparation and that of Palmieri's group (Bisaccia et al., 1990), it can be inferred that Kaplan's preparation is likewise not a degradation product of the preparation obtained in our laboratory.

Polypeptide sequence analysis of the tricarboxylic acid carrier

The fragmentation with Endoproteinase Glu-C, hydroxylamine and CNBr described above, to test the lack of similarity between the different citrate carrier preparations, has yielded peptides which were microsequenced and utilised for constructing DNA probes for screening a cDNA library, in order to obtain a clone for the tricarboxylate carrier. The sequences obtained are reported in Table II.

Table II. Amino acid sequences of the peptides from the citrate carrier

CNBr fragment	SQREL?VGIP VTDEN**GTRLG E**.......
Hydroxylamin fragment	..................................**GTRLG** EGTNAAKQAI..
V8 protease fragment	..LGTAYVSA TTGAVATALG ?NA?T
CNBr fragment	?VTSLE DDLQASIQ?S ?PELR?VYF

5. Conclusions

The preparation of the tricarboxylic acid carrier we have discussed above has the following features. It can be purified from mitoplasts, and in a liposome-reconstituted state, the preparation catalyses citrate transport which is inhibited by benzenetricarboxylic acid. The protein has a molecular mass of 38 kDa and is substantially different from lower molecular mass proteins obtained in other laboratories. These cannot be considered proteolytic degradation products of the 38 kDa protein. Protein sequence information is now being used to obtain a full-length cDNA clone, which allow a more detailed characterisation of the carrier at the molecular level.

6. Acknowledgements

The Swiss National Foundation is acknowledged for financial support.

7. References.

Bisaccia, F., De Palma, A., Prezioso, G. and Palmieri, F. (1990) Kinetic characterisation of the reconstituted tricarboxylate carrier from rat liver mitochondria. Biochim. Biophys. Acta 1019, 250-256.

Claeys, D. and Azzi, A. (1989). Tricarboxylate carrier of bovine liver mitochondria. Purification and reconstitution. J. Biol. Chem. 264, 14627-14630.

Chappell, J.B. and Haarhoff, K.N. (1967) The penetration of the mitochondrial membrane by anions and cations. In: Biochemistry of Mitochondria, (Slater, E.C., et al., eds) Academic Press, London. pp. 75-91.

Coleman, P.S. (1986) Membrane cholesterol and tumour bioenergetics. Ann. N.Y. Acad. Sci. 488, 451-467.

Fleischer, S., McIntyre, J.O. and Vidal, J.C. (1979) Large scale preparation of rat liver mitochondria in high yield. Methods Enzymol. 55, 32-39.

Glerum, D.M., Claeys, D., Mertens, W. and Azzi, A. (1990). The tricarboxylate carrier from rat liver mitochondria. Purification, reconstitution and kinetic characterisation. Eur. J. Biochem. 194, 681-684.

Halperin, M.L., Robinson, B.H. and Fritz, I.B. (1972) Effects of palmitoyl CoA on citrate and malate transport by rat liver mitochondria. Proc. Natl.

Acad. Sci. USA 69, 1003-1007.

Kaplan, R.S., Mayor, J.A., Johnston, N. and Oliveira, D.L. (1990) Purification and characterisation of the reconstitutively active tricarboxylate transporter from rat liver mitochondria. J. Biol. Chem. 265, 13379-13385.

Lüthy, R. and Azzi, A. (1989) Novel reconstitution and enzymatic assay of the mitochondrial tricarboxylate carrier. Anal. Biochem. 177, 323-326.

Markwell, M.A.K., Haas, S.M., Tolbert, N.E. and Bieber, L.L. (1981) Protein determination in membrane and lipoprotein samples: manual and automated procedures. Methods Enzymol. 72, 296-303.

Palmieri, F., Stipani, I., Quagliariello, E. and Klingenberg, M. (1972) Kinetic studies of the tricarboxylate carrier in rat liver mitochondria. Eur. J. Biochem. 26, 587-594.

Ragan, C.I., Wilson, M.T., Darley-Usmar, V.M. and Lowe, P.N. (1987) Sub-fractionation of mitochondria and isolation of the proteins of oxidative phosphorylation. In: Mitochondria, A practical approach. IRL Press, Oxford. pp. 79-112.

Robinson, B.H., Williams, G.R., Halperin, M.L. (1971) Factors affecting the kinetics and equilibrium of exchange reactions of the citrate transporting system of rat liver mitochondria. J. Biol. Chem. 246, 5280-5286.

Robinson, B.H., Williams, G.R., Halperin, M.L., Leznoff, C.C. (1972) Inhibitors of the dicarboxylate and tricarboxylate transporting systems of rat liver mitochondria. J. Membr. Biol. 7, 391-401.

Robinson, B.H. (1972) The role of mitochondrial tricarboxylate anion transport in metabolism. In: Rate control of biological processes., pp. 195-214.

Saris, C.J.M., Van Eenbergen, J., Jenks, B.G. and Bloemers, H.P.J. (1983) Hydroxylamine cleavage of proteins in polyacrylamide gels. Anal. Biochem. 132, 54-67.

Sluse, F.E., Meijer, A.J. and Tager, J.M. (1971) Anion translocators in rat heart mitochondria. FEBS Lett. 18, 149-153.

THE MITOCHONDRIAL PYRUVATE CARRIER: THE MECHANISM OF SUBSTRATE BINDING

Katarzyna A. Nałęcz
Nencki Institute of Experimental Biology
Pasteur str.3
02-093 Warsaw
Poland

Pyruvate, a compound connecting the metabolic pathways in catabolism and anabolism of sugars, fatty acids and amino acids, in most cases must enter mitochondria where its further metabolism takes place. Pyruvate transport, catalysed by a specific carrier (Papa *et al.*, 1971), occurs either due to a $\triangle$pH dependent process (Papa & Paradies, 1974) or in an exchange with other 2-oxo- or halogenated monocarboxylates (Paradies & Papa, 1977), which can have great significance in the removal of ketone bodies *in vivo*.

Identification and purification of the pyruvate carrier

The mitochondrial transport systems have been identified and purified to varying extent (for review see Azzi *et al.*, 1989; Krämer & Palmieri, 1989; Palmieri *et al.*, 1990). As in the case of other carriers, extraction of the pyruvate transporter was performed in the presence of a detergent (Triton X-100 or X-114) (Nałęcz *et al.*, 1986a & 1986b) followed

Abbreviations: DEPC, diethylpyrocarbonate; DIDS, 4,4'-diisothiocyano-2,2'-stilbene disulfonate; DTE, dithioerythritol; HTP, hydroxylapatite; NEM, N-ethylmaleimide; pCMB, p-hydroxymercuribenzoate (p-chloromercuribenzoic acid)

NATO ASI Series, Vol. H 83
Molecular Biology of Mitochondrial Transport Systems
Edited by M. Forte and M. Colombini

by further exclusion chromatography on HTP. The final step in isolation of the carrier was based on the reported inhibition of pyruvate transport by 2-cyanocinnamate and its analogues (Halestrap, 1975). An affinity resin has been synthesized by coupling 2-cyano-4-hydroxycinnamate, through the diazo bond, to a 16-carbon spacer attached to Sepharose 4B (Bolli *et al.*, 1989; Nałęcz *et al.*, 1989).

Elution with 20 mM ligand resulted in purification of two polypeptides of 31.5 and 34 kDa (Bolli *et al.*, 1989; Nałęcz *et al.*, 1989), which have been separated by variation of pH values and addition of salt: 0.2 M NaCl at pH 6.2 eluted a polypeptide with M_r 34,000 and pyruvate carrier activity, while at pH 8, the other protein, with M_r of 31,500 and activity of the 2-oxoglutarate carrier, was released from the affinity resin (Bolli *et al.*, 1989; Nałęcz *et al.*, 1989). The fact that the latter protein was capable of interacting with the immobilized 2-cyano-4-hydroxycinnamate implies that the amount (35-50 pmoles/mg protein) of pyruvate carrier in mitochondria (Paradies, 1984; Paradies & Ruggiero, 1986) inferred from titration with 2-cyano[^{14}C]cinnamate may have been overestimated.

In the case of higher eukaryotes the pyruvate carrier isolated from different organisms and organs, as demonstrated for rat liver, rat brain and bovine heart mitochondria, had the same molecular mass, although differences in abundance of this protein were observed, with a relatively high content of the pyruvate carrier in brain (Nałęcz *et al.*, 1992).

Of proteins extracted with Triton X-114 from mitochondria of wild type *S. cerevisiae* and subjected to affinity chromatography with immobilized 2-cyano-4-hydroxycinnamate, three polypeptides were isolated, different than for higher eukaryotes (Nałęcz *et al.*, 1991). One of these, of apparent molecular mass of 30 kDa, proved to be copurified porin from the outer mitochondrial membrane, as it disappeared in a preparation obtained from the porin-deficient mutant strain HR-2A$^-$ (Nałęcz *et al.*, 1991).

Activity of the pyruvate carrier in the reconstituted system

Incorporation of isolated transporting protein into an artificial membrane, a necessary prerequisite for measurements of its transport activity, in the case of the pyruvate carrier, was achieved either by a freeze-thaw-sonication procedure (Nałęcz *et al.*, 1986a) or by the use of hydrophobic beads (Bolli *et al.*, 1989; Nałęcz *et al.*, 1989; 1991; 1992; Nałęcz & Zambrowicz, 1990), according to a modified procedure originally described by Krämer and Heberger (Krämer & Heberger, 1986). The latter technique, based on detergent removal, allows formation of a △pH in proteo- and liposomes (Nałęcz *et al.*, 1990; 1992) after a pulse, external addition of acid. Although it was postulated by Bakker and Van Dam, 1974, that the accumulation of pyruvate in liposomes could be attributed to diffusion, the △pH-driven pyruvate uptake into phosphatidyl choline vesicles was higher in proteoliposomes than in liposomes, especially at short incubation times (Nałęcz *et al.*, 1992). Moreover, the activity of pyruvate carrier isolated from various tissues was different, for instance twice as high in the case of the protein isolated from heart, as compared to that from liver mitochondria.

The same protein was capable of taking up 2-oxoisocaproate into vesicles in a △pH-dependent way. This branched-chain 2-oxoacid inhibited pyruvate uptake in a competitive way with a K_i of 1.7 ± 1.1 mM, the value comparable with the one reported for intact mitochondria (Paradies & Papa, 1977), whilst 2-oxoisovalerate and 2-oxo-3-methylvalerate were inhibitory with K_i values of 10.5 ± 0.4 mM and 7.0 ± 1.1 mM, respectively. The tissue differences described above, as well as the sensitivity of pyruvate accumulation towards other 2-oxoacids and towards reagents affecting SH groups (Table 1, Nałęcz *et al.*, 1990; 1992) pointed to the specificity of this transport. However, the observations that pyruvate uptake was such a fast and transient process led to the exchange reaction as the method of choice for the pyruvate transport activity assay.

In proteoliposomes, the pyruvate carrier was capable of exchanging pyruvate with pyruvate itself, with acetoacetate (Nałęcz *et al.*, 1992) and with branched-chain 2-oxoacids (Nałęcz *et al.*, 1991; 1992), but not with 2-oxoglutarate (Nałęcz *et al.*, 1992). The exchange reaction was not affected by anion substrates of other inner mitochondrial carriers, namely malate and citrate (Nałęcz *et al.*, 1992). In order to determine properties of the pyruvate carrier in an exchange reaction, the initial velocity was measured as a function of substrate concentration in both internal and external compartments. Fig. 1 presents the results of two-reactant initial velocity studies of the pyruvate homoexchange when the internal and external substrate concentrations varied from 20-50 mM and 0.1-1 mM, respectively. Neither the slopes in the Lineweaver-Burk were found to be parallel, nor was the secondary plot found to be horizontal (Fig. 1B), which led to the exclusion of the possibility of a ping-pong

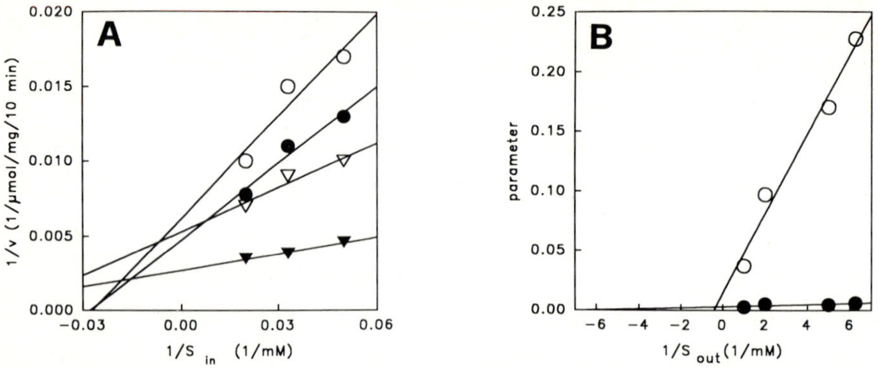

Fig.1.Two-substrate analysis of the pyruvate/pyruvate homologous exchange reactions catalysed by the reconstituted pyruvate carrier. A. Lineweaver-Burk plots presenting the dependence of exchange velocity on internal pyruvate at (○) 0.16 mM, (●) 0.20 mM, (▽) 0.50 mM and (▼) 1.0 mM external substrate. B. Secondary plot replotting slopes (○) and intercepts (●) from plot A, as a function of external substrate.

mechanism. These results would imply that the pyruvate exchange occurs by a sequential mechanism and that the ternary complex has to be formed during this reaction. Similar analyses of the catalytic cycle were reported for the reconstituted aspartate/glutamate (Dierks *et al.*, 1988), dicarboxylate (Indiveri *et al.*, 1989) and 2-oxoglutarate (Indiveri *et al.* 1991) carriers, pointing to a sequential type of mechanism as a possible general intrinsic mechanism of transport through carriers of the inner mitochondrial membrane. A bisubstrate analysis of the pyruvate exchange reaction yielded the following kinetic data: the V_{max} value of the exchange reaction was 33 ± 9 μmol/mg protein/min, whilst K_m values were drastically different for the external (40 ± 2 nM) and the internal (150 ± 10 μM) side of proteoliposomes, indicating an assymmetric insertion of the pyruvate carrier into phosphatidylcholine vesicles.

Functional groups involved in the pyruvate transport

Sensitivity of pyruvate transport towards 2-cyanocinnamate and its derivatives

(Halestrap, 1975), as well as inhibition of translocation of this compound through the inner mitochondrial membrane by several SH group specific reagents implied the involvement of sulphydryl group(s) in binding of the substrate. Although initial attempts at labeling of the isolated protein with eosin-5-maleimide were unsuccessful, a successful reaction was achieved after reduction with tributylphosphine, indicating that SH groups of the isolated pyruvate carrier were all in an oxidized state, and that an even number of SH groups was present in this protein (Nałęcz *et al.*, 1990). Some of the SH groups must change their respective positions upon binding of the substrate, since phenylarsine oxide, a compound reacting with two SH groups in close vicinity, had a much more pronounced inhibitory effect (over 60%) after preincubation with pyruvate (Nałęcz *et al.*, 1990). As presented in Table 1, the uptake of pyruvate was inhibited by either permeable or nonpermeable SH group reagents, whilst the exchange reaction was not affected by even relatively high concentration of NEM, a compound penetrating the phospholipid bilayer. Since the same diverse effect was reported for the pyruvate transport in intact mitochondria (Papa & Paradies, 1974), it would suggest that the sidedness of the incorporated carrier resembled the one occuring *in vivo*.

Even after a prolonged treatment leading to SH group blockage, a full inhibition of

Table 1. Effect of SH group reagents on the activity of the pyruvate carrier. Pyruvate transporting activity was measured as given in Nałęcz *et al.*, 1991 & 1992, after 3 min preincubation with pCMB, NEM or 2-cyano-4-hydroxycinnamate at the indicated concentrations.

Addition	Pyruvate transport (% of control)	
	Uptake (30 s)	Exchange (2 min)
None	100	100
pCMB (0.2 mM)	53	46
NEM (2 mM)	59	93
2-Cyano-4-hydroxy-cinnamate (1 mM)	51	52

pyruvate transport was not achieved. Since a mercurial-induced uniport was reported for other carriers, namely for aspartate/glutamate (Dierks *et al.*, 1990a & 1990b), ADP/ATP (Dierks *et al.*, 1990a) and phosphate (Stappen & Krämer, 1992) transporting proteins, the possibility of a ''leak'' through the pyruvate carrier was considered. As was shown elsewhere (Nałęcz *et al.*, 1992), accumulation of radioactive acetate occuring to the same extent in vesicles without and with reconstituted pyruvate carrier, was only affected by 2-cyano-4-hydroxycinnamate in proteoliposomes, suggesting an opening of the channel in the protein

Table 2. Effect of compounds reacting with different functional groups on the pyruvate uptake by proteoliposomes with the reconstituted pyruvate carrier. Proteoliposomes after reconstitution of the pyruvater carrier were preincubated for 30 min at 37°C in the dark with indicated compound, which was removed by passage of vesicles through Sephadex G-25 (medium) columns equilibrated with 170 mM sucrose. Transport reaction was measured as described by Nałęcz *et al.*, 1990.

Addition	Concentration (mM)	Activity μmol/mg protein/30s	%
None	-	23.1 ± 2.5	100
DIDS	0.02 + 0.06	9.0 ± 0.7	39
Pyridoxal-phosphate	0. 5	21.8 ± 3.3	95
	2.0	21.7 ± 2.6	94
	5.0	14.1 ± 2.1	62
Phenylglyoxal	1.0	20.0 ± 2.2	87
	3.0	11.8 ± 1.4	51
	10.0	9.0 ± 1.1	39
	20.0	3.6 ± 0.8	16
4-Hydroxy-phenylglyoxal	5.0	6.5 ± 0.7	28
	10.0	2.2 ± 0.8	6
	20.0	1.0 ± 0.1	4
4-Nitro-phenylglyoxal	5.0	4.2 ± 0.7	18
	10.0	2.2 ± 0.3	10
	20.0	1.0 ± 0.2	4

moiety. These observations can lead to conclusions that the redox state of SH groups in the inner mitochondrial membrane carrier proteins could be an important regulatory factor in their function in general.

Halestrap, in 1978, proposed a scheme for the pyruvate carrier mechanism, in which the active site contained not only SH group(s) involved in a thiol-addition reaction. Also, due to pH dependence of pyruvate transport, involvement of a positively charged group of pK_a of 8.35 was postulated. Therefore, an effect of compounds reacting with free amino groups and lysyl, histydyl and guanidyl residues was explored in more detail here.

As presented in Table 2, DIDS, the nonpermeable compound reacting with free amino groups, affected the uptake of pyruvate very strongly, whilst the effect of pyridoxal phosphate was visible only at high concentrations of this compound. Due to the fact that in uptake measurements it is difficult to distinguish an effect on ΔpH or on the protein moiety itself, the same compounds were applied in the exchange reaction assay. As shown in Fig. 2, none of the compounds tested affected the pyruvate/pyruvate exchange, pointing to the fact that a free amino group and/or lysyl residue were not involved in binding of the substrate, or they were protected by the presence of pyruvate. It should be emphasized that neither DIDS nor pyridoxal phosphate changed the permeability of the membrane, since accumulation of radioactive pyruvate was not changed in the presence of these compounds, with 2-oxoglutarate used as the internal substrate. As presented in Table 2, the guanidyl-specific reagent -phenylglyoxal (Takahashi, 1968), and its more hydrophilic derivatives, strongly inhibited the pyruvate uptake. However, chemical similarity of these compounds to protonophores suggested the need for checking their effect on the permeability of liposomal and proteoliposomal membranes. An enzymatic assay with a ß-hydroxybutyrate coupled reaction (Capuano *et al.*, 1990; Nałęcz *et al.*, 1991 & 1992) allowed monitoring of an initial leak, as well as efflux of substrate from vesicles. As is shown in Table 3, the $pyruvate_{out}/acetoacetate_{in}$ exchange was strongly inhibited by phenylglyoxal, whilst high efflux of acetoacetate was still observed after treatment with 4-nitro- and 4-hydroxy- derivatives (86 and 52% of the control efflux, respectively). Both more hydrophilic compounds promoted a high (about 30%) substrate leak from proteo- and liposomes (Nałęcz *et al.*, 1992), pointing to changed permeability of the lipid phase as an explanation of the observed effect, rather than the carrier inhibition. On the other hand, a strong inhibition of acetoacetate/pyruvate exchange by phenylglyoxal indicates an involvement of arginine at the active site of the pyruvate carrier.

Table 3. Effect of arginine specific reagents on acetoacetate efflux from vesicles. Vesicles were incubated without any additions or with 5 mM of compounds indicated, at 37°C for 1 h. After removal of the reagents, as described in the legend to Table 2, the exchange reaction was measured with the use of an enzymatic assay (Nałęcz *et al.*, 1992). The values of leak were estimated in liposomes. The amount of lipid was the same in both kind of vesicles, the value of efflux is recalculated per mg protein in proteoliposomes.

Compound	Acetoacetate efflux (μmol/min) Proteoliposomes	Liposomes
None	119	16
Phenylglyoxal	16	16
4-Nitrophenylglyoxal	98	41
4-Hydroxyphenylglyoxal	59	37

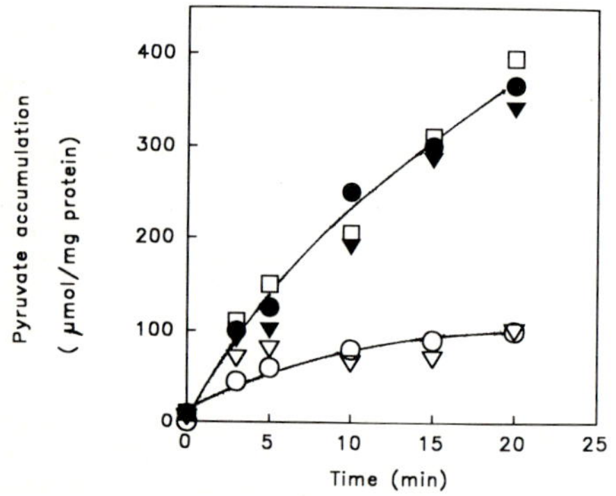

Fig. 2. Effect of compounds reacting with amino groups on the pyruvate exchange. Proteoliposomes with the pyruvate carrier were incubated for 30 min at 37°C in the dark without any additions (circles), with 5 mM pyridoxal phosphate (triangles) or 60 μM DIDS (squares). The substrate entrapped inside the vesicles was either 50 mM pyruvate (●,▼,□) or 50 mM 2-oxo-glutarate (○ , ▽). The excess of each reagent was removed by passing through Sephadex G-50 equilibrated with 170 mM sucrose and the exchange reaction was measured by accumulation of [1-^{14}C]pyruvate (Nałęcz *et al.*, 1992).

One of the putative positively charged groups at the active centre of the pyruvate carrier could be an imidazole ring of histidine. Although DEPC, a histydyl reagent (Miles, 1977), inhibited the pyruvate carrier activity when added to the assay (Fig. 3A), the ability of this compound to react with cysteine was taken into consideration. SH groups of the pyruvate carrier were blocked with pCMB as described by Kassab *et al.*, 1968, and an attempt was undertaken to carbethoxylate histydyl chains. After reactivation of SH groups with DTE the activity of pyruvate carrier was not affected, implying that histidine was

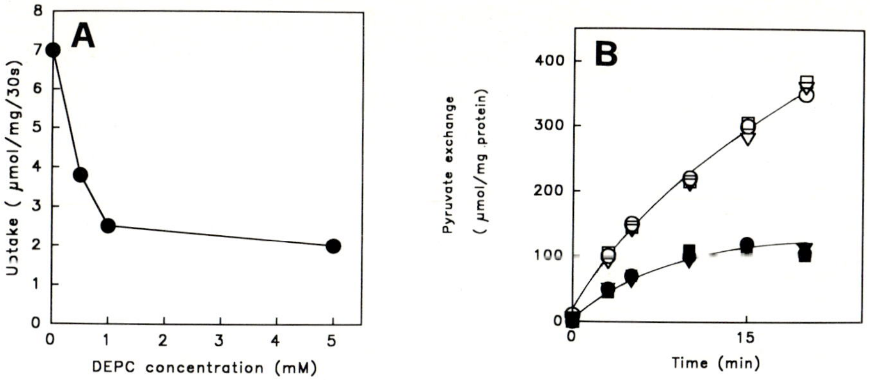

Fig. 3. Effect of DEPC on the pyruvate carrier activity. A. Effect of DEPC on the pyruvate uptake. DEPC at indicated concentration was added 3 min before the uptake measurements, as described by Nałęcz *et al.*, 1990. B. Effect of DEPC on pyruvate exchange. The pyruvate carrier was either incubated without any treatment (circles), or in the presence of 0.2 mM pCMB (triangles) or with 0.2 mM pCMB and 1 mM DEPC (squares), for 1 h at 4°C. After reactivation with 25 mM DTE the carrier protein was subjected to reconstitution with either 50 mm pyruvate (open symbols) or 2-oxoglutarate (closed symbols) and the exchange was measured, as given by Nałęcz *et al.*, 1992.

neither directly involved in substrate binding, nor was it a part of the active site.

In summary, it can be concluded from the labeling and chemical modification experiments with the pyruvate carrier, that this protein needs free SH group(s) and an arginyl residue for its activity and/or substrate binding, whilst such a role for histidine and, most probably for lysine, has been excluded.

A hypothetical mechanism of pyruvate translocation

An involvement of arginine in the activity of the pyruvate carrier led to analysis and comparison of the guanidyl group and pyruvate itself. A crystal structure of methylguanidinium orthophosphate allowed calculation of the C-N distances in the molecule (1.319, 1.326Å), as well as the corresponding angles, which have been almost the same in the strictly planar methylguanidyl group (Cotton *et al.*, 1973). This very rigid structure can interact not only with oxygens of orthophosphate (O_2-P-O_4 angle of 111° or 113°), but also with oxygens of the carboxyl group of pyruvate (126.3°) (Tavale *et al.*, 1961). Moreover, the pyruvate molecule can rotate around the central C_1-C_2 bond (Tavale *et al.*, 1961) which could enable movement of this molecule around a guanidyl group and facilitate transport inside the carrier protein. Studies on a model reaction betwen L-cysteine and pyruvate demonstrated its strong pH dependence, synthesis of 2-methyl-2,4-thiazolidinedicarboxylic acid was reported to have a pH optimum at pH 7-8, whilst degradation of this compound was facilitated by a pH shift towards higher values (8.5-9) (Włodek & Czubak, 1984). These kind of interactions can explain pyruvate uptake as a consequence of following events:

- binding of the substrate outside proteoliposomes or at the outer side of inner mitochondrial membrane, facilitated by a positively charged guanidyl group of arginine and lower pH, promoting an interaction of pyruvate with cysteine;
- movement of the substrate around a guanidyl residue due to the rotation of the pyruvate molecule, followed by interaction with another SH group within the carrier molecule;
- breakage of semimercaptal bond at high pH inside the vesicles or mitochondria.

It seems , however, to be more speculative to explain how the process of exchange could function. For many transporting proteins a dimeric structure has been postulated for the functional unit (Klingenberg, 1981). Such a claim can also be true for the pyruvate carrier, since preincubation of the reconstituted protein with its substrate increased sensitivity towards phenylarsine oxide, suggesting a change in the vicinity of two SH groups, most probably due to a change of conformation or aggregation state. Moreover, when the pyruvate carrier was extracted from proteoliposomes treated with either pyruvate or pyruvate and

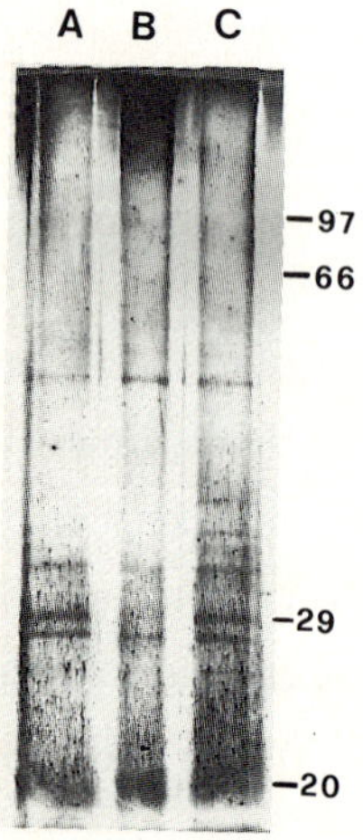

Fig. 4. Sodium dodecyl sulfate polyacrylamide gel electrophoresis pattern representing the pyruvate carriers extracted from proteoliposomes treated in a different way. A. No treatment, B. Preincubation with pyruvate, C. Incubation with pyruvate and next with phenylarsine oxide. The protein was extracted from proteoliposomes with 2% Triton X-100 and after precipitation with precooled acetone subjected to electrophoresis without 2-mercaptoethanol, as described by Nałęcz *et al.*, 1990.

phenylarsine oxide, an additional band of M_r 59,000 on polyacrylamide gel was visible when electrophoresis was performed in the absence of 2-mercaptoethanol (Fig. 4). The putative mechanism of pyruvate exchange could thus be presented in the following way:

- binding of the substrate moves at least one SH group per monomer of the carrier molecule in such a way, that it can interact with an SH group of another monomer; an S-S bridge between them is formed, giving a dimer;
- this conformational change allows exposure of another internal SH group towards a more hydrophilic environment, the group becomes more accessible for the internal substrate and less accessible for NEM, which is then unable to inhibit the process of exchange;
- when SH groups are blocked, binding of the substrate or SH group modifying reagent opens a channel in the carrier protein; an unspecific leak occurs.

References:

Azzi A, Nałęcz KA, Nałęcz MJ, Wojtczak L (eds) (1989) Anion carriers of mitochondrial membranes. Springer Berlin Heidelberg New York

Bakker EP, Van Dam K (1974) The movement of monocarboxylic acids across phospholipid membrane: Evidence for an exchange diffusion between pyruvate and other monocarboxylate ions. Biochim Biophys Acta 339:285-289

Bolli R, Nałęcz KA, Azzi A (1989) Monocarboxylate and alpha-ketoglutarate carriers from bovine heart mitochondria. Purification by affinity chromatography on immobilized 2-cyano-4-hydroxycinnamate. J Biol Chem 264:18024-18030

Cotton FA, Day VW, Hazen EE, Larsen S (1973) Structure of methylguanidinium dihydrogenorthophosphate. A model compound for arginine-phosphate hydrogen bonding. J Am Chem Soc 95:4834-4840

Capuano F, Di Paola M, Azzi A, Papa S (1990) The monocarboxylate carrier from rat liver mitochondria. Purification and kinetic characterization in a reconstituted system. FEBS Lett 261:39-42

Dierks T, Riemer E, Krämer R (1988) Reaction mechanism of the reconstituted aspartate/glutamate carrier from bovine heart mitochondria. Biochim Biophys Acta 943:231-244

Dierks T, Salentin A, Heberger C, Krämer R (1990a) The mitochondrial aspartate/glutamate and ADP/ATP carrier switch from obligate counterexchange to unidirectional transport after modification by SH-reagent. Biochim Biophys Acta 1028:268-280

Dierks T, Salentin A, Krämer R (1990b) Pore-like and carrier-like properties of the mitochondrial aspartate/glutamate carrier after modification by SH-reagents:evidence for a preformed channel as a structural requirement of carrier-mediated transport.

Biochim Biophys Acta 1028:281-288
Halestrap AP (1975) The mitochondrial pyruvate carrier. Kinetics and specificity for substrates and inhibitors. Biochem J 148:85-96
Halestrap AP (1978) Pyruvate and ketone-body transport across the mitochondrial membrane. Biochem J 172:377-787
Indiveri C, Dierks T, Krämer R, Palmieri F (1989) Kinetic discrimination of two substrate binding sites of the reconstituted dicarboxylate carrier from rat liver mitochondria. Biochim Biophys Acta 977:194-199
Indiveri C, Dierks T, Krämer R, Palmieri F (1991) Reaction mechanism of the reconstituted oxoglutarate carrier from bovine heart mitochondria. Eur J Biochem 198:339-347
Kassab R, Roustan C, Pradel L-A (1968) Site actif des ATP: guanidine phosphotransférases. I. Réaction des groupes ϵ-NH_2 lysine essentiels avec le 1-diméthylamino-naphtalene-5-sulfochlorure. Biochim Biophys Acta 167:308-316
Klingenberg M (1981) Membrane protein oligomeric structure and transport function. Nature 290:449-454
Krämer R, Heberger C (1986) Functional reconstitution of carrier proteins by removal of detergent with a hydrophobic ion exchange column. Biochim Biophys Acta 863:289-296
Krämer R, Palmieri F (1989) Molecular aspects of isolated and reconstituted carrier proteins from animal mitochondria. Biochim Biophys Acta 974:1-23
Miles EW (1977) Modification of histidyl residues in proteins by diethylpyrocarbonate. Methods Enzymol 47:431-442
Nałęcz KA, Bolli R, Wojtczak L, Azzi A (1986a) The monocarboxylate carrier from bovine heart mitochondria: partial purification and its substrate-transporting properties in a reconstituted system. Biochim Biophys Acta 851:29-37
Nałęcz KA, Bolli R, Wojtczak L, Azzi A (1989) Purification of the monocarboxylate carrier by affinity chromatography. In Anion carriers of mitochondrial membranes, Azzi A, Nałęcz KA, Nałęcz , Wojtczak L (eds) pp 45-57. Springer Berlin Heidelberg New York
Nałęcz KA, Kamińska J, Nałęcz MJ, Azzi A (1992) The activity of pyruvate carrier in a reconstituted system: Substrate specificity and inhibitor sensitivity. Arch Biochem Biophys 297:162-168
Nałęcz KA, Müller M, Zambrowicz EB, Wojtczak L, Azzi A (1990) Significance and redox state of SH groups in pyruvate carrier isolated from bovine heart mitochondria. Biochim Biophys Acta 1016:272-279
Nałęcz KA, Zambrowicz BE (1990) Monocarboxylate carrier from bovine heart mitochondria: Reconstitution in a functionally active state. Acta Biochim Polonica 37:109-112
Nałęcz MJ, Nałęcz KA, Azzi A (1991) Purification and functional characterisation of the pyruvate (monocarboxylate) carrier from baker's yeast mitochondria (*Saccharomyces cerevisiae*). Biochim Biophys Acta 1079:87-95
Nałęcz MJ, Nałęcz KA, Broger C, Bolli R, Wojtczak L, Azzi A (1986b) Extraction, partial purification and functional reconstitution of two mitochondrial carriers transporting keto acids: 2-oxoglutarate and pyruvate. FEBS Lett 196:331-336
Palmieri F, Bisaccia F, Capobianco L, Iacobazzi V, Indiveri C , Zara V (1990) Structural and functional properties of mitochondrial anion carriers. Biochim Biophys Acta 1018:147-150
Papa S, Francavilla A, Paradies G, Meduri B (1971) The transport of pyruvate in rat liver mitochondria. FEBS Lett 12:285-288

Papa S, Paradies G (1974) On the mechanism of translocation of pyruvate and other monocarboxylic acids in rat-liver mitochondria. Eur J Biochem 49:265-275

Paradies G (1984) Interaction of alpha-cyano [^{14}C]cinnamate with the mitochondrial pyruvate translocator. Biochim Biophys Acta 766:446-450

Paradies G, Papa S (1977) On the kinetics and substrate specificity of the pyruvate translocation in rat liver mitochondria. Biochim Biophys Acta 462:333-346

Paradies G, Ruggiero FM (1986) Characterization of the alpha-cyanocinnamate binding site in rat heart mitochondria and in submitochondrial particles. Biochim Biophys Acta 850:249-255

Stappen R, Krämer R (1992) Kinetic characterization of the reconstituted phosphate carrier from bovine heart mitochondria. EBEC Short Reports Vol 7, p 86

Takahashi K (1968) The reaction of phenylglyoxal with arginine residues in proteins. J Biol Chem 23:6171-6179

Tavale SS, Pant LM, Biswas AB (1961) The crystal structure of sodium pyruvate. Acta Cryst 14:1281-1286

Włodek L, Czubak J (1984) Formation of 2-methyl-2,4-thiazolidinedicarboxylic acid from L-cysteine in rat tissues. Acta Biochim Polonica 31:279-288

MEMBRANE TOPOLOGY OF THE ADP/ATP CARRIER FROM MITOCHONDRIA

G. Brandolin, G.J.M. Lauquin*, R. Brasseur**, P.V. Vignais
DBMS/Biochimie (URA 1130 CNRS alliée à l'INSERM)
Centre d'Etudes Nucléaires, 85X, 38041 Grenoble cedex, France

Introduction

The exchange between ATP generated by oxidative phosphorylation in mitochondria and extramitochondrial ADP is catalysed by a specific carrier protein located in the inner mitochondrial membrane. Although ADP/ATP transport is well characterized on physiological and kinetic grounds, its detailed mechanism at the molecular level remains unknown. Due to severe difficulties encountered in the crystallisation of membrane transport proteins, no direct structural information on the ADP/ATP carrier is presently available. However, through a number of indirect approaches applied to the membrane-bound ADP/ATP carrier, it has been possible to propose a schematic view of the topographical organization of its polypeptide chain within the mitochondrial membrane. These approaches were theoretical and experimental, including prediction of the secondary structure from the amino-acid sequence, probing of binding sites with photochemical derivatives of inhibitors and substrates, and localisation of hydrophilic restricted segments of the peptide chain on one face of the membrane or the other using proteolytic and immunological techniques.

As summarized in this paper, some segments of the membrane-bound ADP/ATP protein were shown to undergo a significant motion during the transition of the carrier between two well-identified conformations stabilized by atractylosides on the one hand and bongkrekic acids on the other. These results allow new insights in the dynamics of the ADP/ATP transport system.

*IBC-CNRS, 1 rue C. Saint-Saëns, 33077 Bordeaux cedex, France
**Faculté des Sciences, CP 206/2, 1050 Bruxelles, Belgique

NATO ASI Series, Vol. H 83
Molecular Biology of Mitochondrial Transport Systems
Edited by M. Forte and M. Colombini

Caracterization of the conformational states of the carrier

The ADP/ATP carrier protein is a key mitochondrial component present in all aerobic eukaryotic cells (for review see Vignais et al, 1985). Comparison of the sequences determined so far for the ADP/ATP carrier from mitochondria of various tissues show that it is remarkably conserved (Wohlrab et al, 1989). The ADP/ATP transport system can be blocked by very specific inhibitors that belong to two distinct families : the atractyloside family, i.e atractyloside (ATR) and carboxyatractyloside (CATR) on the one hand and the bongkrekic acid family, i.e bongkrekic acid (BA) and iso bongkrekic acid (isoBA) on the other. Atractylosides bind to the carrier at sites exposed to the cytosol and bongkrekic acids at sites oriented to the matrix compartment. The binding of CATR and BA to the ADP/ATP carrier are mutually exclusive and it is well established that the inhibitors recognize two distinct conformations of the carrier, referred to as CATR and BA conformations respectively. In fact, as developed later, we have shown that the non-inhibited carrier protein exists in equilibrium between the CATR and the BA conformations. Only in the presence of either CATR or BA, able to trap the carrier in the corresponding conformation, is this equilibrium shifted, as a consequence of the formation of stable CATR-carrier or of BA-carrier complexes. Distinct conformations of the ADP/ATP carrier in these two complexes have been demonstrated by the absence of cross-reactivity to antibodies specific to each conformation (Buchanan et al, 1976, Boulay et al, 1986) as well as by distinct susceptibilities to chemical reagents such as iodine (Brdiczka and Schumacher, 1976) and N-ethylmaleimide (Boulay and Vignais, 1984).

Kinetics of the transition between the CATR conformation and the BA conformation of the ADP/ATP carrier, either in the membrane-bound state or solubilized and isolated in the presence of detergent, have been extensively studied by intrinsic and extrinsic fluorescence techniques. It was found that the tryptophanyl fluorescence of the isolated carrier is significantly modified by addition of ADP or ATP to the preparation (Brandolin at al, 1981). Only the non-chelated forms of ADP and ATP were found to be efficient, demonstrating they are the true substrates for the ADP/ATP carrier ; non-transportable nucleotides were ineffective. The ADP (ATP)-induced conformation changes of the carrier resulted in kinetically complex fluorescence changes that could be analysed in detail if measured at low temperature (Brandolin et al, 1985). A rapid fluorescence increase corresponded to the binding of ADP or ATP to saturable high affinity binding sites, followed by an additional slower fluorescence increase, due to the conformational change of the carrier protein in response to the ADP/ATP binding. The latter increase was further enhanced in the presence of BA. The high fluorescence level attained upon addition of BA was attributed to the carrier-BA complex ; it contrasted with the low fluorescence level of the carrier-CATR complex. To summarize, this study allowed the caracterization of two distinct fluorescence levels corresponding to the CATR conformation and to the BA conformation, and the demonstration that the transition between the two states is triggered exclusively by the transportable nucleotides ADP or ATP. In addition, we were able to titrate the nucleotide as well as the inhibitor (CATR and BA) binding sites on the isolated ADP/ATP carrier in detergent solution. In further studies of the intrinsic fluorescence changes of the ADP/ATP carrier, it was found that the nature of the detergent used to solubilize the mitochondrial membrane influenced the conformational state of the purified unliganded carrier : whereas the CATR conformer was favoured in the presence of laurylamidodimethyl-propyl-aminoxide (LAPAO), the BA conformer was stabilized in the presence of cholamidopropyl-dimethylammoniopropane sulfonic acid (CHAPS) (Block and Vignais, 1986).

The kinetic features of the transition of the membrane-bound ADP/ATP carrier between the CATR and the BA conformations were probed with naphthoyl-ADP (N-ADP), a fluorescent derivative of ADP (Block et al, 1982). N-ADP is not transported and it binds with high affinity to the carrier which results in a decrease of the N-ADP fluorescence.

Conversely, the fluorescence of N-ADP is greatly increased when the probe is released from the carrier upon addition of CATR or of BA. This property was used to titrate the nucleotide and the inhibitor binding sites of the carrier. The bound N-ADP was released rapidly in the presence of CATR or BA. Additivity of fluorescence signals observed upon simultaneous addition of both inhibitors demonstrated that the two conformers coexist in the same membrane preparation. From experiments performed at low temperature, it was shown that the slow transition between the CATR conformation and the BA conformation is considerably accelerated by addition of minute amounts of ADP or ATP. In the presence of ADP or ATP, either BA or CATR was able to induce the rapid and complete release of bound N-ADP (Block et al, 1983). Non transportable nucleotides used instead of ADP or ATP were inefficient. Both fluorescence studies with N-ADP (on the membrane-bound carrier) and with the intrinsic tryptophanyl fluorescence (on the isolated carrier) indicated that the rapid transition between the CATR and the BA conformations depends on the presence of specifically transported nucleotides, i.e ADP or ATP, and probably reflects molecular events inherent to the transport process.

The oligomeric state of the carrier

N-ADP, together with a number of transportable and non-transportable derivatives of ADP and ATP were used to explore the parameters of the nucleotide binding sites of the ADP/ATP carrier either in the membrane-bound state or in the solubilized form in the presence of detergent (Dupont et al, 1982, Brandolin et al, 1982, Block and Vignais, 1984). All nucleotides were able to bind to the same total number of binding sites on the carrier, but only with transportable nucleotides was the occurence of equal amounts of high and low affinity sites evidenced. The presence of more than one binding site per carrier unit was in agreement with the sequential mechanism of transport proposed by Duyckaerts et al (1980).

The results of titration of binding sites for adenine nucleotides and CATR in the membrane-bound carrier led to the concept that the functional carrier is oligomeric. There is an equal amount of high affinity and low affinity nucleotide binding sites in the ADP/ATP carrier and the bound nucleotides are removed upon CATR binding. Since the total number of CATR binding sites is twice that of high affinity ADP binding sites (Block and Vignais, 1984), each carrier unit contains two CATR binding sites. As it was previously established that one mole of CATR is bound per mole of carrier dimer (Klingenberg et al, 1978), one has to conclude that the carrier is tetrameric. The presence of two CATR binding sites per carrier unit would be in accordance with the positive cooperativity binding exhibited by the CATR binding sites in the membrane-bound carrier as well as in the isolated carrier in detergent solution. Binding of CATR to the carrier, resulting in inhibition of transport, would result in the splitting of the ADP/ATP carrier tetramer into two stable carrier-CATR dimeric inactive complexes.

A plausible model for the folded organization of the ADP/ATP carrier protein was first reported by Saraste and Walker (1982) and Runswick et al (1987). This model was essentially theoretical and it presented two main features : 1) the carrier protein exhibited a triplicate structure as if it was made up of three homologous repeats, each about one hundred amino-acids long, 2) the hydropathy plot of the carrier indicated the presence of five clearly hydrophobic segments of sufficient length to span the lipid bilayer. Because of the three-fold symmetry concept, it was suggested that the organization of the carrier protein involved a triply repeated motif containing two transmembrane α helices connected by a hydrophilic loop. To fit with this model, a sixth transmembrane segment was postulated, that was supposed to span the region 66-91 of the beef heart mitochondria ADP/ATP carrier, despite its marked hydrophilic caracter (Runswick et al, 1987, Klingenberg, 1989).

The internal three-fold symmetry is not restricted to the

ADP/ATP carrier protein. It also occurs in other mitochondrial metabolite carriers, including the uncoupling protein from brown adipose tissue (UCP), the phosphate carrier and the oxoglutarate carrier (Runswick et al, 1987, 1990). This suggests that these carriers belong to the same family and exhibit similar foldings possibly related to their transport function, even though the transported ions are very different, ranging from the small cation H_3O^+ for the uncoupling protein to the large anion ATP^{4-}.

Although appealing, the six helix model of the ADP/ATP carrier could not accommodate conclusions deduced from immunological and chemical data. As a consequence, other models of folding of the carrier monomer more in agreement with experimental data have been proposed. They include arrangements with four α helices (Klingenberg, 1985), with five α helices, a β strand and a transmembrane segment (Bogner and Aquila, 1986) or six α helices, a β sheet and another membrane spanning segment (Bogner at al, 1986). These models were based on the results of chemical labeling of lysine groups of the carrier with pyridoxal phosphate. In our group, a five helix model has been proposed, in accordance with the hydropathy plot and with discrete regions, identified from immunochemical and photolabeling experiments, and assigned to one face or the other of the mitochondrial membrane (Dalbon et al, 1988). Recently this model has been refined with the results concerning the proteolytic cleavage of the membrane-bound carrier (Marty et al, 1992, Brasseur et al, unpublished data, 1992).

Experimental approaches to the determination of the membrane topography of the ADP/ATP carrier

Photolabelling, immunochemistry and limited proteolysis have been applied to intact beef heart mitochondria or inside-out submitochondrial particles to determine the presence of accessible well-defined sequences of amino-acids of the ADP/ATP carrier on one side or the other of the mitochondrial membrane.

1 - Photolabeling

Photolabeling experiments were carried out with azido derivatives of ADP and atractyloside (Boulay et al, 1983, Dalbon et al, 1988). Both azido derivatives satisfied the prerequisites for photoactivable probes used in the determination of the topography of membrane proteins; they showed a high affinity, a high specificity and they were non-permeant. Moreover, the membrane sidedness of the binding sites for azido-ADP and azido-atractyloside was unambiguously assessed since both reagents were demonstrated to bind only to sites located on the cytosolic face of the membrane. Therefore, it could be safely concluded that the regions covalently labelled by these photoactivable ligands are accessible from the cytosol. The two regions labelled by azido-ADP and azido-atractyloside spanned Cys 159 to Met 200 and Tyr 250 to Met 281, with Lys 162, 165 and Ile 183 preferentially labelled in the ADP/ATP carrier from beef heart mitochondria. In the yeast ADP/ATP carrier the region labelled by 2-azido-ADP extended from residue 172 to residue 210 (Mayinger et al, 1989); interestingly, the segment 172-210 of the yeast carrier shows strong homology with the segment 159-200 of the beef heart mitochondrial carrier. These labelled homologous regions of the yeast carrier and the beef heart carrier correspond to segments connecting the two C-terminal predicted transmembrane helices. Surprisingly, no photolabeling with 2-azido-ADP was detected in the ADP/ATP carrier of yeast mitochondria in the region corresponding to the segment Tyr 250-Met 281 of the ADP/ATP carrier from beef heart mitochondria, whereas an homologous region was labelled in the UCP (Winkler and Klingenberg, 1992).

2 - Immunochemistry

Experimental assignment of the extra-membrane segments of the peptide chain of the ADP/ATP carrier was essentially based on the combined use of immunochemical and limited proteolysis techniques. In a first approach, antibodies were directed against a ten residue long synthetic peptide whose sequence corresponded to the N-terminal sequence of the beef heart ADP/ATP carrier protein. From ELISA, performed either with mitoplasts coated onto microtiter plates or used in back-titration experiments after the antibodies had reacted with mitoplasts in suspension, it could be concluded unambigously that the N-terminal region of the membrane-bound carrier is exposed to cytosol (Brandolin et al, 1989). Detailed examination of the immunological reactivity of the N-terminal region of the membrane-bound beef heart ADP/ATP carrier revealed that the accessibility of the N-terminal region of the carrier to peptide-directed antibodies is modulated by the conformational state of the carrier : it was very high for the carrier stabilized in the CATR conformation, and, in contrast, it was considerably decreased when the carrier was in the BA conformation. However, no direct involvement of the N-terminal region of the carrier in the binding of specific substates or inhibitors could be demonstrated and the precise role of this region of the carrier has still to be established.These results demonstrated that the N-terminal segment of the ADP/ATP carrier is very flexible and is able to sense structural changes occurring in the peptide chain when the carrier shifts between the CATR conformation and the BA conformation, and probably when adenine nucleotide transport is initiated. It has been observed that the conformation-sensitive masking or unmasking of the N-terminal segment localized on the cytosolic face of the mitochondrial inner membrane is correlated with conformational changes that take place on the opposite face of the membrane (the matrix face) (Brandolin et al, 1989). For example, the shift of the ADP/ATP carrier to the BA conformation results not only in a decreased accessibility of the N-terminal region to specific anti N-terminal antibodies, but also to an enhanced chemical reactivity of Cys 56 to alkylating reagents (Boulay and Vignais, 1984).

3 - Limited proteolysis

The topography of the membrane-bound ADP/ATP carrier protein was further explored with proteolytic enzymes under controlled experimental conditions. Site-specific endoproteases were used and the fragments of the carrier that were generated were first identified by immunoreaction with antipeptide antibodies, and the identification was corroborated after purification, by microsequencing (Marty et al, 1992, Brasseur et al, 1992).

To assess the cytosolic or matrix exposure of the extramembrane fragments of the ADP/ATP carrier, proteolytic digestion of mitoplasts and inside-out submitochondrial particles (SMP) was carried out. As SMP are usually contaminated by some right-side out particles, only proteolytic cleavage of the membrane-bound carrier in SMP, but not in mitoplasts, was considered to be relevant. In addition, both types of particles were pretreated with CATR or with BA in order to stabilize the carrier either in the CATR or in the BA conformation, respectively.

Only in SMP pretreated by BA (SMP-BA) was the carrier susceptible to cleavage by proteases. In other words, it is only in the BA conformation that some peptide bonds of the membrane-bound carrier are exposed to added proteases. This result is a further illustration of the topological changes undergone by the carrier during the transition between the CATR and the BA conformations and it rules out any possible pitfall that could be a consequence of particle permeabilisation. Three site-specific proteases, able to cleave the membrane-bound carrier in SMP-BA, were selected in the course of this study, namely an arginine-specific endoprotease, a lysine-specific endoprotease and trypsin. Analysis of the cleavage products allowed the identification of a limited number of bonds exposed to the matrix compartment and accessible to the proteases when the carrier is in the BA conformation : Arg 30-Val 31 and/or Arg 59-Ile 60, Lys 42-Glu 43, Lys 146-Gly 147 and Lys 244-Gly 245. No cleavage of the carrier is observed on regions exposed to the cytosol. It is likely that these cleavage sites are located on, or close to, regions of the carrier flexible enough either to be masked or to be unmasked during the shift of the carrier between the CATR conformation and the BA conformation.

Five vs six transmembrane helix models of the ADP/ATP carrier

Theoretical and experimental conclusions that have accumulated concerning the topography of the ADP/ATP carrier from beef heart mitochondria have been interpreted in terms of two main models taking into account several possible arrangements of the peptide chain of the carrier within the mitochondrial membrane. These models differ by the number of the transmembrane segments, assumed to be five or six. The six helix model is essentially based on the triplicate organization of the sequence of the ADP/ATP carrier, which appears to be an attribute of the mitochondrial carrier family (Runswick et al, 1987). This consists of the juxtaposition of three segments of about one hundred amino-acids each. Assuming that sequence homologies should result in structural homologies, each element is proposed to constitute a building block made up of two membrane spanning α helices, connected by a hydrophilic loop. However, in the case of the ADP/ATP carrier, clearly only one of the two segments belonging to the N-terminal block seems to be hydrophobic enough to span the membrane as an α helix (see Figure 1). The second segment of this block contains 40 % hydrophilic amino acid residues, and its organization as a transmembrane α helix can hardly be postulated.

A direct consequence of the six helix model proposed for the mitochondrial carriers would be the exposure of the N- and C-terminal regions of their peptide chains on the same side of the membrane. Therefore, the determination of the orientation of these segments relative to the membrane plane would provide a key assessment for the validity of the models. Results concerning the localisation of the terminal segments of the ADP/ATP carrier, the phosphate carrier and the UCP by immunological and proteolytic methods are summarized in

Table I.

Carrier	N-terminus exposure	C-terminus exposure	Reference
ADP/ATP	cytosolic	unknown	Brandolin et al, 1989
UCP	unknown	cytosolic	Eckerskorn et al, 1987
Pi	cytosolic	cytosolic	Capobianco et al, 1991

Table I. Orientation of the N- and C-terminal segments of mitochondrial carriers

The C-terminal segment of the UCP and the N-terminal segment of the ADP/ATP carrier are exposed to cytosol, but there is no data concerning the orientation of the N-terminus of UCP. Attempts to localize the C-terminal segment of the ADP/ATP carrier by an immunological approach, using anti C-terminal peptide antibodies, were not conclusive : ELISA with mitoplasts coated onto microtiter plates indicated a cytosolic exposure of the C-terminus whereas back-titration ELISA showed a greater reactivity of the C-terminus on SMP. Only in the case of the phosphate carrier was the orientation of both the N- and C-terminal regions shown to be cytosolic ; in addition the hydropathy plot of the phosphate carrier (see Figure 1) as well as data obtained from limited proteolysis experiments support a six transmembrane helix model. Because of the homologies that were found between the sequences of the mitochondrial carriers, a membrane arrangement of the ADP/ATP carrier similar to that of the phosphate carrier, i.e. in which the N- and the C-terminal segments are exposed to cytosol, has been suggested (Klingenberg, 1989). However, at present, the only experimental data that might support the six helix model of the ADP/ATP carrier come from the demonstration of three equidistant cleavable bonds, Lys 42-Glu 43, Lys 146-Gly 147 and Lys 244-Gly 245, on the matrix face of the mitochondrial membrane when the carrier adopts the BA conformation.

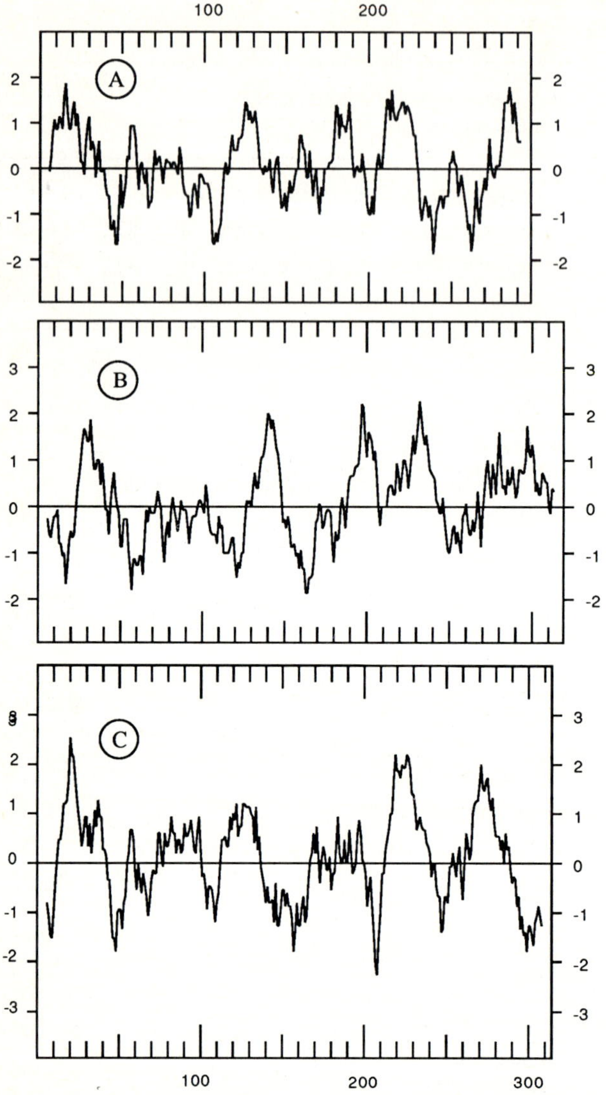

Figure 1. Hydropathy plot of the ADP/ATP carrier protein from (A) beef heart mitochondria, and (B) *S. cerevisiae* mitochondria and (C) of the phosphate carrier protein from beef heart mitochondria. Only five hydrophobic segments can be clearly deduced from the hydropathy plots concerning the ADP/ATP carriers.

The proposal of a five helix model for the ADP/ATP carrier is based essentially on the hydrophobic/hydrophilic analysis of the sequence of the carrier and is consistent with photolabelling data. According to the five helix model, the cytosolic orientation of the N-terminal segment would result in the cytosolic exposure of regions extending from residues

137 to 170 and 234 to 266, that correspond precisely to the hydrophilic parts of the segments labelled by 2-azido-ADP and by azido-atractyloside and that are therefore accessible from cytosol. However, as mentioned above, investigations concerning the orientation of the C-terminus of the ADP/ATP carrier, predicted in this model to face the matrix compartment, has led to ambiguous conclusions. It should be added that attempts to model the ADP/ATP carrier as a monomer seem to be of limited interest since the functional carrier unit is obviously oligomeric.

The translocation pathway for nucleotides

Two domains of the ADP/ATP carrier, located between residues 137-170 and 234-266 are likely to be involved in the building of the nucleotide translocation path since they contain both : 1) the segments accessible from the cytosol and photolabelled by azido derivatives of ADP and of atractyloside 2) two of the protease-sensitive bonds exposed to the matrix compartment in a conformation-sensitive manner. As a consequence, one should consider that restricted peptide segments, adjacent in the sequence of the ADP/ATP carrier, could be alternatively accessible from the matrix or from the cytosol depending of the conformational state of the carrier. This peculiar feature is probably inherent in the functional dynamics of the carrier.

The regions 137-170 and 234-266 display an amphipathic caracter, with an overall positive charge consistent with their nucleotide binding capacity. The fact that these loops could be accessible either from the cytosol or from the matrix compartment indicated they were inserted in the membrane, possibly as hairpin structures. This is only conceivable if the charges of polar residues are neutralized by ion-pairing or shielded from the hydrophobic core of the bilayer. To fulfill this condition it should be assumed that the segments 140-151 and 153-167 on one hand and 237-249 and 253-267 on the other are organized in α helices. As shown on Figure 2, this structure is expected to result in the clustering of polar residues on one face of the amphipathic helices.

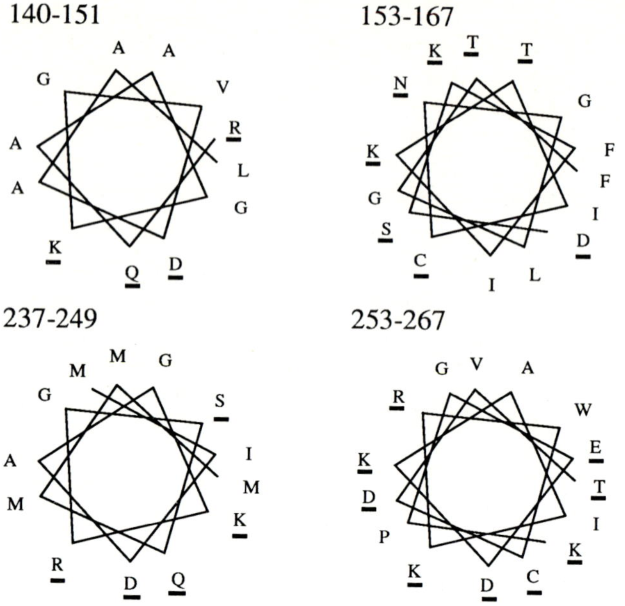

Figure 2. Possible helical structure of the short amphiphilic segments of the ADP/ATP carrier showing the clustering of polar amino acid residues (underlined) on one face of the helices. *From Marty et al., 1992. With permission.*

The existence, in multispanning membrane-inserted carriers, of short amphipathic segments within the membrane has been postulated by Lodish (1988) on the basis of steric considerations. Such an organization would result in shortening of the transmembrane paths for hydrophilic molecules because of locally reduced thickness of the membrane, therefore facilitating transport. A dynamic model of the ADP/ATP carrier has been proposed, in which the association of monomers in functional carrier dimers or tetramers takes place in order to delineate a hydrophilic translocation channel by the juxtaposition of the short amphipathic α helices (Figure 3). The gating of the channel and the translocation process itself might be the results of conformational changes undergone by the amphipathic helices during the functioning of the carrier.

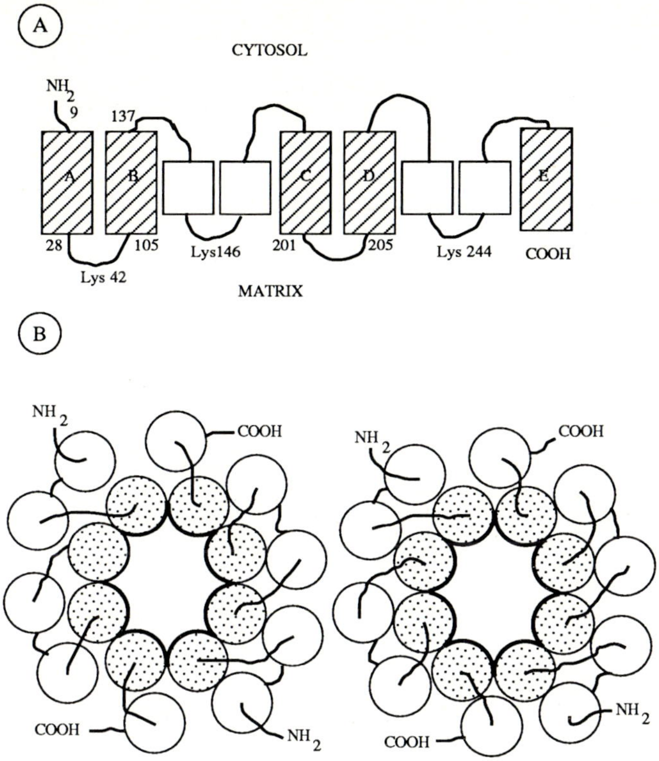

Figure 3. Postulated arrangement of the ADP/ATP carrier in the mitochondrial membrane. (A) front view of the monomeric carrier with five transmembrane helices and four short amphipathic helices. (B) top view of a possible tetrameric organization of the carrier showing the hydrophilic path delineated by the polar faces of the short amphipathic helices (highlighted). *Adapted from Marty et al., 1992. With permission.*

Bibliography

Block MR, Lauquin GJM, Vignais PV (1982) Interaction of 3'-O(1-naphthoyl) adenosine 5'-diphosphate, a fluorescent adenosine 5'-diphosphate analog, with the adenosine 5'-diphosphate/adenosine 5'-triphosphate carrier protein in the mitochondrial membrane. Biochemistry 21:5451-5457

Block MR, Lauquin GJM, Vignais PV (1983) Use of 3'-O(1-naphthoyl) membrane-bound adenosine 5'-diphosphate/adenosine 5'-triphosphate carrier. Biochemistry 22:2202-2208

Block MR, Vignais PV (1984) Substrate-site interactions in the membrane-bound adenine nucleotide carrier as disclosed by ADP and ATP analogs. Biochim Biophys Acta 767:369-376

Bogner W, Aquila H (1986) ADP/ATP carrier : analysis of transmembrane folding using pyridoxal phosphate. Methods Enzymol 125:650-658

Bogner W, Aquila H, Klingenberg M (1986) The transmembrane arrangement of ADP/ATP carrier as elucidated by the lysine reagent pyridoxal 5-phosphate. Eur J Biochem 165:611-620

Boulay F, Lauquin GJM, Tsugita A, Vignais PV (1983) Photolabeling approach to the study of the topography of the atractyloside 5'-triphosphate carrier protein. Biochemistry 22:477-484

Boulay F, Lauquin GJM, Vignais PV (1986) Localization of immunoreactive regions in the beef heart adenine nucleotide carrier using rabbit antisera against the carboxyatractyloside-liganded and the sodium dodecyl sulfate denatured carrier forms. Biochemistry 25:7567-7571

Boulay F, Vignais PV (1984) Localization of the N-ethylmaleimide reactive cystein in beef heart mitochondrial ADP/ATP carrier protein. Biochemistry 23:4807-4812

Brandolin G, Boulay F, Dalbon P, Vignais PV (1989) Orientation of the N-terminal region of the membrane-bound ADP/ATP carrier protein explored by antipeptide antibodies and an arginine-specific endoprotease. Evidence that the accessibility of the N-terminal residues depends on the conformational state of the carrier. Biochemistry 28:1093-1100

Brandolin G, Dupont Y, Vignais PV (1981) Substrate-induced fluorescence of the isolated ADP/ATP carrier protein in solution. Biochem Biophys Res Commun 98:28-35

Brandolin G, Dupont Y, Vignais PV (1982) Exploration of the nucleotide binding sites of the isoplated ADP/ATP carrier protein from beef heart mitochondria. II. Probing of the nucleotide sites by formycin triphosphate, a fluorescent transportable analogue of ATP. Biochemistry 21:6348-6353

Brandolin G, Dupont Y, Vignais PV (1985) Substrate-induced modifications of the intrinsic fluorescence of the isolated adenine nucleotide carrier protein : demonstration of distinct conformational states. Biochemistry 24:1991-1997

Brasseur R, Lins L, Ruysschaert JM, Brandolin G, Vignais PV (1992) submitted

Brdiczka D, Schumacher D (1976) Iodination of peripheral mitochondrial membrane proteins in correlation to the functional state of the ADP/ATP carrier. Biochem Biophys Res Commun 73:823-832

Buchanan BB, Eiermann W, Riccio P, Aquila H, Klingenberg M (1976) Antibody evidencefor different conformational states of ADP/ATP translocator protein isolated from mitochondria. Proc Natl Sci Acad USA 73: 280-2284

Capobianco L, Brandolin G, Palmieri F (1991) Transmembrane topography of the mitochondrial phosphate carrier explored by peptide-specific antibodies and enzymatic digestion. Biochemistry 39:4963-4969

Dalbon P, Brandolin G, Boulay F, Hoppe J, Vignais PV (1988) Mapping of the nucleotide-binding sites in the ADP/ATP carrier of beef heart mitochondria by photolabeling with 2-azido(α-^{32}P)adenosine diphosphate. Biochemistry 27:5141-5149

Dupont Y, Brandolin G, Vignais PV (1982) Exploration of the nucleotide binding sites of the isolated ADP/ATP carrier protein from beef heart mitochondria. I. Probing of the nucleotide sites by naphthoyl-ATP, a fluorescent non transportable analogue of ATP. Biochemistry 21: 6343-6347

Duyckaerts C, Sluse-Goffart CM, Fux JP, Sluse FE, Liebecq C (1980) Kinetic mechanism of the exchange catalyzed by the adenine nucleotide carrier. Eur J Biochem 106:1-6

Eckerskorn C, Klingenberg M (1987) In the uncoupling protein from brown adipose tissue the C-terminus protrudes to the c-side of the membrane as shown by tryptic cleavage. FEBS Letters 226:166-170

Klingenberg M (1985) The ADP/ATP carrier in mitochondrial membrane. In : Martonosi A (ed) The Enzymes of Mitochondrial Membranes : Membrane Transport, vol 4, John Wiley New York pp 511-553

Klingenberg M (1989) Molecular aspects of the adenine nucleotide carrier from mitochondria. Arch Biochem Biophys 270: 1-14

Klingenberg M, Riccio P, Aquila H (1978) Isolation of the ADP,ATP carrier as the carboxyatractylate complex from mitochondria. Biochim Biophys Acta 503:193-210

Lodish HF (1988) Multi-spanning membrane proteins : how accurate are the models? Trends Biochem Sci 13:332-334

Marty I, Brandolin G, Vignais PV (1991) Topography of the membrane-bound ADP/ATP carrier assessed by enzymatic proteolysis. Biochemistry 31:4058-4065

Mayinger P, Winkler E, Klingenberg M (1989) The ADP/ATP carrier from yeast (AAC-2) is uniquely suited for the assignment of the binding center by photoaffinity labeling. FEBS Lett 244:421-426

Runswick MJ, Powell SJ, Nyren P, Walker JE (1987) Sequence of the bovine mitochondrial phosphate carrier protein : structural relationship to ADP/ATP translocase and the brown fat mitochondria uncoupling protein. EMBO J 6:1367-1373

Runswick MJ, Walker JE, Bisaccia F, Iacobazzi V, Palmieri F. (1990) Sequence of the bovine 2-oxoglutarate/malate carrier protein : structural relationship to other mitochondrial transport proteins. Biochemistry 29:11033-11040

Saraste M, Walker JE (1982) Internal sequence repeats and the path of polypeptide in mitochondrial ADP/ATP translocase. FEBS Lett 144:250-254

Vignais PV, Block MR, Boulay F, Brandolin G, Lauquin GJM (1985) Molecular aspects of structure-function relationships in mitochondrial adenine nucleotide carrier. In : Bengha G (ed) Structure and Properties of Cell Membranes, vol II, CRC Press, pp 139-179

Winkler E. and Klingenberg, M (1992) Photoaffinity labeling of the nucleotide-binding site of the uncoupling protein from hamster brown adipose tissue. Eur J Biochem 203:295-304

Wohlrab H, Bukusoglu C, Defoe H (1989) Recent studies on the mitochondrial phosphate transport protein (PTP) and on its relationship to the ADP/ATP translocase (AAC) and the uncoupling. In : Azzi A, Nalecz KA, Nalecz MJ, Wojtczak L (eds) Anion Carriers of Mitochondrial Membranes. Springer-Verlag, Berlin Heidelberg New York, pp 113-121.

THE USE OF LIGAND BINDING TECHNIQUES TO IDENTIFY TRANSPORT PROTEINS: THE CHARACTERIZATION OF THE [^{125}I]-IODOPHENYLARSINE OXIDE RECEPTOR IN RAT MITOCHONDRIAL MEMBRANES.

Maureen W. McEnery*
Department of Physiology and Biophysics
Case Western Reserve University School of Medicine
10900 Euclid Street
Cleveland, OH 44106

and

Robert D. Hoffman
Institute of Pathology
Case Western Reserve University School of Medicine and
University Hospitals of Cleveland
2085 Adelbert Road
Cleveland, OH 44106

Introduction

The use of radioactive ligands that bind selectively and with high affinity to specific proteins has long been a valuable tool for the identification and purification of membrane proteins and receptors (Cuatrecasas and Hollenberg, 1975; Enna, 1985; Snyder, 1986). This report introduces a novel radioligand, [^{125}I]-4-iodophenylarsine oxide ([^{125}I]PAO) and its application to the study of mitochondrial proteins. The broader goal of this study is to emphasize the utility of radioligand binding assays as an approach which complements standard biochemical and biophysical methods. Indeed, the functional assay for mitochondrial transport proteins often necessitates detergent solubilization, biochemical purification, and reconstitution into phospholipid vesicles. The use of specific radioligands for the purpose of defining a membrane-bound protein by its receptor properties has permitted the purification of numerous ion channels and transport proteins. Moreover, by using covalent radioligand, the specific binding protein can be purified under denatured conditions. Examples of the application of radiolabelled amino acid-selective covalent modifying reagents (e.g. [^{14}C]

NATO ASI Series, Vol. H 83
Molecular Biology of Mitochondrial Transport Systems
Edited by M. Forte and M. Colombini

dicyclohexylcarbodiimide ([^{14}C]DCCD) (Cattell, et al. 1970; 1971) and [^{3}H]N-ethylmaleimide ([^{3}H]NEM) (Coty and Pedersen, 1975)) presented valuable information about the structure of the DCCD-binding protein and the phosphate/proton transporter (PiT), respectively. Also, the synthesis of photoactivatable radiolabelled nucleotide analogs (e.g. 8-azido-ADP (Lauguin, et al., 1978), 8-azido-ATP (Winkler and Klingenberg, 1992), and substrate analogs (e.g. 2-azido-4-nitrophenylphosphate (Tommassino, et al. 1987); 5-(1-hydroxy-4-azidophenylazo)-1,2,3-benzetricarboxylic acid (Szewczyk, et al. 1991) have also been of great ultility in the study of specific mitochondrial carriers.

Drugs and pharmacologically active compounds (e.g. benzodiazepines, isoquinoline carboxamides, 1,4-dihydropyridines, phenylalkylamines, and cyclosporin A) comprise a third class of ligands (Table 1) that identify high affinity "alloreceptors" on mitochondrial membranes (see McEnery, 1992; Zernig, 1990; Ballestra, et al., 1990). Although these radioligands bind with nanomolar affinity to their receptor(s) permitting purification based upon recovery of radioligand binding sites, often the mechanism and site of binding cannot be predetermined. On the other hand, the nanomolar affinities of these compounds offers the opportunity to study the effects of potent modulators of receptor function (Kinnally, et al., 1993).

This study focuses upon the binding of radiolabelled organic arsenicals to membrane proteins. Until recently, general methods for the demonstration of arsenical-protein interactions have been lacking. This is particularly surprising since the concept of specific binding sites for arsenical drugs ("Arsenozeptoren") was explicitly formulated by Paul Ehrlich as early as his 1909 Nobel address, and, as such, numbers among the first receptors of any kind to be considered. The recent synthesis of [^{125}I]PAO and a structurally related affinity chromatography resin (Hoffman and Lane, 1992) has permitted the identification of distinct [^{125}I]PAO binding proteins in 3T3-L1 adipocytes. The basis for the chemical interaction of PAO with biological targets is the formation of a cyclic covalent adduct (Figure 1) that can be reversed by the addition of dithiol reagents such as 2,3-dimercaptopropanol (British anti-lewisite (BAL)). Monothiol reagents such as β–mercaptoethanol are less potent in their ability to reverse this adduct. The utility derived from the use of organic arsenicals is their apparent selectively for a vicinal dithiol binding site. The thiol groups which comprise the binding site in proteins are believed to be closely spaced in the target molecule. Among numerous cellular protein targets, PAO has been demonstrated at micromolar concentrations affect numerous mitochondrial enzymes including the soluble lipoic acid-containing α-ketoacid dehydrogenase complexes (Webb, 1967), subunits of the mitochondrial ATP synthase (Joshi and Hughes, 1981; Yagi and Hatefi, 1984; Lippe, et al. 1988) and affect mitochondrial

Table 1

Radioligands that identify mitochondrial receptors.

Ligand	Kd	tissue(s)	major target
[^{125}I]PAO[1]	150 nM	rat kidney, liver,brain, adipocytes	phosphate/proton transporter (PiT)
[^{3}H]Nitrendipine[2]	593 nM	guinea pig heart and liver	inner memb. anion channel[3] (IMAC)
[^{3}H]Cyclosporin A[4]	8 nM	rat liver	permeability transition pore (PTP)/ cyclophilin
[^{3}H]Diltiazem[5]	120 nM	guinea pig heart and liver	Na^{+}-dependent-Ca^{++} channel
[^{3}H]Diazepam[6]	80 nM	rat kidney	mitochondrial benzodiazepine receptor (mBzR)
[^{3}H]Ro 5-4864[7]	10 nM	rat kidney	mBzR
[^{3}H]Ro 5-4864[8]	5.2 nM	vas deferens	unknown
[^{3}H]PK11195[7]	10 nM	rat adrenal, kidney, testis, ovary	mBzR
[^{3}H]Idazoxan[9]	6 nM	human liver rat liver	imidazoline-guanidinium receptive site (IGRS)

1 This study
2 Zernig, et al. (1988)
3 Zernig, et al. (1990)
4 McGuinness, et al. (1990)
5 Chiesi and Carafoli 1987)
6 Braestrup and Squires (1977)
7 Hirsch, et al. (1989)
8 Escubedo, et al. (1992)
9 Tesson, et al. (1991)

transporters (Naez, et al., 1990) and ion channels (Bernardi, 1992; Bernardi, et al., 1992; Diwan, et al., 1986, 1988; Novgorodov, et al. 1987; Sanadi, et al., 1981). Using radioligand binding assays, we extend the previous observations, and attempt to define pharmacologically, biochemically and immunologically the major [^{125}I]PAO binding protein in mammalian mitochondrial membranes (Hoffman, et al., 1992).

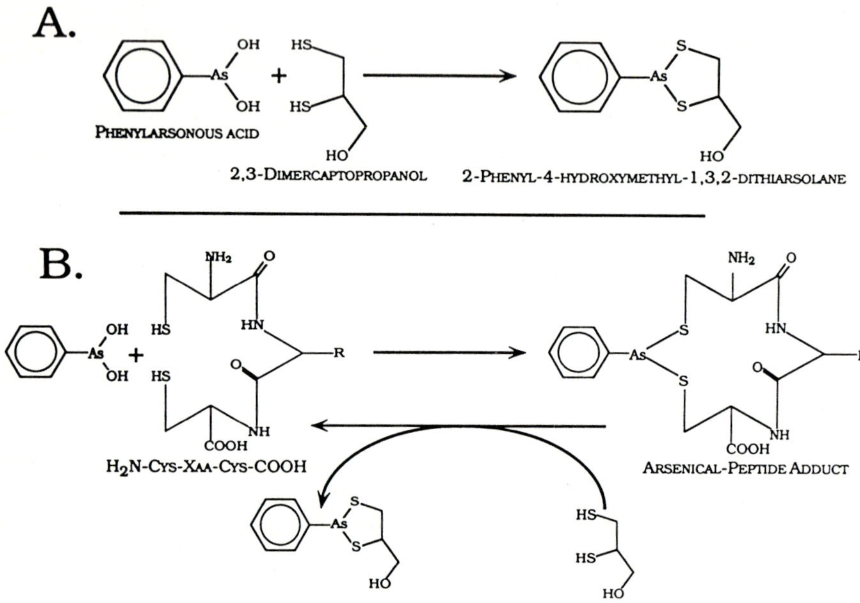

Figure 1: Mechanism of action of phenylarsonous acid with dithiols.
A. Phenylarsonous acid (the hydrated form of phenylarsine oxide) reacts with vicinal sulfydryl groups of the antidote 2,3-dimercaptopropanol to form a stable cyclic adduct. B. A similiar reaction is proposed for the interaction of phenylarsonous acid and two cysteine residues.

Preparation of [^{125}I]PAO and radioligand filter binding assays.

[^{125}I]PAO was synthesized from arsanilic acid according to the method of Hoffman and Lane (1992) by diazotization, iodination and reduction. The specific activity of [^{125}I]PAO was routinely within the range of 5 to 10 mCi/umol. The sample was dissolved in dimethylsulfoxide and stored at -80°C. [^{125}I]PAO was stable for several months when stored as the concentrated stock solution; however, when stored diltuted in aqueous buffers, [^{125}I]PAO became inactive. There is no evidence for the catastrophic decay of [^{125}I]PAO due to radioinactivation (Halpern and Stocklin, 1977), thus, the appropriate corrections for radioactive decay are required.

Radioligand filter binding conditions were optimized to measure the binding of [^{125}I]PAO to receptor fractions. The standard assays included in a 200 uL final volume: 20,000 cpm of [^{125}I]PAO (corresponding to 1-5 nM); protein in the range of 0.5 to 50 ug, and Binding Buffer (25 mM Tris, pH 7.7, 5 mM EDTA, 40 mM potassium phosphate, 2.5% ethylene glycol). The binding of [^{3}H]PK11195 and [^{3}H]Ro5-4864 to tissue samples was carried out in 200 uL final volume in the presence of Binding Buffer and 1 nM radioligand (approximately 20,000 cpm/sample). Displaceable binding was determined under the identical conditions with the addition of 10 uM unlabeled PK11195, 10 uM unlabeled Ro 5-4864 or 25 uM unlabeled PAO. Nonspecific binding was determined in the absence of protein. The samples were incubated on ice for 60 min, and then rapidly filtered on a Brandel cell harvester manifold (model N-24) onto pre-soaked glass fiber filters (Whatman GF/B) by washing twice with 3 ml of ice-cold 50 mM NaCl. Bound radioactivity was quantitated by scintillation counting.

Standard radioligand binding assays use the polycation polyethyleneime (PEI) to coat glass fiber filters, thus trappping the filtered samples. In the absence of added protein (data not shown) or in the presence of kidney mitochondria preincubated with 25 uM unlabeled PAO, [^{125}I]PAO adheres to PEI-treated glass fiber filters (Table 2 part A). Pretreament of the filters with PEG (mw 8 kDa), however, dramatically decreased non-specific filter binding, and promoted the recovery of [^{125}I]PAO-receptor complexes. The recovery of [^{125}I]PAO binding using PEG-soaked filters was comparable under the same conditions to the specific cpm binding (total cpm bound - displaceable cpm bound) of both [^{3}H]PK11195 and [^{3}H]Ro 5-4864, ligands diagnostic for the mitochondrial benzodiazepine receptor also present in kidney mitochondria (Table 2 part B). Scatchard analysis of [^{125}I]PAO binding sites was determined in the presence of a constant concentration of [^{125}I]PAO (as above) with unlabelled IPAO in the range of 1 nM to 25 uM.

Preparation of mitochondria, F_1F_0-ATPase, F_0-proton channel, F_1-ATPase and the mitochondrial benzodiazepine receptor (mBzR):

Mitochondria from different peripheral rat tissues were were prepared by the method of Wehrle et. al. (1985) in the presence of H-medium plus 1 mM EGTA. Triton X-100 solubilization of rat liver and kidney mitochondria and hydroxylapatite chromatography in the presence of 3% Triton X-100 was according to De Pinto et al. (1985). F_1F_0-ATPase and F_0-proton channel were prepared by CHAPS solubilization of extensively washed rat liver inner membrane vesicles as reported (McEnery, et al. 1984 and 1989, respectively). F_1-ATPase was prepared by the method of Catterall and Pedersen (1971). The mitochondrial benzodiazepine receptor (mBzR) was solubilized from rat kidney mitochondrial membranes in the presence of dodecyl-β-maltoside (DM) and purified by hydroxyapatite chromatography in the presence of DM according to the method of McEnery et al. (1992).

Table 2

Determining the optimal conditions for [125]PAO radioligand filter binding in the presence of 1 ug kidney mitochondrial protein.

A.

Filter Treatment	$[^{125}I]$PAO bound to filter total	$[^{125}I]$PAO bound to filter plus 25 uM unlabeled PAO
1.0 % PEG	4,000	250
0.3 % PEI	18,000	18,000

B.

Filter Treatment	$[^{3}H]$PK11195 specific cpm bound	$[^{3}H]$Ro 5-4864 specific cpm bound
1.0 % PEG	1173	800
0.3 % PEI	1098	850

PAO-affinity column chromatography and immunological detection of mitochondrial proteins:

Solubilized 4-aminophenylarsine oxide was coupled to an Affi-gel 10 support as described (Hoffman and Lane, 1992). Triton X-100 extracts of rat liver mitochondria (De Pinto, et al., 1985) were incubated with 1 ml resin for 1 hr. The flow-through was collected and the column sequentially washed with increasing concentrations of β-mercaptoethanol. The eluted fractions were resolved by SDS-PAGE (Laemmli, 1970) on a 12% polyacrylamide gel in the presence of β-mercaptoethanol and SDS. The gels were transferred to PVDF membranes overnight and blocked with powdered milk according to standard procedures. Primary antibodies against the voltage-dependent anion channel (McEnery, et al, 1992), and the phosphate carrier (a gift from H. Wohlrab; Rasmussen and Wohlrab, 1986) were used at optimal dilutions. The blots were incubated with secondary antibody (horseradish peroxidase-goat anti-rabbit IgG conjugate) and subsequently developed using the Amersham ECL chemiluminescence detection method according to manufacture's instructions.

Results

$[^{125}I]$PAO binds in a protein-dependent manner to rat kidney mitochondria (linear range 0.1 to 20 ug protein/assay), and did not bind to bovine serum albumin at 10-100X greater concentration (data not shown). The binding of $[^{125}I]$PAO to rat kidney homogenate and subcellular fractions was characterized by Scatchard analysis (Table 3). High affinity $[^{125}I]$PAO binding is present in rat kidney crude homogenate and is enriched 7 to 10-fold in purified inner membranes compared to kidney mitochondria. There is no apparent enrichment of the $[^{125}I]$PAO binding site in purified outer membranes, although there is a slight decrease in the affinity for $[^{125}I]$PAO.

Table 3

Submitochondrial distribution of $[^{125}I]$PAO binding site in kidney

Fraction	Kd (nM)	Bmax (pmol/ug
Total kidney homogenate	345	1.7
Kidney mitochondria	127	2.1
Inner membranes	157	15.6
Outer membrane	517	1.5

Table 4

Pharmacology of $[^{125}I]$PAO binding to rat mitochondria

Compound	IC_{50} (uM)	Compound	IC_{50} (uM)
PAO	0.3	IPAO	0.3
Eosin-5-maleimide	0.1	Mercuric chloride	0.25
2,3-Dimercaptopropanol	0.25	N-ethylmaleimide	5.0
Dithiothreitol	2.0	Glutathione	2.0
Diamide	10.0	β-mercaptoethanol	1000

The specificity of the [^{125}I]PAO binding site in rat kidney mitochondria was investigated using a panel of monothiol and dithiol modifying reagents (Table 4). [^{125}I]PAO binding was inhibited by nanomolar concentration of unlabeled PAO and IPAO which would compete for the binding site on the receptor. Also the impermeant sulfhydryl reagent eosin-5-maleimide (Houstek and Pedersen, 1985) and mercuric chloride inhibited [^{125}I]PAO binding with IC_{50} values of 100 nM and 250 nM, respectively. N-ethylmaleimide at higher concentrations was also inhibitory. The potency of 2,3-dimercaptopropanol reflects the interaction of this dithiol containing compound and [^{125}I]PAO itself, leading to the neutralization of the radioligand. Dithiothreitol would be expected to interact with [^{125}I]PAO via a similiar mechanism, while the monothiols β-mercaptoethanol and glutathione would be less potent. The inhibitory effect of diamide which oxidatively forms disulfides from dithiols on [^{125}I]PAO binding strongly suggests that the mechanism for the interaction of [^{125}I]PAO and its binding site is via an interaction with vicinal thiols.

The localization of [^{125}I]PAO binding site to the inner membrane led us to examine the binding of [^{125}I]PAO to purified ATPase components which constitute 1% of the mitochondrial inner membrane protein (Table 5). Although the affinity for [^{125}I]PAO binding was almost identical among the preparations tested, there was no apparent enrichment in the [^{125}I]PAO binding site in either purified F_1F_0-ATPase, purified F_0-proton channel, or purified F_1-ATPase. These results suggest that the [^{125}I]PAO binding protein is a minor contaminant of the F_1F_0-ATPase and F_0 preparations (see Figure 3).

Table 5

Identification of [^{125}I]PAO binding sites in liver mitochondrial preparations.

Fraction	Kd (nM)	Bmax (pmol/ug)
Liver mitochondria	80	1.5
Inner membranes	150	15.0
Purified F_1F_0-ATPase	125	12.0
Purified F_0-proton channel	130	14.0
Purified F_1-ATPase	500	1.0
HA flow-thru fraction	200	32.0

In contrast, the HA flow-through fraction showed an enrichment in the recovery of [^{125}I]PAO binding sites relative to inner membranes. Moreover, [^{125}I]PAO binding site in purified in HA flow-thru fraction demonstrates the same sensitivity to eosin-5-maleimide, mercuric chloride and diamide as the native binding site in mitochondria (data not shown). The HA flow-thru fraction contains very few proteins and is enriched in the voltage-dependent anion channel (VDAC), the phosphate/proton transporter (PiT), and the adenine nucleotide tranporter (ANT) (De Pinto, et al., 1985).

Covalently modified [^{125}I]PAO binding proteins in intact mitochondria and purified F_1F_0-ATPase was detected even after SDS-PAGE in the absence of reducing agents. Upon exposure to film, a major 33-35 kDa and a minor 50 kDa and 75 kDa proteins were observed to incorporate [^{125}I]PAO in rat liver mitochondria (Figure 2 lane 1). The 33-35 kDa protein was also observed in the purified F_1F_0-ATPase preparation (**Figure 2 lane 2**). When aligned with the Coommassie stained preparation of the F_1F_0-ATPase (Figure 2 lane 3), the

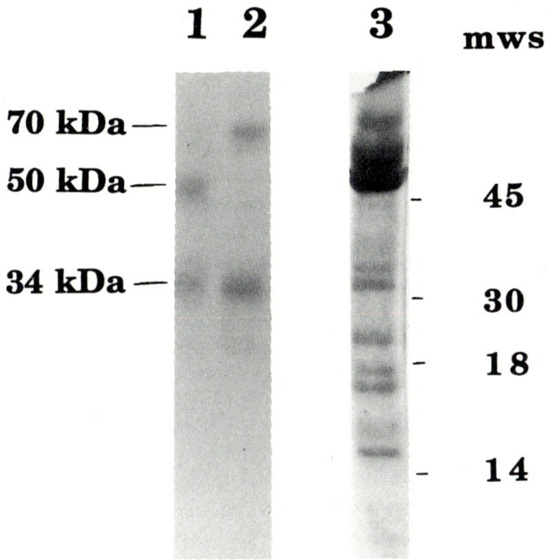

Figure 2: [^{125}I]PAO covalently modifies a 33-35 kDa protein in intact rat liver mitochondria and purified F_1F_0-ATPase.
Samples were incubated with [^{125}I]PAO as stated in methods and resolved by SDS-PAGE. Lane 1: autoradiogram of [^{125}I]PAO labelled rat liver mitochondria; lane 2: autoradiogram of [^{125}I]PAO labelled purified F_1F_0-ATPase; lane 3: Coomassie blue stain of purified F_1F_0-ATPase. Mobility of molecular weight standards (mws) and apparent molecular weight of [^{125}I]PAO labelled proteins are indicated.

[125I]PAO binding protein has a mobility similiar to the gamma subunit of the F_1 portion of the complex (Catterall and Pedersen, 1971; McEnery, et al. 1989). As [125I]PAO did not label the gamma subunit in the purified F_1-ATPase preparation, the [125I]PAO binding protein was considered to be a contaminant of the F_1F_0-ATPase preparation.

The effect of PAO on the uptake of [32P]orthophosphate, [3H]ADP, and [45Ca] into intact rat kidney mitochondria was examined. As shown in **Figure 3**, PAO inhibits the inital rate of K+-induced [32Pi] uptake at 1 uM. At higher concentrations of PAO (10 uM), both initial rate and V_{max} are inhibited. PAO did not effect either the uptake of [3H]ADP or [45Ca] at 10 uM (data not shown). [32P] orthophosphate uptake is also inhibited by the uncoupler dinitrophenol (DNP) and the sulfhydryl reagent eosin-5-maleimide (E-5-M). These results indicate that PAO inhibits [32P] orthophosphate uptake and suggests that PAO directly effects the PiT.

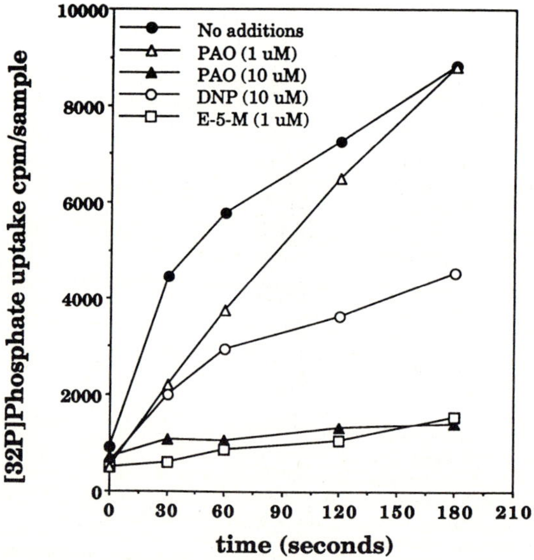

Figure 3: PAO inhibits [32P]orthophosphate uptake into mitochondria. Rat kidney mitochondria were incubated with 1 uM and 10 uM PAO, 10 uM dinitrophenol (DNP), and 1 uM eosin-5-maleimide (E-5-M) for 15 minutes prior to phosphate uptake assay. At specific times, uptake was rapidly terminated by filtration.

The direct association of PAO with the PiT was further examined by the use of a phenylarsine oxide affinity column. Triton X-100 solubilized rat liver mitochondria were chromatographed on the PAO-Affigel 10 column (see methods) and the column was washed batch-wise with increasing concentrations (0-1.0 M) of b-merceptoethanol. All samples were analyzed for the presence of PiT and VDAC, by immunoblotting with specific antisera. The PiT was present in both the column flow-through and the 10 mM β-mercaptoethanol elutions (**Figure 4**). When the PAO-Affigel column was pretreated with 2,3-dimercaptopropanol to neutralize the reactive groups, the PiT was found exclusively in the column flow-through. These results indicate a direct, high affinity association between reactive PAO and the PiT. In contrast, the VDAC was found exclusively in the column flow-indicating that VDAC does not possess highly reactive vicinal cysteine residues.

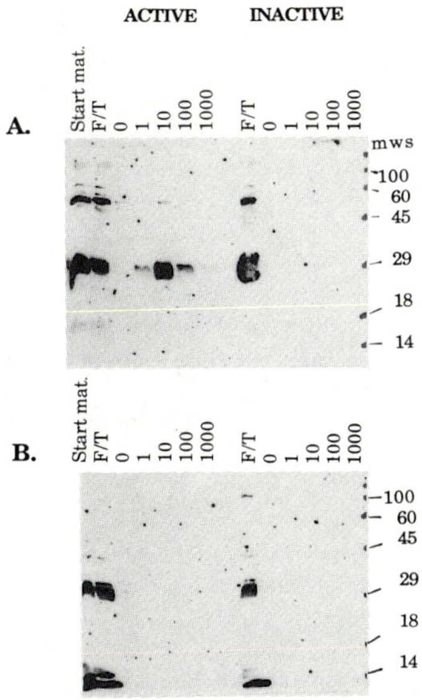

Figure 4: PAO-affinity column chromatography of Triton-X 100 solubilized rat liver mitochondria.
Rat liver mitochondria were solubilized with Triton-X 100, chromatographed on an active PAO-Affigel 10 affinity column or a similiar column neutralized with 2,3-dimercaptopropanol (inactive) as described in Methods. The columns were eluted with increasing concentrations (1-1000 mM) β-mercaptoethanol. Panel A: immunodectection of the PiT by Western blot analysis; Panel B: immunodection of VDAC.

Additional evidence for the direct interaction of PAO with the PiT comes from the effect of sulfhydryl reagents on the apparent mobility of the PiT in SDS-PAGE (Kolbe and Wohlrab, 1985; Pratt et al., 1991). As shown in **Figure 5,** N-ethylmaleimide (NEM) modification of the PiT leads to a decrease in the mobility of the PiT (PiT-3). Pretreatment with increasing concentrations of PAO (0.5, 5 and 50 uM) protects against subsequent alkylation by NEM with the major isoforms of the PiT corresponding to PiT 1 and PiT 2 (Figure 5A). Pretreatment of samples with high concentrations of mercuric chloride (Figure 5B) protects completely against NEM alkylation with only PiT 1 evident. These results suggest that the conversion of PiT 1 to PiT 2 and PiT 3 reflects multiple sites of alkylation by NEM. Furthermore, the observation that PAO at micromolar concentrations prevents the formation of PiT in the presence of NEM, suggests a specific mode of binding for PAO.

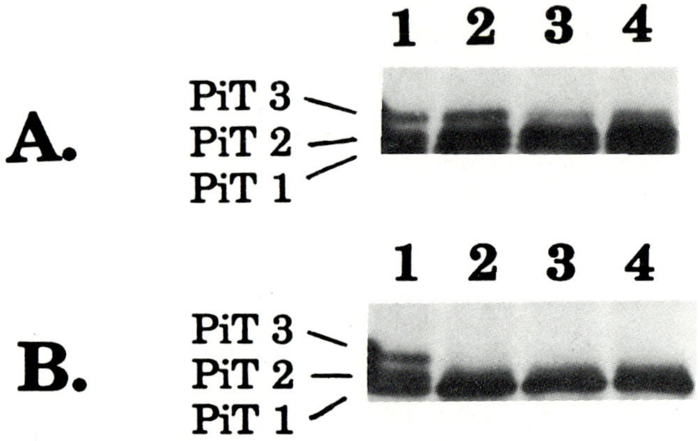

Figure 5: PAO protects PiT against NEM-induced mobility shift.
Triton X-100 solubilized rat kidney mitochondria (25 ug) were incubated with the following: Panel A lane 1, no additions; lane 2, 0.5 uM PAO; lane 3, 5 uM PAO; lane 4, 50 uM PAO; or in Panel B lane 1, no additions; lane 2, 50 uM mercuric chloride; lane 3, 500 uM mercuric chloride; lane 4, 5000 uM mercuric chloride for 30 minutes at room temperature. 5X Laemmli sample buffer plus 20 mM NEM and 3% SDS was added to quench the reaction and the samples were resolved by SDS-PAGE on a 10% gel. The PiT was detected with anti-PiT antibody (see Methods). Apparent molecular weights of PiT 1: 33 kDa; PiT 2: 34 kDa; PiT 3, 35 kDa.

Discussion

In this report, we introduce ^{125}I-4-iodophenylarsine oxide (^{125}I-PAO), a novel radioligand for the detection of vicinal dithiols and, for the first time, demonstrate the presence of a high affinity [^{125}I]PAO binding motif in the PiT. The primary structure of the rat liver PiT does not contain vicinal cysteines, or cysteines separated by one or two amino acid residues. The most closestly spaced cysteines in the rat liver PiT protein corresponding to cysteines 18 and cysteine 27. It is therefore likely that the cysteines which comprise the [^{125}I]PAO binding site are brought together only in the native, folded structure.

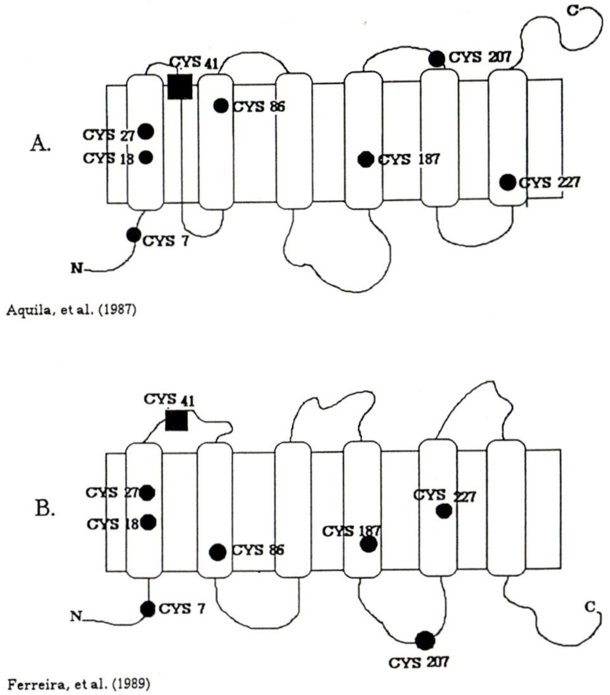

Figure 6: Membrane orientation of Phosphate/Proton Transporter (PiT): topographic orientation of cysteines (redrawn from Ferreira, et al, 1990)
Two possible models for the topological disposition of the PiT are presented. The data to support the role of CYS 41 in the vicinal dithiol binding domain of [^{125}I]PAO is presented in Discussion section.

Which cysteine residues might participate in [^{125}I]PAO binding to the PiT? These results demonstrate the impermeant sulfhydryl reagent eosin-5-maleimide protects against [^{125}I]PAO binding to intact mitochondria, implicating as one component a cysteine oriented toward the cytoplasm. The ability of 50 uM PAO to protect against NEM-alkylation and the associated change in apparent mobility strongly suggests that the highly reactive cysteine 41 is involved in PAO binding. It is significant to note the study by Ferreira et al. (1990) which reports the effect of NEM-alkylation on the trypsin-sensitivity of the PiT C-terminal amino acids. It was suggested that the interaction with NEM confers a significant conformation change in the protein rendering the C-terminal amino acids (aa 302-312) of the PiT less accessible to proteases. It might therefore be possible that cysteine residue 41 and a C-terminal cysteine residue are allosterically coupled.

These hypotheses should be considered in the context of the membrane orientation of the PiT and the topological orientation of cysteines (**Figure 6**). The consequence of these models is the relative disposition of the cysteines in the membrane. In Model A, cysteine 41 may be in close proximity to cysteine 86 and cysteine 207, while in model B cysteine 41 would be more isolated. As a third alternative, a study by Palmieri et al. (1990), clearly establishes the orientation of the N-terminal as being disposed to the cytoplasmic side. Although the projected position of cysteine 41 is imbedded in the lipid bilayer, the relative interactions between cysteines 41, 86 and 207 would be conserved as in Model A. Clearly, PAO binding to site-directed cysteine mutants of the PiT would help to sort out these issues.

Lastly, [^{125}I]PAO may be a useful tool for the identification of other high affinity PAO-binding proteins in mitochondria. The effect of PAO on the K^+ channel of the inner membrane (Diwan, et al. 1986; 1988) and the sensitivity of the pyruvate carrier to sulfhydryl modifying reagents (Naez, et al., 1990) may indicate that these proteins might also be labelled by [^{125}I]PAO and bind with high affinity to the PAO-affinity resin. Lastly, the use of IPAO may be extended to the introduction of heavy atoms in PAO-binding proteins for X-ray diffraction studies.

* To whom comments should be addressed

References:

Aquila, H., Link, T.A., and Klingenberg, M. (1987) Solute carriers involved in energy transfer of mitochondria form a heterologous protein family. FEBS Lett. 212: 1-9.

Ballestra, J.J., Garcia, A.G., Gutierrez, L.M., Hildalgo, M.J., Pamero, M., Reig, J.A., and Viniegra, S. (1990) Separate [^{3}H]nitrendipine binding sites in mitochondria and plasma membrane of bovine adrenal medulla. Brit. J. Pharmacol. 101: 21-26.

Bernardi, P. (1992) Modulation of the mitochondrial cyclosporin A-sensitive permeability transition pore by the proton electrochemical gradient. Evidence that the pore can be opened by membrane depolarization. J. Biol. Chem. 267: 8834-8839.

Bernardi, P., Vassanelli, S., Veronese, P., Colonna, R., Szabo, I., and Zoratti, M. (1992) Modulators of the mitochondrial permeability transition pore. Effect of protons and divalent ions. J. Biol. Chem. 267: 2934-2939.

Braestrup, C. and Squires, R.F. (1977) Specific benzodiazepine receptors in rat brain characterized by high-affinity [^{3}H]diazepam binding. Proc. Nat. Acad. Sci. 74: 3805-3809.

Catterall, W.A. and Pedersen, P.L. (1971) Adenosine triphosphatase from rat liver mitochondria. I. Purification, homogeneity, and physical properties. J. Biol. Chem. 246: 4987-4994.

Cattell, K.J., Knight, I.G., Lindop, C.R., and Beechey, R.B. (1970) The isolation of dicyclohexylcarbodiimide-binding proteins from mitochondrial membranes. Biochem. J. 117: 1011-1013.

Cattell, K.J., Lindop, C.R., Knight, I.G., and Beechey, R.B. (1971) The identification of the site of action of dicyclohexylcarbodiimide as a proteolipid in mitochondrial membranes. Biochem. J. 125: 169-177.

Chiesi, M., Rogg, H., Eichenberger, K., Gazzotti, P., and Carafoli, E. (1987) Stereospecific action of diltiazem on the mitochondrial Na/Ca^{++} exchange system and on sarcolemmal Ca^{++} channels. Biochem. Pharmacol. 36: 2735-2740.

Coty, W.A. and Pedersen, P.L. (1975) Phosphate transport in rat liver mitochondrial. Membrane components labeled by N-ethylmaleimide during inhibition of transport. J. Biol. Chem. 250: 3515-3521.

Cuatrecasas, P. and Hollenberg, M. (1976) Membrane receptors and hormone action. Adv. Protein Chem. 30: 251-445.

De Pinto, V., Tommasino, M., Benz, R., and Palmieri, F. (1985) The 35 kDa DCCD-binding protein from pig heart mitchondria is the mitochondrial porin. Biochim. Biophys. Acta 813: 230-242.

Diwan, J.J., Strivastava, J., Moore, C., and Haley, T. (1986) Stimulation of K+ flux into mitochondria by phenylarsine oxide. J. Bioener. Biomembr. 18: 123-134.

Diwan, J.J., Haley, T., Moore, C. (1988) Sensitivity of mitochondrial Mg++ flux to reagents which effect K+ flux. J. Bioener. Biomembr. 20: 261-271.

Enna, S.J. (1985) Radioreceptor assays. In Neurotransmitter receptor binding. 2nd edition, editied by H.I. Yamamura et al., Raven Press, pp 177-188.

Escubedo, E., Camins, A., Talaveron, C., and Camarasa, J. (1992) Characterization of [^{3}H]Ro5-4864 binding sites in rat vas deferens. J. Neurochem. 58: 39-45.

Ferreira, G.C., Pratt, R.D., and Pedersen, P.L. (1989) Energy-linked anion transport. Cloning sequencing and characterization of a full length cDNA encoding the rat liver mitochondrial proton/phosphate symporter. J. Biol. Chem. 264: 15628-15633.

Ferreira, G., Pratt, R.D., and Pedersen, P.L. (1990) Mitochondrial Proton/Phosphate Transporter. An antibody directed against the COOH terminus and proteolytic cleavage experiments provide new insights about its membrane topology. J. Biol. Chem. 265: 21202-21206.

Halpern, A. and Stocklin, G. (1977) Chemical and biological consequences of b-decay. Radiat. Environ. Biophysics 14: 167-183.

Hirsch, J.D., Beyer, C.F., Malkowitz, L., Loullis, C.C., and Blume, A.J. (1989) Characterization of ligand binding to mitochondrial benzodiazepine receptors. Mol. Pharm. 34: 164-172.

Hoffman, R.D. and Lane, M.D. (1992) Iodophenylarsine oxide and arsenical affinity chromatography: new probes for dithiol proteins. Application to tubulins and to components of the insulin receptor-glucose transporter signal transduction pathway. J. Biol. Chem. 267: 14005-14011

Hoffman, R.D. Wohlrab, H. and McEnery, M.W. (1992) manuscript in preparation.

Houstek, J. and Pedersen, P.L. (1985) Adenine nucleotide and phosphate transport systems of mitochondria. Relative localization of sulfhydryl groups based on the use of the novel fluorescent probe eosin-5-maleimide. J. Biol. Chem. 260: 6288-6295.

Joshi, S. and Hughes, J.B. (1981) Inhibition of uncoupling factor B activity by cadium ion, arsenite-2,3-dimercaptopronanol, and phenylarsine oxide, and preferential activation by dithiols. J. Biol. Chem. 256: 11112-11116.

Kinnally, K.W., Zorov, D., Antonenko, N., Snyder, S.H., McEnery, M.W., and Tedeschi, H., (1993) Mitochondrial benzodiazepine receptor linked to specific ion channels by nanomolar actions of ligands. Proc. Natl. Acad. Sci. in press.

Kolbe, H.V. and Wohlrab, H. (1985) Sequence of the N-terminal formic acid peptide and location of the N-ethylmaleimide binding site of the phosphate transport protein from beef heart mitochondria. J. Biol. Chem. 260: 15899-18906.

Laemmli, U.K. (1970) Cleavage of structural proteins during the assembly of the head of bacteriophage T4. Nature 227: 465-477.

Lauguin, G.J., Brandolin, G., Lunardi, J., and Vignais, P.V. (1978) Photoaffinity labelling of the adenine nucleotide carrier in heart and yeast mitochondria by an arylazido ADP analog. Biochim. Biophys. Acta 501: 10-19.

Lippe, G., Deni-Sala, F., and Sorgato, M.C. (1988) ATP synthase complex from beef heart mitochondria. Role of the thiol groups of the 25 kDa subunit of F_o in the coupling mechanism between F_o and F_1. J. Biol. Chem. 263: 18627-18634.

McEnery, M.W. (1992) The mitochondrial benzodiazepine receptor: evidence for association with the voltage-dependent anion channel (VDAC). J. Bioener. Biomembr. 24: 63-69.

McEnery, M.W., Buhle, E.L., Aebi, U., and Pedersen, P.L. (1984) Proton ATPase of rat liver mitochondria. J. Biol. Chem. 259: 4642-4651.

McEnery, M.W., Hullihen, J., and Pedersen, P.L. (1989) F_o proton channel of rat liver mitochondria. Rapid purification of a functional complex and a study of its interaction with the unique probe diethylstilbestrol. J. Biol. Chem. 264: 12029-12036.

McEnery, M.W., Snowman, A.M., Triffiletti, R.R., and Snyder, S.H. (1992) Isolation of the mitochondrial benzodiazepine receptor: association with the voltage-dependent anion channel and the adenine nucleotide carrier. Proc. Natl. Acad. Sci. 89: 3170-3174.

McGuinness, O., Yafei, N., Costi, A., and Crompton, M. (1990) The presence of two classes of high affinity cyclosporin A binding sites in mitochondria. Eur. J. Biochem. 194: 671-679

Naez, K.A., Muller, M., Zambrowicz., E.B., Wojtczak, L., and Azzi, A. (1990) Significance and redox state of the SH-groups in pyruvate carrier isolated from bovine heart mitochondria. Biochim. Biophys. Acta 1016: 272-279.

Novgorodov, S.A., Kultayeva, E.V., Yaguzhinsky, L.S., Lemshko, V.V. (1987) Ion permeability induction by the SH crossling reagents in rat liver mitochondria is inhibited by the free radical scavenger butylhydroxytoluene. J. Bioener. Biomembr. 19: 191-202.

Palmieri, F., Bisaccia, F., Capobianco, L., Iacobazzi, V., Indiveri, C., and Zara, V. (1990) Biochim. Biophys. Acta 1018: 147-140.

Pratt, R.D., Ferreira, G., and Pedersen, P.L. (1991) Mitochondrial phosphate transport. Import of the H+/Pi symporter and role of the presequence. J. Biol. Chem.. 266: 1276-1284.

Rasmussen, U.B. and Wohlrab, H. (1986) Conserved structural domains among species and tissue-specific differences in the mitochondrial phosphate-transport protein and the ADP/ATP carrier. Biochim. Biophys. Acta 852: 306-314.

Sanadi, D.R., Hughes, J.B., and Joshi, S. (1981) Activation of potassium dependent H+ efflux from mitochondria by cadium and phenylarsine oxide. J. Bioener. Biomembr. 13: 425-431.

Snyder, S.H. (1986) Brain receptors: the emergence of a new pharmacology. Trends Neurosci. 455-459.

Szewczyk, A., Claeys, D., Mertens, W., and Azzi, A. (1991) Azido derivative of tricarboxylic acid for photoaffinity labelling. Biochim. Biophys. Acta 1073: 209-212.

Tesson, F., Prip-Buus, C., Lamoine, A., Pegorier, J.P., Parini, A. (1991) Subcellular distribution of imidazoline-guanidinium receptive sites in human and rabbit liver. Arch. Biochem. Biophys. Major localization to the mitochondrial outer membrane. J. Biol. Chem. 266: 155-160.

Tommasino, M., Prezioso, G., and Palmieri, F. (1987) Photoaffinity labeling of the mitochondrial phosphate carrier by 4-azido-2-nitrophenyl phosphate. Biochim. Biophys. Acta 890: 39-46.

Webb, J.L. (1963) Enzyme and Metabolic Inhibitors. Vol. III, pp. 595-793, Academic Press, N.Y.

Wehrle, J.P., Cintron, N.M., and Pedersen, P.L. (1978) Phosphate transport in rat liver mitochondria,. Energy dependent accumulation of phosphate by inverted inner membrane vesicles. J. Biol. Chem. 253: 8598-8603.

Winkler, E. and Klingenberg, M. (1992) Photoaffinity labeling of the nucleotide-binding site of the uncoupling protein from hampster brown adipose tissue. Eur. J. Biochem. 203: 295-304.

Yagi, T. and Hatefi, Y. (1984) Thiols in oxidative phosphorylation: inhibition and energy potentiated uncoupling by monothiols and dithiol modifiers. Biochem. 23: 2449-2455.

Zernig, G. (1990) Widening potential for Ca++ antagonists: non-L-type Ca++ channel interaction. Trends Pharmacol. Sci 11: 38-44.

Zernig, G. and Glossmann, H. (1988) A novel 1,4-dihydropyridine-binding site on mitochondrial membranes from guinea pig heart, liver and kidney. Biochem. J. 253: 49-58.

Zernig, G., Graziadei, I., Moshammer, T., Zech, C., Reider, N., and Glossmann, H. (1990) Mitochondrial Ca++ antagonist binding sites are associated with an inner mitochondrial membrane ion channel. Mol. Pharmacol. 38: 363-369.

CHANNEL PROPERTIES OF MITOCHONDRIAL CARRIERS

Thomas Dierks, Reiner Stappen and Reinhard Krämer
Institut für Biotechnologie
Forschungszentrum Jülich
5170 Jülich
F.R.G.

Solutes cross lipid bilayers by various mechanisms including diffusion, flux through pores, and carrier mediated processes. Although distinct examples can be well classified, the discrimination between carrier mechanisms and channel (pore) processes is not always unequivocal.

Carrier systems of the inner mitochondrial membrane are in general regarded as explicit examples for the carrier type of solute transport, catalyzing movement of substrates and ions by uniport, symport and antiport mechanisms [Krämer & Palmieri, 1992]. By protein and DNA sequencing, the pattern of a structural family of these carriers has evolved [Aquila et al., 1987; Runswick et al., 1990; Krämer & Palmieri, 1992]. This pattern was extended by studies on kinetic mechanisms in reconstituted systems, leading to the definition of a corresponding functional family [Krämer & Palmieri, 1992]. This means that mitochondrial carriers, the kinetic mechanism of which has been elucidated in detail so far, function according to a sequential type of mechanism. This has been shown for the oxoglutarate, the aspartate/glutamate, the citrate, and the dicarboxlyate carrier [Sluse et al, 1979; Dierks et al., 1988; Indiveri et al., 1991; Bisaccia et al., 1992]. The sequential mechanism involves a ternary complex of the carrier protein and two substrate molecules which are bound before the translocation step in the catalytic cycle of transport occurs [Dierks et al., 1988]. The only exception of this functional family of mitochondrial carrier proteins, studied in detail so far, seems to be the carnitine carrier, which presumably functions according to a ping-pong mechanism, thus only accepting

NATO ASI Series, Vol. H 83
Molecular Biology of Mitochondrial Transport Systems
Edited by M. Forte and M. Colombini

one substrate at a given time [C.Indiveri, personal communication].

Recent studies in our laboratory, using functionally reconstituted mitochondrial carrier proteins, showed that besides the kinetic properties as discussed above, which are fully in agreement with the mechanism of typical carrier-type transport proteins, channel-type properties can be revealed after specific modifications, at least in the case of some of the mitochondrial carriers. This has been especially well established for the aspartate/glutamate carrier [Dierks et al, 1990a,b], however, also other mitochondrial carriers could be converted to uniport function to variable extent (see below).

Antiport and uniport function of the aspartate/glutamate carrier

Substrate transport by the aspartate/glutamate carrier, as in the case of other mitochondrial carriers too, is characterized by high substrate specificity, high activation energy (indicating significant conformational changes of the carrier protein during the catalytic cycle), and a "carrier-like" (i.e. relatively low) transport rate. Additionally, similar to the other antiport carriers, movement of one solute (aspartate or glutamate) is strictly coupled to the movement of the counter solute (aspartate or glutamate). A number of (unspecific) inhibitors, which in general means amino acid modifying reagents, have been tested in order to characterize essential amino acid residues of the aspartate/glutamate carrier in functionally reconstituted state (see below). In the course of these studies, we detected that upon addition of certain cysteine-specific inhibitors, the strict antiport function was converted into solute uniport [Dierks et al., 1990a,b]. The uniport mode of mitochondrial carrier proteins is usually measured as substrate efflux from proteoliposomes.

There are a number of arguments which prove that this particular function was not due to unspecific damage of the liposomal membrane by the applied reagents, but was in fact catalyzed by the reconstituted aspartate/glutamate carrier. (i) The functional conversion from antiport to uniport, as induced by cysteine modification, was fully reversible upon addition of SH-reagents like DTE. (ii) The activity of the induced uniport was linearily dependent on the amount of inserted protein. (iii) If the protein was in denatured state, no efflux could be observed. (iv) The induction of substrate efflux could be specifically prevented by addition of aspartate and glutamate, the correct ligands of the aspartate/glutamate carrier (see below).

After conversion to the uniport mode, the reconstituted aspartate/glutamate carrier functionally combines properties of a typical carrier protein with some of that related to a channel (Tab. I).

Tab. I. Comparison of kinetic properties of the reconstituted aspartate/glutamate carrier in antiport and in uniport mode.

Property compared	antiport mode	uniport mode
transport mechanism	strictly coupled antiport	uniport (efflux)
K_M (substrate affinity)	50 - 100 μM	≥ 200 mM
substrate specificty	very high (only asp and glu)	very low (many ions, size selection)
activation energy (15 - 30 ^oC)	77 kJ/mol	86 kJ/mol
first order rate constant of transport	0.3 - 0.6 min^{-1}	up to 0.4 min^{-1}
substrate interaction	+ (binding)	(+) (protection against inhibition)

The original ("carrier-type") rate was conserved, as well as the original activation energy, indicating comparable conformational changes within the carrier protein during the translocation step. On the other hand, substrate affinity as well as substrate specificity were more or less completely lost. Under these conditions, the K_M ("transport affinity") became unmeasurably high, and transport of other compounds such as small inorganic ions (chloride, sufate), some carboxylic acids as well as uncharged molecules up to the size of glucose was observed. Positively charged (lysine) and larger molecules (ATP) were excluded. These results can be summarized by the statement, that the reconstituted aspartate/glutamate carrier under conditions mentioned above functioned as a more or less unspecific channel with the energetic and some kinetic properties of a carrier.

Table II. Comparison of transport acitivity of several mitochondrial carriers when functioning in the transport modes of antiport, physiological uniport (CAC) and symport (PIC), respectively, as well as mercurial induced uniport (efflux). Carrier proteins: AGC, aspartate/glutamate carrier; AAC, ADP/ATP carrier; CAC, carnitine carrier; PIC, phosphate carrier. The antiport activity of the reconstituted carriers was set to 100% in all cases, since the different systems are not directly comparable (various reconstitution methods and efficiencies). Because mercurial induced uniport was measured under significantly different conditions as compared to the physiological transport activities, the relative rates are estimates only.

Carrier	relative activity [%] of: physiological antiport	physiological uniport/ symport	mercurial induced uniport
AGC	100	≤ 2	up to 80
AAC	100	< 1	up to 10
CAC	100	10	≈ 10
PIC	100	35	up to 100

Uniport function of other mitochondrial carriers.

The functional interconversion between antiport and uniport has not only been observed in the case of the reconstituted aspartate/glutamate carrier, but has been detected with respect to at least three other mitochondrial transport proteins, i.e. the ATP/ADP carrier [Dierks et al, 1990a], the carnitine carrier [Indiveri et al., 1992a], and the phosphate carrier [R.Stappen and R.Krämer, unpublished results]. Also the oxoglutarate carrier in principle could be converted to uniport function [C.Indiveri & T.Dierks, unpublished results].

The extent of uniport induction was different in the different carriers studied so far (Table II). When comparing these activities it has to be taken into account that two of the carriers (AGC and AAC) only function in antiport mode under physiological conditions, whereas the carnitine and the phosphate carrier both function in antiport and in uniport or symport mode, respectively, already in intact mitochondria.

Molecular basis for uniport induction of the aspartate/glutamate carrier

A long list of reagents has been tested on their influence on the activity of the reconstituted aspartate/glutamate carrier. These reagents could be classified according to their functional effects and their specificity of reaction. A number of reagents had to be critically tested with respect to their actual target amino acid which, in some cases, proved to be different from generally acknowledged reaction schemes. We applied the following cysteine reagents *p*-(chloromercuri) benzoate (PCMB), *p*-(chloromercuri)benzenesulfonate (PCMBS), 5,5'-dithiobis(2-nitrobenzoic acid) (DTNB), mersalylic acid, $HgCl_2$, and N-ethyl maleimide (NEM). Furthermore, 7-chloro-4-nitrobenzo-2-oxa-1,3-diazole (NBD-Cl), which is reported to modify cysteine, lysine as well as tyrosine residues, was shown to react solely with SH-groups of the reconstituted aspartate/glutamate carrier [Stappen et al., 1992]. *p*-Bromophenacyl

Tab.III. Classification of specific amino acid reagents (see text for explanations and reagent abbreviations). IC_{50} (AC_{50}): inhibitor (activator) concentration leading to 50% inhibition (activation). Substrate protection by NEM is weak (+), reagent classification by competition with PCMB. EDC was applied in the presence of 50 mM acetate [Dierks et al., 1992].

reagent class	reagent	modified residue	IC_{50} (µM)		AC_{50} (µM)	substrate protection against		
			antiport	uniport	uniport induction	antiport inhibit.	uniport inhibit.	uniport induction
A: inhibition of antiport, no effect on uniport								
	PCMB	cys(a)	2	-	-	+		
	BPAB	cys(a)	3	-	-	+		
	DTNB	cys(a)	10	-	-	+		
	NBD-Cl	cys(a)	50	-	-	+		
	NEM	cys(a)	1000	-	-	(+)		
B: low reagent concentration: inhibition of antiport high reagent concentration: inhibition of antiport and induction of uniport								
	$HgCl_2$	cys(a)+(b)	2	-	2	-		+
	Mersalyl	cys(a)+(b)	2	-	100	-		+
	PCMBS	cys(a)+(b)	2	-	40	-		+
	DIDS	cys(a)+lys	10	-	70	-		+
C: inhibition of both antiport and uniport								
	DEPC	his	20	80	-	+	+	
	EDC/acet.	lys	500	4000	-	+	+	
	PLP	lys	440	>1000	-	+	+	
	SITS	lys	300	1000	-	+	+	

bromide (BPAB), generally acknowledged as histidine reagent, also modified cysteine(s) in this case. In addition to the well known lysine reagents pyridoxal 5'-phosphate (PLP) and 4-acetamido-4'-isothiocyanostilbene-2,2'-disulfonate (SITS), the carbodiimide 1-ethyl-3-(3-dimethylaminopropyl)-carbodiimide (EDC) did not react with with carboxylic side chains of the carrier protein but led to acylation of lysine in the presence of carboxylic acids [Dierks et al., 1992]. The bifunctional stilbene 4,4'-diisothiocyanostilbene-2,2'-disulfonate (DIDS) was shown to react with both a lysine and a cysteine residue of the aspartate/glutamate carrier [Stappen et al., 1992]. Finally, diethyl pyrocarbonate (DEPC) was used as histidine-specific reagent.

In the case of several of these reagents we analyzed the time course of inhibition in presence of various inhibitor concentrations. Thereby we could show that reaction with only one single amino acid led to complete inhibition of the carrier function. This could be demonstrated for EDC/acetate (reaction with one lysine), SITS (cysteine), NBD-Cl (cysteine) and DEPC (histidine). By consecutive modification using different reagents and by competition experiments we could further show that under conditions of uniport induction, two (classes of) cysteines were modified by that type of reagents which converted the aspartate/glutamate carrier into uniport function. Further characterization was obtained by experiments testing the protection against inhibition by the presence of the substrate ligands of the carrier (i.e. aspartate or glutamate). This means that the specific modification of a particular amino acid of the carrier protein by these inhibitors was prevented by addition of carrier substrates. It turned out that several of the identified amino acids could fully and specifically be protected by the presence of substrate. Even more important was the fact that the concentration of substrate necessary to protect 50% of the carrier's activity was in all cases more or less identical with the observed apparent substrate affinity of the aspartate/glutamate carrier. This was taken as indication that the amino acid residues, which were characterized by this type of specific interaction with the transport substrate as well as with the inhibiting ligand, reside within or near the active site of the carrier.

On the basis of these experiments, we classified the inhibitor reagents in the following way (Tab. III). Reagents of class A solely react with cysteine (a) in the active site thus blocking the (physiological) antiport function of the aspartate/glutamate carrier. Reagents of class B react with two cysteines (a) and (b), or with cys (a) and lysine in the case of DIDS. Modification of either of these amino acids leads to antiport inhibition, modification of both residues together additionally induces the above described uniport function of the protein. Finally, reagents of class C modify the active site lysine (or histidine) thereby completely blocking all activities of the carrier. The interconversion of the various functional states of the carrier by these types of reagents and the reversion of these events by dithioerythitol is summarized in Fig. 1.

As described above, we obtained functional evidence that one cysteine, one lysine and one histidine is located in or near the active site. Since most of the reagents applied are impermeable and since the aspartate/glutamate carrier has been shown to be oriented right-side out in proteoliposomes [Sluse et al., 1991], these inhibitors affect the exofacial binding site of the carrier protein. By consecutive application of various reagents it could be shown that after modification of the active site lysine, only small SH-reagents (e.g. $HgCl_2$) could bind to cys(a), whereas bulky inhibitors (e.g. $HgCl_2$/EDTA) were excluded [Stappen et al., 1992]. No substrate protection could be detected for cys(b), furthermore, reagents leading to S-S bridge formation did not inhibit carrier function, i.e. cys(a) and cys(b) are not located closely enough to be crosslinked by these reagents. These and other experiments led to the suggestion of a topological model of the exofacial active site of the reconstituted aspartate/glutamate carrier (Fig. 2).

Fig. 1. Interconversion of functional states of the aspartate/ glutamate carrier.

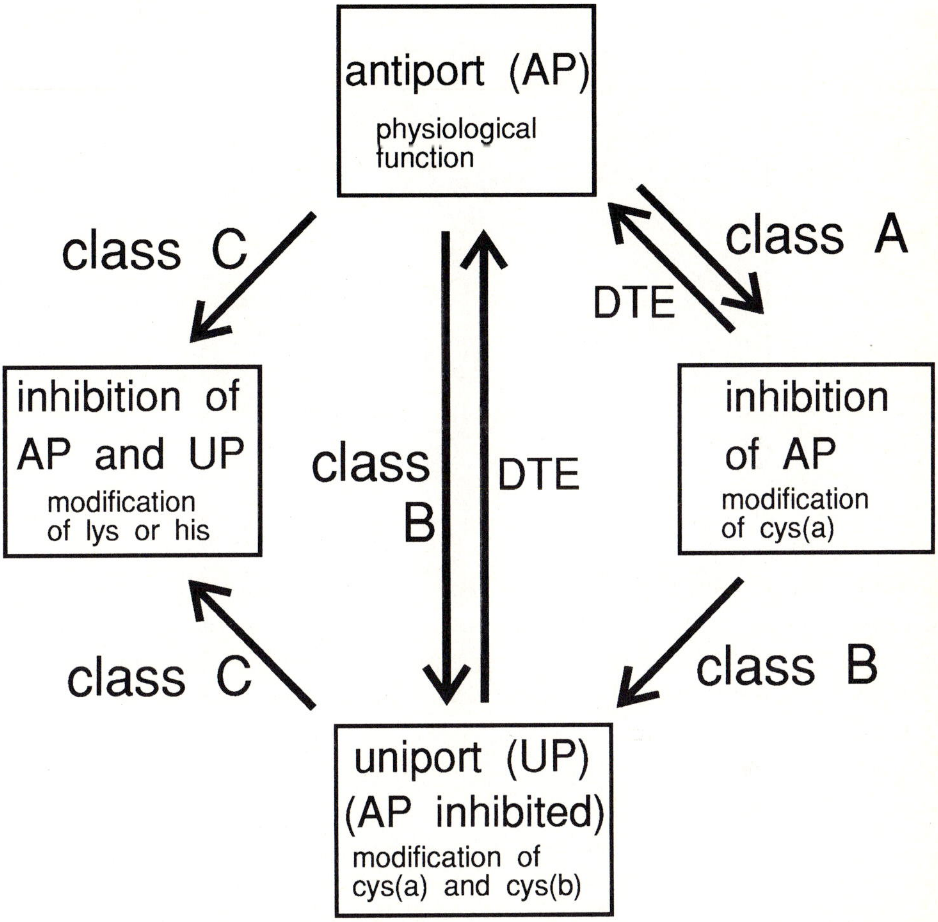

Interconversion of four different functional states (boxed) of the aspartate/glutamate carrier by the three classes of reagents (cf. Tab.III) and dithioerythritol (DTE). In the physiological antiport mode all relevant amino acid residues are unmodified. Reversible modification of cys(a) by class A reagents (for classification of reagents see Tab. III) leads to inhibition of antiport. Modification of cys(a) and (b) together by class B reagents leads to reversible in-duction of uniport. Class C reagents react at the active site lysine or histidine and inhibit both antiport and uniport.

Fig. 2. Putative topological relationship of functionally important amino acid residues at the exofacial binding site of the aspartate/glutamate carrier.

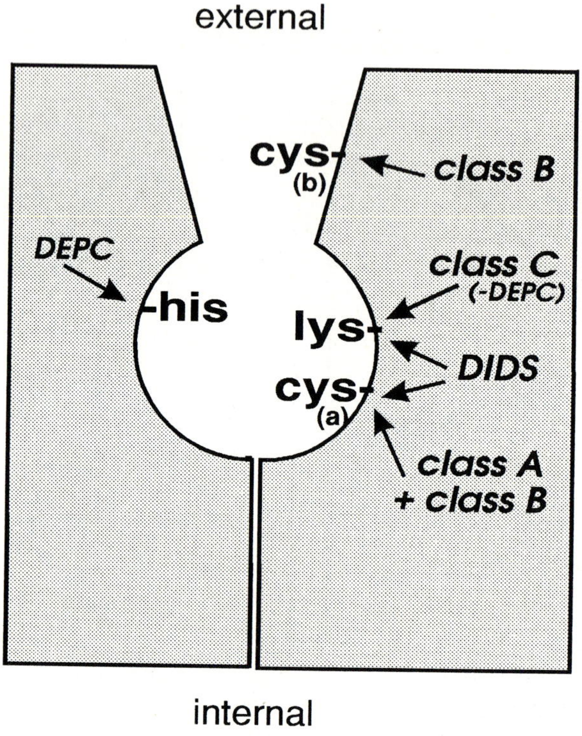

Functional significance of uniport activity

It is very likely that the observed uniport (efflux) function of mitochondrial carrier proteins, as induced by treatment with particular (mercurial) reagents, has no physiological significance in intact mitochondria. Besides the fact that this result may explain a lot of old observation made with isolated mitochondria, where efflux of solutes was observed after treatment with mercurials [for review see Dierks et al., 1990a], important conclusions can be drawn regarding the molecular function of mitochondrial transporters which may also have implications for understanding carrier function in general.

It is obvious, that the chemical modifications introduced by the reagents discussed above change the binding site of the carrier protein. This can be concluded on the basis of the observed loss of binding affinity and substrate specificity. Simultaneously we observed that the functional coupling of the movement of two solute molecules (i.e. antiport mode in a sequential type of transport mechanism) was lost. In other words, the antiport carrier was functionally converted into a uniporter. On the other hand, this conversion did not result in formation of a typical channel, since the overall translocation rate was not enhanced and even the activation energy of the transport event remained constant. These results can be interpreted by assuming that the actual conformational change of the carrier protein within the transport cycle, as indicated by the size of the activation energy, remained essentially the same, irrespective whether functioning in the (physiological) antiport mode or in the (artefactual) uniport mode induced by reversible chemical modification.

We conclude that by this specific modification we can functionally discriminate between gating functions of the carrier protein and channel functions. The gating functions are severely affected by the reagents applied and are lost to large extent after modification of two particular cysteine residues at or near the active site in the case of the aspartate/glutamate carrier. Modification of the two cysteines, besides knocking out substrate recognition, obviously interferes with the tight, so-called "intrinsic" binding of the substrate [Jencks, 1980; Klingenberg, 1991], which is considered to be the crucial step during transport catalysis. The channel functions became apparent only after chemical modification of the gates. The intrinsic channel obviously does not discriminate substrates within a wide range with respect to molecular weight (size) and even charge. Passage through this channel, however, is obviously always correlated to the same conformational change of the carrier protein.

Thus we interpret our data as a functional indication of a permanent intrinsic channel present within at least the mitochondrial antiport carriers and possibly within other carrier proteins too. In this context it is tempting to speculate that this intrinsic channel may be an evolutionary conserved element already present in the most simple transporters (pores, facilitators) up to the highly sophisticated membrane bound protein machineries coupling two and more solute fluxes in a useful way [Saier, 1991].

Acknowledgement: Continuous support of Prof. H. Sahm, Jülich, is greatfully acknowledged. This work was supported by grants from the Deutsche Forschungsgemeinschaft (SFB 189) and the Fonds der Chemischen Industrie.

References:

Aquila H, Link TA, Klingenberg M (1987) Solute carriers involved in energy transfer of mitochondria form a homologous protein family. FEBS Lett 212:1-9

Bisaccia F, de Palma A, Dierks T, Krämer R, Palmieri F (1992) Reaction mechanism of the tricarboxylate carrier from rat liver mitochondria. Biochim Biophys Acta, in press

Dierks T, Riemer E, Krämer R (1988) Reaction mechanism of the isolated and reconstituted aspartate/glutamate carrier. Biochim Biophys Acta 943:231-244

Dierks T, Salentin A, Heberger C, Krämer R (1990a) The mitochondrial aspartate/glutamate and ADP/ATP carrier switch from obligate counterexchange to unidirectional transport after modification by SH-reagents. Biochim Biophys Acta 1028:268-280

Dierks T, Salentin A, Krämer R (1990b) Pore-like and carrier-like properties of the mitochondrial aspartate/glutamate carrier after modification by SH-reagents: evidence for a preformed channel as astructural requirement of carrier-mediated transport. Biochim Biophys Acta 1028:281-288

Dierks T, Stappen R, Salentin A, Krämer R (1992) Probing the active site of the reconstituted aspartate/glutamate carrier from bovine heart mitochondria: carbodiimide-catalyzed acylation of a functional lysine residue. Biochim Biophys Acta 1103:13-24

Indiveri C, Dierks T, Krämer R, Palmieri F (1991) Reaction mechanism of the reconstituted oxoglutarate carrier from bovine heart mitochondria. Eur J Biochem 198:339-347

Indiveri C, Tonazzi T, Dierks T, Krämer R, Palmieri F (1992a) The mitochondrial carnitine carrier: Characterization of SH-groups relevant for its transport function. Biochim Biophys Acta, in press

Jencks WP (1980) The utilization of binding energy in coupled vectorial processes. Adv Enzymol 51:75-106

Klingenberg M (1991) Mechanistic and energetic aspects of carrier catalysis - exemplified with mitochondrial translocator. In A study of Enzymes Vol 2 (Kuby SA, ed.) pp. 367-389

Krämer R, Palmieri F (1992) Structure and function of metabolite carriers in mitochondria. In Comprehensive Biochemistry (Ernster, L., ed.), in press

Runswick MJ, Walker JE, Bisaccia, F, Iacobazzi, V, Palmieri, F. (1990) Sequence of the bovine 2-oxoglutarate/malate carrier protein - Structural relationship to other mitochondrial transport proteins. Biochemistry 29:11033-11040

Saier M (1990) Evolution of permease diversity and energy-coupling mechanisms. Res Microbiol 141:281-286

Sluse FE, Duyckaerts C, Liebecq S, Sluse-Goffart M (1979) Kinetic and binding properties of the oxoglutarate translocator of rat-heart mitochondria. Eur J Biochem 100:3-17

Sluse FE, Evens A, Dierks T, Duyckaerts C, Sluse-Goffart CM, Krämer R (1991) Kinetic study of the aspartate/glutamate carrier in intact rat heart mitochondria and comparison with a reconstituted system. Biochim Biophys Acta 1058:329-338

Stappen R, Dierks T, Salentin A, Krämer R (1992) Probing the active site of the reconstituted aspartate/glutamate carrier from bovine heart mitochondria: Structure-function relationship involving one lysine and two cysteine residues. Eur J Biochem, in press

A MINIMAL KINETIC MODEL OF THE ACTIVITY OF THE 107 pS CHANNEL OF THE INNER MEMBRANE OF MITOCHONDRIA

C. Ballarin, M. C. Sorgato, Dipartimento di Chimica Biologica, Via Trieste 75, 35121 Padova (Italy) and O. Moran, Istituto CNR di Cibernetica e Biofisica, Via Dodecaneso 33, 16146 Genova (Italy)

Introduction

Large conductances of different kind have been found in the inner membrane of mitochondria by direct observation of the membrane with a patch clamp electrode (Moran and Sorgato, 1992; Szabò and Zoratti, 1992; Kinnally et al., 1992). The role of any of these is, however, still obscure. The first high conductance pathway to be detected has been a slightly anionic 107 pS channel (mCS, mitochondrial Centum pico-Siemens), observed *in situ* (i.e., in mitoplasts, which are mitochondria largely deprived of the outer membrane) as well as after reconstitution in large liposomes of both mitochondrial membranes, or of a fraction of the inner one (Sorgato et al., 1987, 1989; Moran et al., 1990; Moran and Sorgato, 1992). Interestingly, the mCS channel possesses a strong and peculiar voltage sensitivity. At high positive or low negative values of the applied potential difference the channel is active, whereas at high negative potentials the channel is in the non-conductive state. As mitochondria sustain a membrane potential negative inside (on the matrix side), the voltage dependence displayed by the channel appears rather intriguing so as to raise the question of whether the voltage gating occurs as such *in vivo* or whether it is consequent to the experimental conditions used. In fact, cytosolic or mitochondrial matrix components, which might be involved in regulating the channel, are likely to get lost during the preparation of mitoplasts. On the other hand, the presence of the mCS channel in all tissues so far tested (liver, heart and adipose tissue) (Sorgato et al., 1987, 1989; Klitsch and Siemen, 1991) argues strongly in favour of a role of physiological importance.

Here, a description of the kinetic behaviour of the mCS channel is presented, which casts light on the molecular mechanism of the channel.

NATO ASI Series, Vol. H 83
Molecular Biology of Mitochondrial Transport Systems
Edited by M. Forte and M. Colombini

Kinetics of the 107 pS channel

Fig. 1A shows the voltage sensitivity displayed by a single mCS channel, which opens at positive potential differences (+40 mV) and closes when the voltage is negative (-40 mV). Interestingly, also the current-voltage relation obtained from a whole mitoplast-patch is strongly asymmetrical (Fig. 1B).

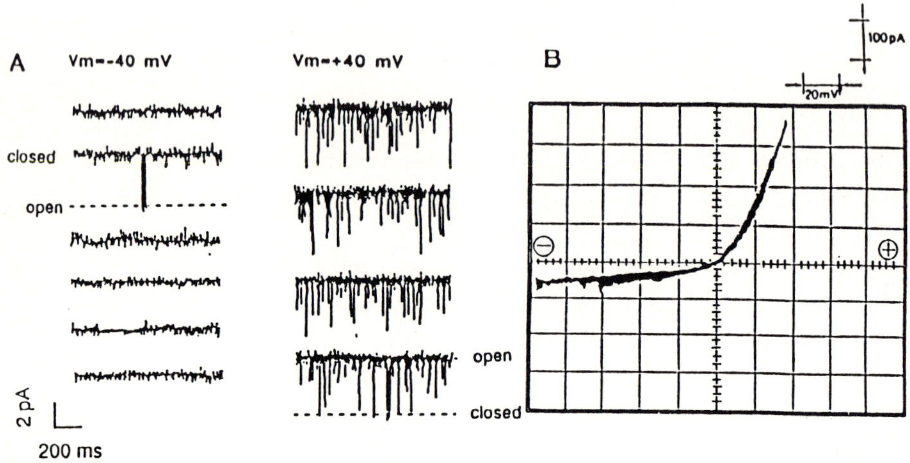

Fig. 1. Current-voltage relationship of the inner membrane of mouse liver mitochondria. Currents flowing through a single 107 pS channel (**A**) or the entire inner membrane (**B**) are recorded from a patch in the mitoplast attached- and whole mitoplast-configuration, respectively. Symmetrical 150 mM KCl solutions are used in both cases. Potential values refer to those found inside the mitoplast. Fig. 1B is taken from Moran and Sorgato (1992), with permission.

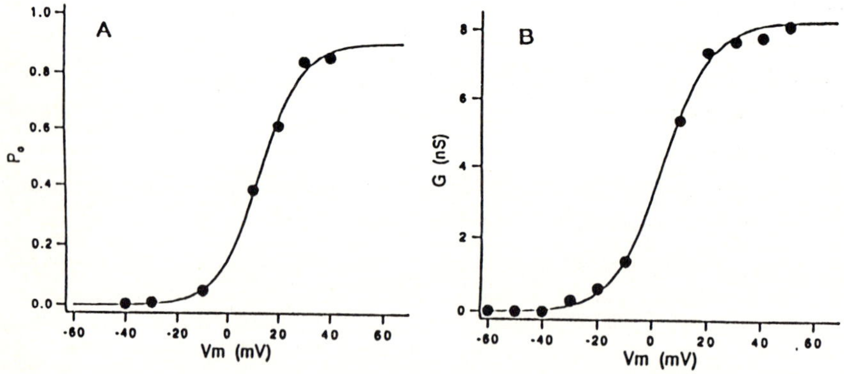

Fig. 2. Variation with voltage of **A**, the open probability (P_O) of a single 107 pS channel; **B**, the integral conductance (G). Experimental conditions are as in Fig. 1.

Therefore, in order to ascertain whether a number of mCS channels is the source of the macroscopic current, the open probability (P_O) of a single channel at different voltages is compared with the macroscopic conductance of the inner membrane (G). This operation is justified by the proportionality between G and P_O in the equation $G=N\gamma P_O$, under the assumption that the number of channels (N) and the conductance of each channel (γ) remains unaltered throughout the potential change. Fig. 2 shows that in either case a similar sigmoidal curve is obtained and that the P_O for a single channel can be close, but not equal, to one.

From the slope of the curves, the number of equivalent charges (z) of the voltage sensor, being moved from one side to the other of the membrane during the closing (or opening) event, can be obtained (Hille, 1984). Again similar values are found, 3.04 for the single channel and 2.83 for the whole membrane. These results thus strongly support the hypothesis that the responsible for the macroscopic current flowing through the entire inner membrane is indeed the mCS channel. Moreover, from data of Fig. 1B and of similar experiments, it has been calculated that the value of N ranges approximately from 50 to 300.

From data of Fig. 2, the value of V_O, i.e. the potential at which there is an equal probability for a channel to be open or closed, can also be estimated (Hille, 1984). A slight difference is found in the two cases, the V_O for the single channel, 12.94, being greater than that for the whole membrane, 3.96. This result does not necessarily invalidate the conclusion reached above. Indeed, the found difference can be explained in that, contrary to the mitoplast attached-configuration used to record the single channel P_O, the matrix content is dialysed away in a whole mitoplast-patch. Alternatively, as the two sets of data of Fig. 2 have been obtained with different mitoplast preparations, differences within the membrane stocks might be responsible for the found variation in the V_O value.

Fig. 3 A reports a typical mCS single channel behaviour at + 30 mV. It is evident that the channel can either flicker rapidly between the open and closed state, or remain in the closed state for a more prolonged time. Analysis of the distribution of the open and closed times (Fig. 3B) demonstrates that the former can be fitted with only one exponential. On the contrary, the distribution of the closed times clearly shows two components of different magnitude, which well represent the flickering and longer closures evident in the trace of Fig. 3A, i.e., a very fast process with a time constant (τ) of less than 1 msec and a slower event with a τ of much higher value.

The time constants for the distribution of the open and closed events were then analysed at different voltages of opposite sign (Fig. 4). The time constant for the distribution of the open event is clearly dependent on the sign and magnitude of the applied voltage, varying from around 50 msec at + 40 mV to around 4.5 msec at -30 mV (Fig. 4A).

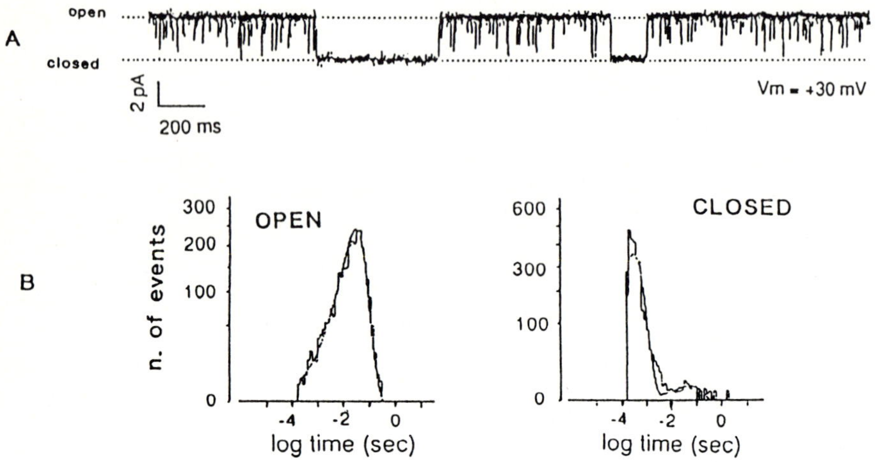

Fig. 3. Fast and slow closure events of the 107 pS channel. **A**. Trace recorded from a mouse liver mitoplast attached-patch at 30 mV. **B**. Distribution of open and closed times at 30 mV.

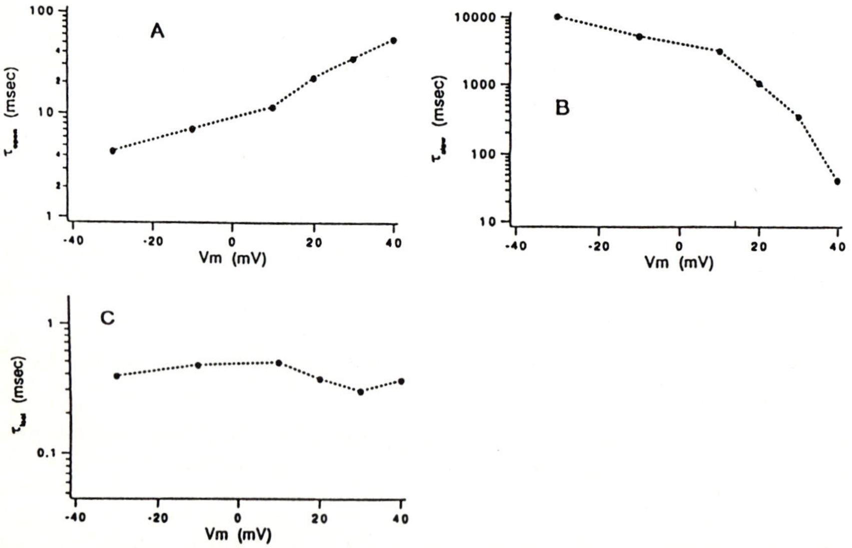

Fig. 4. Effect of voltage on the time constants of the open (**A**) and closed events (**B, C**) of the 107 pS channel.

As expected, the time constant of the slow closure event behaves in the opposite fashion: it increases in value as the voltage is made more negative (10 sec at - 30 mV) and decreases with increasing of the positive potentials (40 msec at + 40 mV) (Fig. 4B). Conversely, the fast component of the closing event of around 0.4 msec is largely independent of voltage (Fig. 4C).

Fig. 5 reports examples of the activation and deactivation phenomenon encountered by the channel at changes of the sign of the applied potential. In Fig. 5A it is shown that if the starting potential is negative (- 40 mV, in correspondence to which the channel is closed) there is a time latency of several seconds before the channel opens at positive potentials. Viceversa, if the potential is switched from positive to negative (Fig. 5 B and C), full closure is delayed the more the lower the value of the negative pulse, after which there is no change of the closed state.

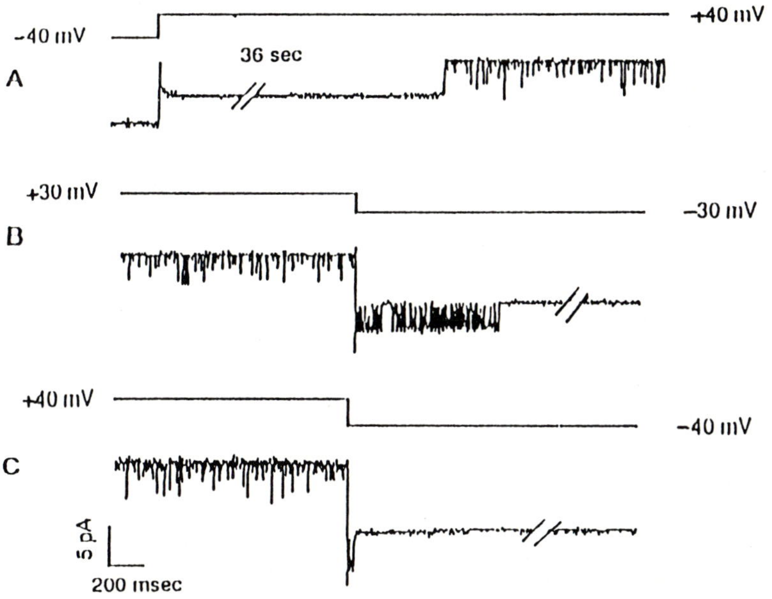

Fig. 5. Time delay of the activation (**A**) and deactivation (**B,C**) processes of the 107 pS channel. Interruptions in the middle and bottom traces indicate a time interval of minutes.

Minimal model of the mCS channel

The data presented above enable the construction of a linear scheme for the mCS channel activity (Fig. 6). It is hypothesized one open state (O), predominant at positive potentials (Fig. 4A), which is in very rapid equilibrium with the closed conformation C_1. C_1 is representative of the voltage-independent closed state (Fig. 4C) reached in the flickering events. Moreover the existence of such state explains the impossibility for the mCS channel to have a P_o equal to 1 (Fig. 2A). Another closed conformation assumed by the channel, C_2, is more probable at negative potentials. C_2 is the slow closure event and the transition from C_1 is clearly voltage-dependent (Fig. 4B). Finally, the assumption of a third closed state, defined quiescent (Q), is justified in order to explain the long period of time necessary to activate the channel (Fig. 5).

$$Q \overset{*}{\leftrightarrow} C_2 \overset{*}{\leftrightarrow} C_1 \leftrightarrow O$$

Fig. 6. A minimal model of the activity of the 107 pS channel at different voltages. Asterisks indicate where rate constants are voltage dependent. For other symbols see text.

Acknowledgements

The work was funded in part by grants of the Ministero dell' Università e della Ricerca Scientifica e Tecnologica of Italy and of the Consiglio Nazionale delle Ricerche to the Unit for the Study of Mitochondrial Physiology.

References

Hille B (1984) Ionic channels of excitable membranes. Sinauer Associates Inc., Sunderland, Ma.

Kinnally KW, Antonenko YN, Zorov DB (1992) Modulation of the inner mitochondrial channel activity. J Bioenerg Biomembr 24: 99-110

Klitsch T, Siemen D (1991) Inner mitochondrial membrane anion channel is present in brown adipocytes but is not identical with the uncoupling protein. J Membr Biol 122: 69-75

Moran O, Sorgato MC (1992) High-conductance pathways in mitochondrial membranes. J Bioenerg Biomembr 24: 91-98

Moran O, Sandri G, Panfili E, Stühmer W, Sorgato MC (1990) Electrophysiological characterization of contact sites in brain mitochondria. J Biol Chem 265: 908-913.

Sorgato MC, Keller BU, Stühmer W. (1987) Patch-clamping of the inner mitochondrial membrane reveals a voltage-dependent ion channel. Nature 330: 498-500

Sorgato MC, Moran O, De Pinto V, Keller BU, Stühmer W (1989) Further investigation on the high-conductance ion channel of the inner membrane of mitochondria. J Bioenerg Biomembr 21: 485-496

Szabo' I, Zoratti M (1992) The mitochondrial megachannel is the permeability transition pore. J Bioenerg Biomembr 24: 111-117

A CHANNEL MODEL TO EXPLAIN REGULATION OF THE MITOCHONDRIAL INNER MEMBRANE ANION CHANNEL (IMAC)

Andrew D. Beavis
Department of Pharmacology
Medical College of Ohio
P.O. Box 10008
Toledo, Ohio 43699-0008
USA

INTRODUCTION

The mitochondrial inner membrane anion channel (IMAC) is a transport pathway in the inner mitochondrial membrane which is able to transport a wide variety of anions ranging from Cl^- to ATP^{4-} (see Beavis (1992) and references therein for a review). In addition to its very broad selectivity, it differs from all the other anion transporters of the inner membrane in that it mediates rapid electrophoretic uniport. On the basis of these properties, we call it a channel. In all other respects, however, its properties are indistinguishable from those of a uniport carrier. Its turnover number is not known, since it has not yet been identified or purified and therefore its abundance is not known.

In respiring mitochondria, IMAC should normally mediate anion efflux driven by the membrane potential generated by the H^+ pumps of the respiratory chain. The resulting pH gradient then provides the driving force for electroneutral K^+ efflux via the K^+/H^+ antiporter. The net effect is respiration driven salt efflux. For volume homeostasis, this efflux must balance the net salt influx, which is also driven by respiration and mediated via the K^+ channel, K^+ leaks and the electroneutral anion transport pathways. Regulation of these processes is believed to be the mechanism by which mitochondrial volume homeostasis is achieved

Abbreviations: IMAC, Inner membrane anion channel; NEM, N-ethylmaleimide; DCCD, N,N'dicyclohexylcarbodiimide; TBT, Tributyltin

NATO ASI Series, Vol. H 83
Molecular Biology of Mitochondrial Transport Systems
Edited by M. Forte and M. Colombini

(Beavis (1992)). Consistent with this proposed function, IMAC is found to be highly regulated. Most of the regulatory properties of IMAC have, however, been established not by assaying respiration driven anion efflux but by assaying anion influx driven by a concentration gradient in the absence of a membrane potential.

PROPERTIES OF IMAC

Introduction - The first evidence for the existence of IMAC was obtained by Azzi and Azzone (1966,1967) who showed that mitochondria become permeable to anions at alkaline pH. Brierley (1970) subsequently showed that alkalinization of the mitochondrial matrix is sufficient to increase anion permeability. More recently, Beavis and Garlid (1987) demonstrated that depletion of mitochondrial divalent cations, specifically matrix Mg^{2+} (Beavis and Powers (1989)), also increases anion permeability and proposed that activation of an anion channel is responsible. Thus, in freshly isolated mitochondria, IMAC is blocked by matrix H^+ and matrix Mg^{2+}. Consequently, in most studies of the regulation of IMAC, mitochondria are either depleted of divalent cations using the ionophore A23187 or assayed under conditions which elevate matrix pH to very high values (see Beavis (1992) for review).

Regulation by Matrix H^+ - Activation by alkaline pH is reversible and therefore H^+ may be considered as an inhibitor of IMAC. A Dixon plot permits the value of the flux at $[H^+] = 0$ to be determined and, this has enabled a value of 7.8 to be determined for the proton pIC_{50} for Cl^- transport (Beavis and Garlid (1987)).

Regulation by Matrix Mg^{2+} - Mg^{2+} inhibits IMAC in a straightforward manner with an IC_{50} of 38 μM and a Hill coefficient between 1 - 1.5. The IC_{50} for Mg^{2+} is, however, dependent on pH, and the relationship between them suggests that Mg^{2+} may bind to a protonated form of IMAC. Brierley's (1970) observation that matrix alkalinization is sufficient to activate IMAC and other evidence presented by Beavis and Powers (1989) suggests that this regulatory protonation site, like the inhibitory site, is located in the matrix. Estimates of the pK of this site, however, suggest that it is too low to be the same one responsible for inhibition.

Regulation by Cationic Amphiphiles - Although many cationic amphiphilic drugs have been shown to inhibit IMAC (Beavis (1989a); Beavis (1992)) most studies have been carried out with propranolol. At pH 7.4, an IC_{50} of 28 μM is obtained with a Hill coefficient of 1.3 - 1.4. As with Mg^{2+}, the IC_{50} is found to increase as the pH is raised but in a more complex manner. The Hill coefficient rises to a value close to 2.0 at pH 8.4. Other drugs such as clonazepam which lack charge also inhibit IMAC in a pH-sensitive manner. Thus, the charge does not seem necessary for inhibition. As with Mg^{2+}, the protonation site regulating propranolol inhibition also appears to be located in the matrix. However, the wide variety of drugs which inhibit IMAC and their relatively high IC_{50} values suggest that the amphiphilic amines interact with IMAC at a hydrophobic site located in the bilayer.

Effects of Mercurials and N-Ethylmaleimide - Transport via IMAC is inhibited by mercurials such as mersalyl and p-chloromercuribenzenesulfonate. In this respect IMAC is similar to other mitochondrial anion transporters. In fact, the dose required to inhibit, e.g. 20 nmol mersalyl/mg, is the same as that required to block the dicarboxylate carrier. Also, like the dicarboxylate carrier, IMAC is not blocked by NEM. Unlike the dicarboxylate carrier, the maximum extent of inhibition of IMAC observed is dependent on the anion species being transported and the size of the R-group of the mercurial. These findings suggest that inhibition results from steric hindrance, and it has been proposed that the mercurial binds to a non-essential -SH group located near the mouth of the channel or anion binding site (Beavis, 1989b, 1991).

Use of mercurials to inhibit IMAC is further complicated by the fact that they have a second effect, which is shared with NEM; they increase the IC_{50} for protons (Beavis (1989b,1991)). Thus, for anions whose transport is not blocked completely by mercurials, it is possible to see stimulation of transport by mercurials at low pH. Comparison of the effect of pH on the activity of IMAC in control and NEM-treated mitochondria suggests that NEM-binding to this regulatory site does not affect the value of J_o, the flux obtained by extrapolating to $[H^+] = 0$. Further studies have revealed that treatment with NEM and mercurials also increases the IC_{50} values for matrix Mg^{2+}, external Mg^{2+} and propranolol (Beavis (1991)).

Regulation by Nucleotide Analogs - Selwyn's group has investigated the effects of palmitoyl CoA (Halle-Smith et al (1988)), agaric acid, bonkrekic acid and Cibacron blue 3GA (Murray et al (1988)) and found that all inhibit Cl^- transport; however, inhibition is incomplete. Powers and Beavis (1990) confirmed this finding but also showed that inhibition of malonate transport approaches 100%. Thus, as with the mercurials, the extent of inhibition appears to depend on the anion. Other nucleotide analogs such as the fluorescein derivative erythrosin B, bromcresol green and alizarin red S also inhibit. Erythrosin is the most effective in inhibiting Cl^- transport. In addition to inhibition being incomplete, this class of inhibitors was also found to differ from H^+, Mg^{2+} and cationic amphiphiles in that the IC_{50} values are not increased but decreased by NEM.

Other Inhibitors - N,N dicyclohexylcarbodiimide (DCCD), which is a hydrophobic alkylating agent, inhibits IMAC irreversibly and completely (Beavis and Garlid (1988)). This reagent reacts with a number of mitochondrial proteins and is usually thought to react with carboxyl groups (Azzi et al (1984)), consequently, we have proposed that it blocks IMAC by reacting with a carboxyl group located in the lipid bilayer. One of the most potent and specific inhibitors of IMAC is tributyltin (TBT) (Powers and Beavis (1991)). This reagent completely blocks the transport of all anions through IMAC at doses around 1 nmol/mg. Inhibition can be prevented or reversed by addition of mercurials at the same dose required to block IMAC. Interestingly, however, while the inhibition by mercurials can be reversed by thiols such as thioglycolate, the protective effect cannot be reversed. It should be noted that thiols by themselves have no effect on inhibition by TBT. The mechanism of this protective effect is unknown; however, it has been shown that TBT can still block IMAC in the presence of mercurials but 10-fold higher doses are required. It has also been shown that NEM has no effect on inhibition by TBT; consequently, the effect of mercurials on inhibition by TBT appears to be effected via a third mercurial binding site (Powers and Beavis (1991)).

The Current Model of IMAC - From the summary of some of the properties of IMAC, presented above, it is evident that this is a complex transport pathway, capable of being modulated by a vast array of different agents. To rationalize many of these observations a model was introduced (Beavis (1991,1992)), in which two regulatory sites modulate the binding of H^+, Mg^{2+} and cationic amphiphiles. The first is a protonation site within the matrix which causes a decrease in IC_{50} for matrix Mg^{2+}, external Mg^{2+}, and propranolol. This could be achieved if the inhibitors were only able to bind to IMAC when the protonation site was occupied. The second is a putative -SH group located on the external surface of IMAC. Reaction of mercurials or NEM at this site induces a conformational change which causes an increase in the IC_{50} for external Mg^{2+}, internal Mg^{2+}, propranolol and protons. This model, however, says nothing about the mechanism of inhibition by these agents, nor does it provide an explanation for why reaction with NEM should decrease the binding constants for four different inhibitory sites. Thus, I shall now present an hypothesis which provides a possible explanation for these phenomena.

A CHANNEL HYPOTHESIS FOR IMAC

Recently Antonenko et al (1991) and Kinnally et al (1992) have obtained evidence for the existence of IMAC from patch-clamp studies. Channels are usually characterized by their ability to undergo spontaneous openings and closings, so that an open state is in equilibrium with a closed state. Thus, channel activity can be regulated not only by agents which block the channel or interfere with the transport mechanism directly, but also by any agent which can bind selectively to the open or closed state of the channel. This type of model also provides a mechanism for both activation and inhibition of the channel.

Thus, I hypothesize that IMAC exists in two states, an open state (C_o) and a completely closed state (C_c). The net flux through IMAC in intact mitochondria, would therefore depend on the number of channels in the open state, which is dependent on the equilibrium constant for the process, K_1 (see Fig. 1). Moreover, I hypothesize that H^+, Mg^{2+} and propranolol inhibit transport by selectively binding to the closed state, i.e. binding sites for these agents are only exposed when the channel is in the closed

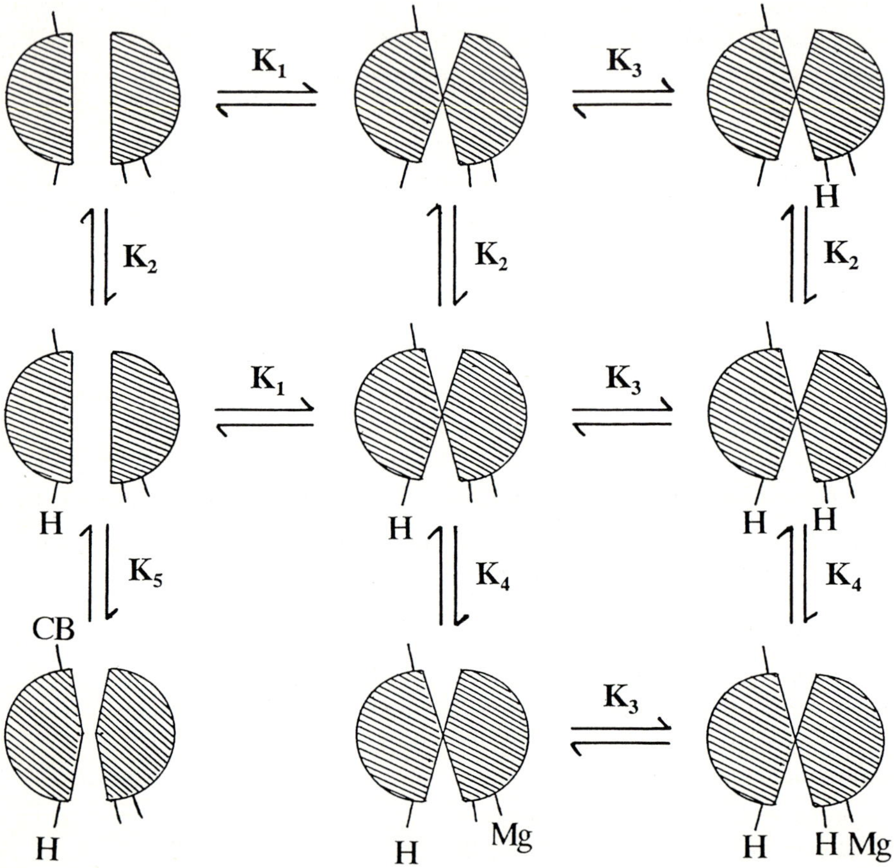

Fig. 1 A channel model to explain the regulation of IMAC by protons, Mg^{2+}, Cibacron blue and N-ethylmaleimide - K_1 is the equilibrium constant describing the open/closed equilibrium. K_2 is the dissociation constant for the binding of a H^+ which must be bound before Mg^{2+} or Cibacron blue (CB) can bind. K_3, K_4 and K_5 are the dissociation constants for the inhibitory H^+, Mg^{2+} and CB binding sites respectively.

state. On the basis of this model, NEM could affect the IC_{50} values for all three of these inhibitors by shifting the open/closed equilibrium toward the open state, i.e. by increasing K_1.

Inhibition by Protons - In order to determine whether this model can explain previous experimental observations, we shall begin by arbitrarily assigning K_1 a value of 1 (i.e. $C_o = C_c$), and we shall assume that the rate of transport $V = kC_o$ where k is a constant and C_o is the number of channels in the open state. Thus, it may be shown that in the presence of an inhibitor such as H^+,

$$V = k.C_o = k.C_T/(1 + (1 + [H^+]/K_3)/K_1)$$

$$V = \frac{V_o}{1 + [H^+]/[H^+]_{50}}$$

$$[H^+]_{50} = IC_{50} = K_3(K_1 + 1)$$

where C_T is the total number of channels, K_3 is the dissociation constant for the "inhibitory" proton, V_o is the rate when $[H^+] \rightarrow 0$, and $[H^+]_{50}$ is $[H^+]$ at which $V = V_o/2$. Note that the model predicts a linear Hill plot with a slope of 1. As shown by the theoretical curves in Fig. 2, if $K_1 = 1$, pK_3 must equal 8.1 to yield a pIC_{50} of 7.8. Moreover, the decrease in pIC_{50} induced by NEM can be explained by an increase in K_1 from 1 to 4. One finding which does not fit the model is that the J_o ($[H^+] \rightarrow 0$), which is predicted to increase 1.6-fold, appears to be unaffected by NEM. It is possible, however, that this increase is masked by a decrease in the conductance of the channel.

Inhibition by Mg^{2+} - If Mg^{2+} also inhibits by selectively binding to the closed state, and if H^+ and Mg^{2+} bind non-competitively, a linear Hill plot with a slope of unity is predicted for inhibition by Mg^{2+}. In this case, the IC_{50} for Mg^{2+} is given by

$$IC_{50} = K_4(K_1/(1 + [H^+]/K_3) + 1)$$

where K_4 is the dissociation constant for Mg^{2+}. This equation

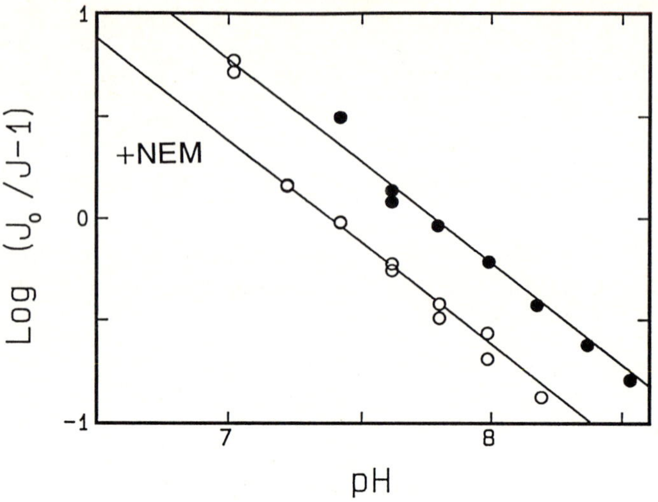

Fig. 2 Regulation of IMAC by protons - Hill plots for the regulation of Cl^- flux (J) by pH in control (●) and NEM-treated (○) mitochondria are shown. J_o is the flux when $[H^+] \rightarrow 0$ determined from a Dixon plot. The curves are drawn on the basis of the model shown in Fig. 1 with $pK_3 = 8.08$ and $K_1 = 1$ and 4 for control and NEM-treated mitochondria respectively. (Data are taken from Beavis (1991)).

predicts that decreasing the pH would decrease the IC_{50} for Mg^{2+}; however, it does not describe the observed pH dependence; moreover, the maximum possible effect is limited to a $(K_1 + 1)$-fold change. Consequently, as in the original model, a second protonation site must be introduced (see Fig. 1). We shall assume that this site must be occupied before Mg^{2+} can bind, and, for simplicity, that all states of the channel can bind a proton at this regulatory site with equal affinity. Thus, in the absence of Mg^{2+}, protonation of this site has no effect on the pH dependence of transport and $V = k(C_o + C_oH)$. This model also predicts a linear Hill plot with a slope of unity. In this case, however, the IC_{50} is given by

$$IC_{50} = K_4(K_2/[H^+] + 1)(K_1/(1 + [H^+]/K_3) + 1)$$

where K_2 is the dissociation constant of the regulatory proton.

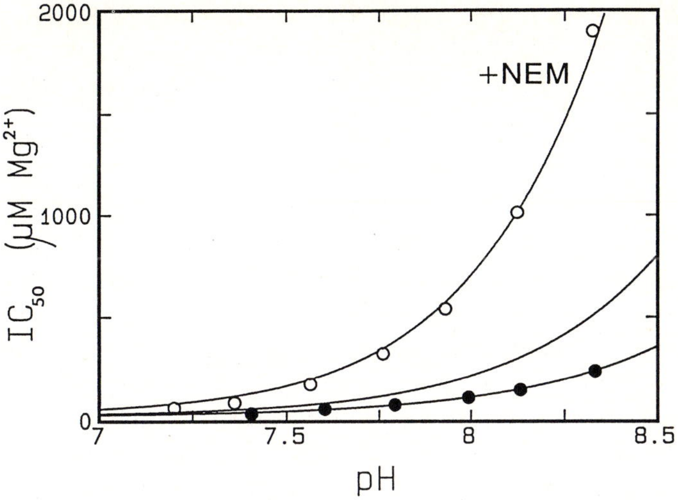

Fig. 3 Dependence of Mg^{2+} IC_{50} on pH - The Mg^{2+} IC_{50} for Cl^- transport is plotted versus the pH of assay medium for control (●) and NEM-treated (○) mitochondria. The curves are drawn on the basis of the model shown in Fig. 1 with $K_1 = 1$ and $pK_2 = 7.5$ for control mitochondria; $K_1 = 4$ and $pK_2 = 7.5$ for the intermediate curve; and $K_1 = 4$ and $pK_2 = 6.9$ for NEM-treated mitochondria. $pK_3 = 8.08$ and $K_4 = 19$ μM for all curves. (Data are taken from Beavis (1991)).

As shown by the data and theoretical curve in Fig. 3, this model describes the data very well using values of $K_4 = 19$ μM and $pK_2 = 7.5$.

We are now in a position to predict the effect of NEM on the relationship between the Mg^{2+} IC_{50} and pH. The intermediate curve in Fig. 3 shows the effect of increasing K_1 from 1 to 4. Clearly, it does not predict a large enough effect. In order to fit the data, pK_2 must also be changed from 7.5 to 6.9. Note that changing values of K_3 and K_1 simultaneously does not allow the curve to be fitted to the data.

Inhibition by Propranolol - Like Mg^{2+}, propranolol blocks anion transport completely and the IC_{50} is dependent on matrix pH; however, the relationship between the IC_{50} and pH is more complex. One way to explain the relationship is to assume that propranolol not only binds to the protonated form of the channel but also, with lower affinity, to the non-protonated form of the channel.

In this case, a linear Hill plot with a slope of unity is predicted with the IC_{50} being given by

$$IC_{50} = \frac{K_7(K_2/[H^+] + 1)(K_1/(1 + [H^+]/K_3) + 1)}{\left(\dfrac{K_2K_7}{[H^+]K_6} + 1\right)}$$

where K_7 and K_6 are the propranolol dissociation constants for the protonated and non-protonated channel respectively. Thus, if $K_6 \rightarrow \infty$, the equation becomes analogous to that for Mg^{2+}. As shown in Fig. 4, the data fit this model quite well with $pK_2 = 6.7$, $K_7 =$ 8 μM and $K_6 = 115$ μM. Moreover, using these values, increasing K_1 from 1 to 4 predicts the effect of NEM. Since, the data do not fit well with $pK_2 = 7.5$, and it is unnecessary to change this constant to account for the effect of NEM, different protonation sites may be involved in propranolol and Mg^{2+} binding.

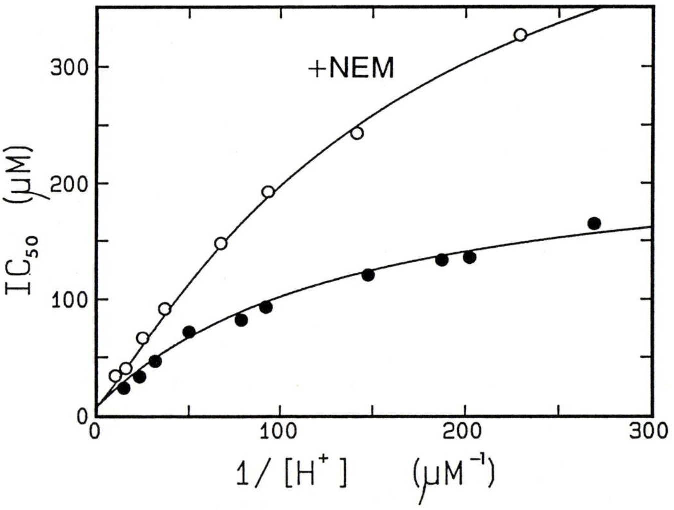

Fig. 4 Dependence of propranolol IC_{50} on pH - The propranolol IC_{50} for Cl^- transport is plotted versus reciprocal H^+ concentration for control (●) and NEM-treated (○) mitochondria. The curves are drawn on the basis of the model in which $K_1 = 1$ and 4 for control and NEM-treated mitochondria respectively and $pK_3 = 8.08$. Propranolol is assumed to bind to all closed states; however, the dissociation constant decreases from 115 μM to 8 μM upon protonation of a site with a pK = 6.7 (Data are taken from Beavis (1991)).

Inhibition by Nucleotide Analogs - According to the model, any agent which only binds to the closed state should be able to inhibit transport 100% (provided the conductance for the closed state is zero). However, if the agent binds to both states, then partial inhibition or even stimulation could be observed depending on the relative affinities for the two states. In either case, the maximum percent effect should be the same for all substrates of the channel. Thus, the partial inhibition of Cl^- transport induced by Cibacron blue, agaric acid and palmitoyl CoA cannot be explained in this way, since the maximum percent inhibition is dependent on the anion species being transported. Thus, we must conclude that these agents bind to the open conformation of the channel and that they inhibit by decreasing its conductance. However, we cannot rule out that they may also bind to the closed state. For simplicity, we shall assume that Cibacron blue only binds to the open state and shall only consider the kinetics for anions whose transport is completely inhibited. Since the IC_{50} appears to increase with pH, we shall also assume that Cibacron blue binds to a protonated form of the channel. On the basis of this model (see Fig. 1), it can be shown that the Hill plot should be linear with a slope of 1 and that the IC_{50} for inhibition by Cibacron blue is given by

$$IC_{50} = K_5(K_2/[H^+] + 1)((1 + [H^+]/K_3)/K_1 + 1)$$

From this equation, it is evident that while H^+ binding to the "regulatory site" (K_2) will decrease the IC_{50}, H^+ binding to the inhibitory site (K_3) will increase the IC_{50}. Thus, the relationship between the IC_{50} and pH is rather complex. Using the IC_{50} values of 1.51 and 2.37 μM measured at pH 7.5 and 8.5 respectively, values of 0.19 μM and 7.83 can be calculated for K_5 and pK_2 respectively. The theoretical relationship between the IC_{50} and pH, calculated on the basis of these numbers, is shown in Fig. 5. In addition, the predicted effect of NEM, i.e. increasing K_1 from 1 to 4 is also shown. Further experiments are required to verify these relationships. The value calculated for pK_2 differs from that for inhibition by Mg^{2+}. However, since the site modulating Cibacron blue binding does not appear to be on the inside but on the

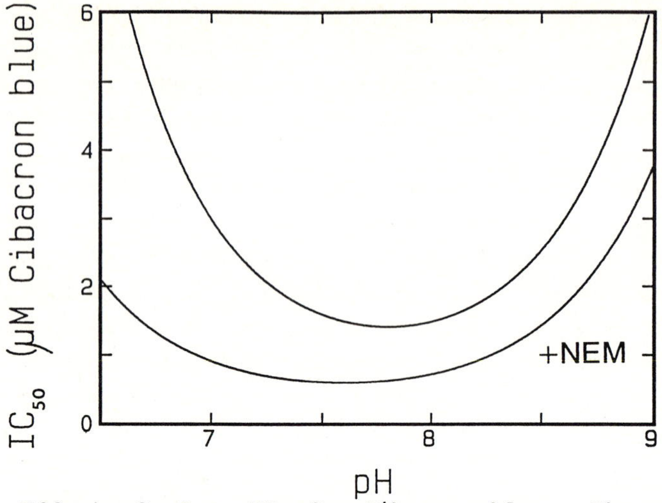

Fig. 5 <u>Effect of pH on IC_{50} for Cibacron blue</u> - The relationship between the IC_{50} and pH predicted by the equation presented in the text is shown. For the upper trace $K_1 = 1$ and for the lower trace (+NEM) $K_1 = 4$, all other constants were the same for both; $pK_2 = 7.83$, $pK_3 = 8.08$ and $K_5 = 0.186\ \mu M$.

outside of the mitochondrion, it is most likely a different protonation site and, consequently, need not have the same value. The model predicts that NEM should decrease the IC_{50} at all pH values. We have confirmed this prediction at pH 7.4 and pH 8.4. Moreover, as shown by the data in Fig. 6, at pH 7.4 the magnitude of the effect is predicted by simply assuming that NEM increases K_1 from 1 to 4. The finding that no change in K_2 needs to be postulated is consistent with the conclusion that two separate protonation sites are involved in binding of Cibacron blue and Mg^{2+}.

Further evidence for the involvement of H^+ in Cibacron blue binding is the finding that the pIC_{50} for H^+ is increased by the addition of Cibacron blue. From the model, the following relationship may be derived:

$$V = \frac{V_o}{1 + [H^+]/[H^+]_{50,o} + K_1[CB]/(K_5(K_2/[H^+] + 1)(1 + K_1))}$$

where $[H^+]_{50,o}$ is the IC_{50} for H^+ in the absence of Cibacron blue.

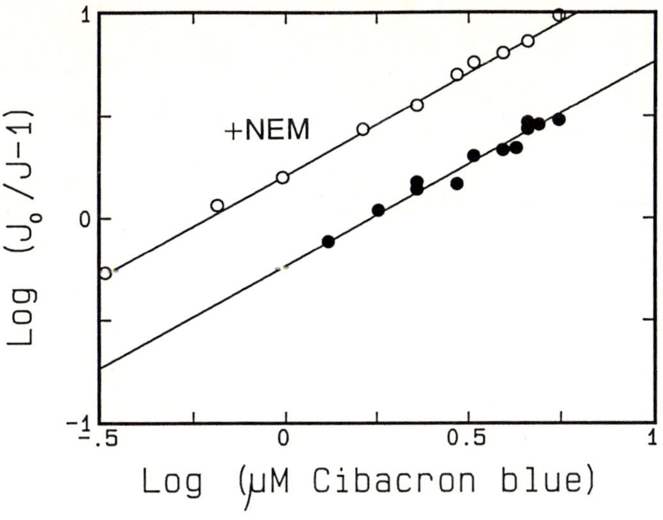

Fig. 6 Inhibition of IMAC by Cibacron blue - Hill plots for inhibition of malonate flux (J) by Cibacron blue in control (●) and NEM-treated (○) mitochondria are shown. J_o ([CB] → 0) was determined from a Dixon plot which was linear. The curves are drawn on the basis of the model shown in Fig. 1 with K_1 = 1 and 4 for control and NEM-treated mitochondria respectively and pK_2 = 7.83, pK_3 = 8.08, K_5 = 0.186 μM and pH = 7.4.

Note that the model predicts a non-linear Hill plot for inhibition by H^+ in the presence of a constant dose of Cibacron blue. The data contained in Fig. 7 show the effect of 1.3 μM Cibacron blue on the pH profile. Except for the data points obtained at low pH values, the data are quite closely described by the model. The deviation observed at low pH, even in the absence of Cibacron blue, has been previously noted (Beavis (1991)) and remains unexplained.

Similar effects are observed with agaric acid and palmitoyl CoA. We believe that these agents all bind at a common site, probably located on the outer surface of the inner membrane. Evidence supporting this is the finding that the inhibitors are competitive with respect to each other, i.e. the IC_{50} for Cibacron blue is increased 2-fold by the addition of an IC_{50} dose of agaric acid (Beavis (unpublished data)).

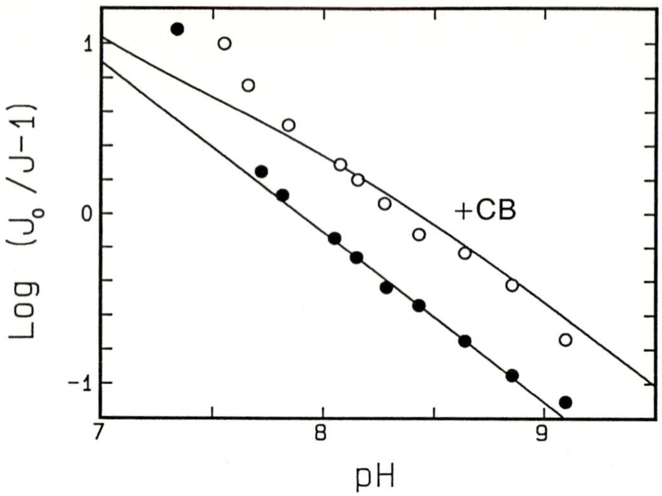

Fig. 7 Effect of Cibacron blue on the pH dependence of IMAC - Hill plots for regulation of malonate flux (J) by pH in the absence (●) and presence (○) of 1.3 μM Cibacron blue are shown. For each curve J_o is the flux when $[H^+] \rightarrow 0$, determined from a Dixon plot. The curves are drawn on the basis of the model shown in Fig. 1 with K_1 = 1, pK_2 = 7.83, K_5 = 0.186 μM and a value of pIC_{50} for H^+ in the absence of Cibacron blue of 7.89.

CONCLUSIONS

In this paper, a channel model to explain some of the regulatory features of IMAC has been presented. In order to determine whether this model is capable of explaining the behavior of IMAC a value for K_1, the open/closed state equilibrium constant was arbitrarily selected, and then the values of the other parameters were chosen to best fit the data. Thus, the values of the other constants are all dependent on the selected value of K_1. Previously, it was suggested (Beavis (1991)) that the protonation sites regulating binding of Mg^{2+} and propranolol may be the same; however, quantitative application of the model suggests not only that the pK values differ but also that one is modulated by NEM while the other is not. Thus, the protonation site K_2 in the model actually represents three different sites and, therefore, any analysis of the model in which Mg^{2+}, propranolol and/or Cibacron blue are present simultaneously must take this into account. The model does, however, provide a simple explanation for the many effects of NEM and mercurials. Agents such as TBT, whose IC_{50} is unaffected by NEM are predicted to bind with equal affinity to the open and closed state, and, therefore, must inhibit by directly blocking the channel or inducing a transition to a different closed state.

ACKNOWLEDGEMENTS

This work was supported by the National Institutes of Health Grant HL 36573 awarded by the National Heart, Lung and Blood Institute, United States Public Health Service, Department of Health and Human Services.

REFERENCES

Antonenko, Y. N., Kinnally, K. W. and Tedeschi, H. (1991) J. Membr. Biol. 124, 151-158
Azzi, A. and Azzone, G. F. (1966) Biochim. Biophys. Acta. 120, 466-468
Azzi, A. and Azzone, G. F. (1967) Biochim. Biophys. Acta. 131, 468-478
Azzi, A., Casey, R. P. and Nalecz, M. J. (1984) Biochim. Biophys. Acta 768, 209-226
Beavis, A. D. (1989a) J. Biol. Chem. 264, 1508-1515
Beavis, A. D. (1989b) Eur. J. Biochem. 185, 511-519
Beavis, A. D. (1991) Biochim. Biophys. Acta 1063, 111-119
Beavis, A. D. (1992) J. Bioenerg. Biomembr. 24, 77-90
Beavis, A. D. and Garlid, K. D. (1987) J. Biol. Chem. 262, 15085-15093
Beavis, A. D. and Garlid, K. D. (1988) J. Biol. Chem. 263, 7574-7580
Beavis, A. D. and Powers, M. F. (1989) J. Biol. Chem. 264, 17148-17155
Brierley, G. P. (1970) Biochem. 9, 697-707
Halle-Smith, S. C., Murray, A. G. and Selwyn, M. J. (1988) FEBS Lett. 236, 155-158
Kinnally, K. W., Antonenko, Y. N. and Zorov, D. B. (1992) J. Bioenerg. Biomembr. 24, 99-110
Murray, A. G., Halle-Smith, S. C. and Selwyn, M. J. (1988) E.B.E.C. Reports 5, 206
Powers, M. F. and Beavis, A. D. (1990) Biophys. J. 57, 179a
Powers, M. F. and Beavis, A. D. (1991) J. Biol. Chem. 266, 17250-17256

THE MITOCHONDRIAL PERMEABILIZATION PORE

Mario Zoratti, Ildiko' Szabo' and Vito De Pinto[1]

CNR Unit for the Physiology of Mitochondria, Dept. of Biomedical Sciences, Via Trieste 75, 35121 Padova, Italy

The megachannel of the permeability transition

The inner membrane of liver or heart mitochondria loaded with more than 30-50 nmol. (mg.prot.)$^{-1}$ of Ca^{2+} becomes permeable to solutes of molecular weight up to 1.5 KDa. This "permeability transition" (PT) is now known to be due to the operation of an aspecific pore (the "PTP") with a diameter of at least 2.8 nm (review: Gunter and Pfeiffer, 1990). The discovery of a high affinity inhibitor, Cyclosporin A, (Fournier et al., 1987; Crompton et al., 1988; Broekemeier et al., 1989) and the possibility of a role of the permeability transition in ischemic damage (Broekemeier et al., 1992; Crompton et al., 1992) have attracted renewed attention towards this phenomenon, at about the same time at which patch-clamp was beginning to be used in mitochondrial research.

Among other findings, the application of this technique to mitochondria has led to the discovery of a high-conductance (1.3 nS in 150 mM KCl), substate-rich "mitochondrial megachannel" (MMC) (Kinnally et al., 1989; Petronilli et al., 1989). We have recently shown that the MMC has a number of properties in common with the permeability transition, so that it may be identified as the PTP (Szabo' and Zoratti, 1991, 1992; Szabo' et al., 1992; Bernardi et al., 1992). Table I summarizes the relevant information.

We wish to emphasize here that the behavior of the MMC suggests a binary structure: the 1.3 nS transitions may well be due to the cooperative gating of two channels with a conductance of about 650 pS. As shown in Fig. 1A, the MMC

[1]Dept. of Pharmaco-Biology, CNR Unit for Mitochondria and Bioenergetics, Bari, Italy

NATO ASI Series, Vol. H 83
Molecular Biology of Mitochondrial Transport Systems
Edited by M. Forte and M. Colombini

Table I. *A comparison of the properties of the MMC (MMC, patch-clamp experiments) and of the permeability transition (PT, swelling experiments).*

MMC	PT
Activation by sub-mM Ca^{2+} at a matrix site[a,b]	Activation by Ca^{2+} binding at an inner site (Km 10μM[c])
Ca^{2+}-competitive inhibition by Mg^{2+}, Mn^{2+}, Sr^{2+}, Ba^{2+}[a]	Ca^{2+}-competitive inhibition with similar parameters by the same divalent cations[c]
Ca2+-competitive inhibition by sub-μM Cyclosporin A[d]	Ca2+-competitive inhibition by sub-μM Cyclosporin A[c]
Ca^{2+}-reversible inhibition by H^+ (matrix side site) in the physiological range[a]	Inhibition by H^+ at an inner face site other than the Me^{2+} site[c]
Inhibition by sub-mM ADP[b]	Inhibition by sub-mM ADP[e]
Inhibition by Amiodarone in the μM range[f]	Inhibition by Amiodarone in the μM range[g]
Data consistent with two cooperating "halves"	Ca2+ and ADP have two sites per channel[e,h]

[a]Szabo' et al., 1992; [b]Szabo' and Zoratti,1992; [c]Bernardi et al., 1992; [d]Szabo' and Zoratti, 1991; [e]Hunter and Haworth, 1980; [f]Antonenko et al., 1991; [g]Bernardi et al., unpublished; [h]Hunter et al., 1976.

often visits a major substate with a conductance of about one-half the maximal size; when in this substate, it exhibits fast flickering to (the) closed state(s). This same conductance level often appears as an intermediate stage during opening and closing events (Fig. 1B). Again, it is characterized by a high level of "open channel noise". Note that the "noise" is much lower in the highest conductance state. We interpret this to mean that when both MMC "halves" are fully open, they reciprocally stabilize this conformation.

The molecular identity of the pore is still unknown. The clues available so far consist mainly in the effects on the PT by compounds (atractyloside, bongkrekate etc.) known to bind with high affinity to the adenine nucleotide carrier (ADC) (Gunter and Pfeiffer, 1990; Hunter and Haworth 1979; Le Quoc and Le Quoc, 1988; Novgorodov et al. 1990, 1992). These effects have led some authors to propose that the PTP may be

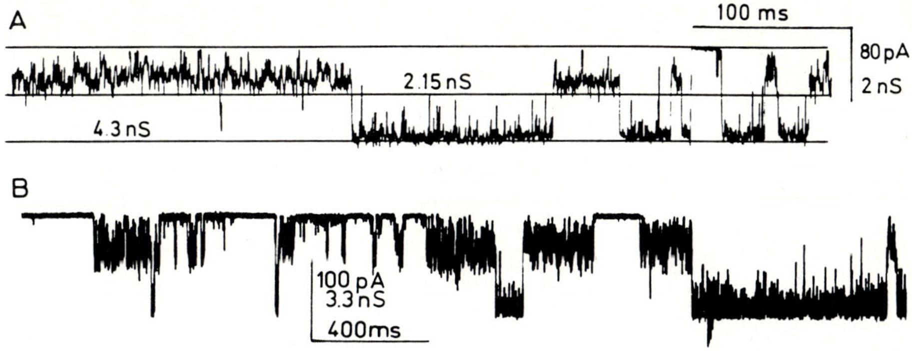

Fig. 1. *Examples of MMC activity showing the "half"-conductance level.* Medium: symmetrical 0.5 M KCl. The flickering "half-size" conductance generally appears less frequently than in these examples. A) V: -40 mV. B) V: -30 mV.

formed by the ADC itself (Halestrap and Davidson, 1990). Even though this hypothesis seems now unlikely (see below and Novgorodov et al., 1992), an interaction between the pore-forming peptide and the ADC represents a definite possibility.

The behavior of the MMC (see Fig. 1) suggested that it might be a dimer, possibly of VDAC in view of its conductance. We found however arguments against such an identification: the PTP/MMC resides in the matrix-enclosing envelope, since its operation can lead to matrix volume changes and it appears in electrophysiological traces together with the 107pS channel (Sorgato et al., 1987), believed to be a marker of the inner membrane. The voltage-dependence and selectivity properties of VDAC, as determined in lipid bilayer studies (Colombini, 1989; Liu and Colombini, 1992; Benz et al., 1990), seemed not to match our observations on the MMC, obtained using a voltage pulse protocol (see below, Fig. 2A). The MMC was found to be essentially non-selective (Szabo' and Zoratti, 1992), while the porin is either cation- or anion-selective, depending on the conductance state (Benz et al., 1990).

On the other hand, it has been suggested that contact sites may contain complexes formed by the ADC and VDAC (Brdiczka, 1991). The peripheral benzodiazepine receptor (Anholt et al.,

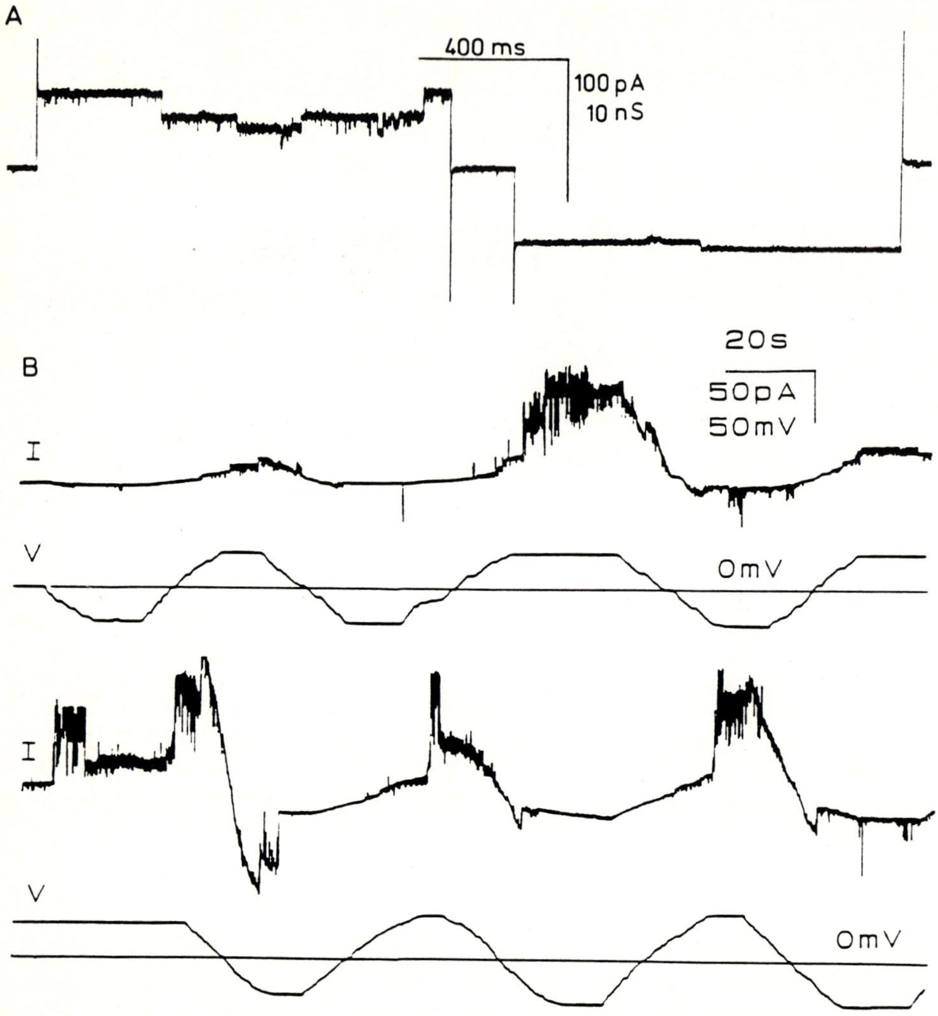

Fig. 2. *Voltage dependence of the MMC*. Medium: symmetrical 150 mM KCl. A) The typical behavior of the MMC subjected to voltage pulses. The protocol consisted in 1-sec. pulses of alternate sign, separated by 100-ms intervals at zero potential. The channels gate predominantly during the positive pulses. V: +/-10 mV. B) The electrical behavior of a mitoplast patch subjected to "slow" potential changes. Current (I) and voltage (V) traces. At positive potentials the megachannels tend to remain open, exhibiting fast gating. Negative potentials induce long-lasting closures. Activity by 107pS channels is also visible in this example.

Electrophysiology of VDAC in artificial membranes

Fig. 3A-B presents examples of activity by VDAC inserted in planar phosphatidylethanolamine bilayers. The properties of the pore conformed to the literature data: transmembrane voltages induced closures, in most cases without regard for the sign of the applied potential. The closures were incomplete, and the kinetics were slow (Fig. 3A). In some cases we observed rapid transitions between states of different conductance (Fig. 3B).

Using the same VDAC preparations, we have performed patch-clamp experiments on VDAC-doped patches of azolectin liposomes. Reconstitution of the porin was achieved by filling the patch pipette with experimental medium containing VDAC (1-10 nM). This approach in principle allows reconstitution of the channel in any membrane amenable to patch-clamp studies, including that of mitoplasts (see below).

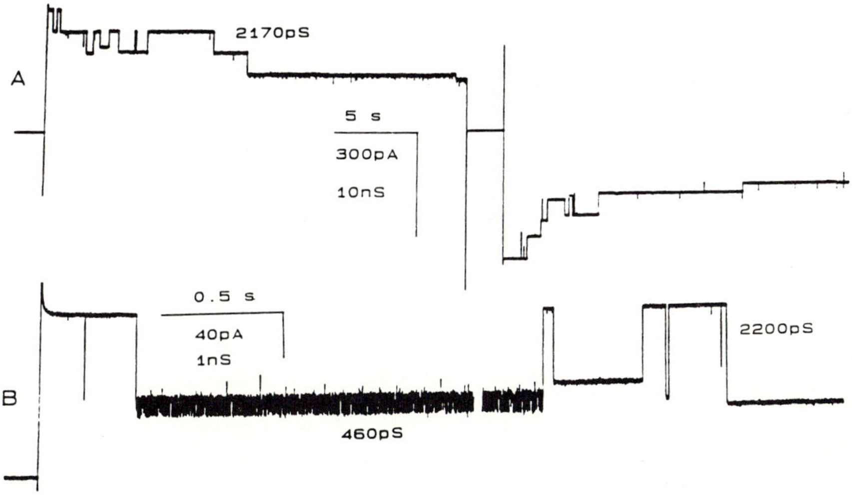

Fig. 3. *Examples of activity by VDAC in phosphatidylethanolamine planar lipid bilayers.* VDAC was obtained following published procedures (De Pinto et al., 1987) as a solution (4.5-6 μg prot./ml) in 3% Triton. Addition of μl volumes to the 3-ml cis compartment of the bilayer apparatus resulted in the incorporation of several channels. Artefacts by Triton were excluded by control experiments. Medium: symmetrical 1 M KCl. A) V(trans): 0/+30/0/-30 mV. B) V(trans): 0/+40 mV. The segment of flickering activity, which lasted several seconds, is shown only in part.

1986) has been reported to consist of VDAC, the ADC and an 18-KDa protein (McEnery et al., 1992). Furthermore, Wunder and Colombini (1991) have attributed to VDAC the channel activity they observed by patch-clamping proteoliposomes containing whole mitochondrial membranes, which is very reminescent of MMC activity. It has been recently reported (Novgorodov et al., 1991; Bernardi, 1992) that the PT is unleashed by depolarization, a suggestive finding since VDAC in planar membranes adopts the open state at potentials close to zero. The properties of VDAC may depend on those of the membrane harboring it (Moran et al., 1992). Bacterial porins have been reported to exhibit different behaviors in planar bilayers and in liposomes under patch-clamp (Berrier, 1992).

These findings have prompted us to investigate the issue, by carrying out a comparative study of the electrophysiological properties of the VDAC and of the MMC. The results suggest that VDAC, or a pore similar to VDAC, is the channel-forming component of the mitochondrial permeability transition pore.

The voltage dependence of the MMC

In experiments utilizing the voltage pulse protocol (Fig. 2A and Petronilli et al., 1989), positive potentials favoured the occupancy of lower-than-maximal conductance states, while the MMC remained essentially in the fully open state at negative voltages. We have now obtained data on the response of the MMC to slowly varying or long-lasting voltages. The dominant behavior is exemplified in Fig. 2B. In agreement with the pulse experiments, the channels exhibited relatively fast gating to lower conductance states at positive potentials ("noisy" trace), while at negative voltages the maximal conductance state is preferred, and substates appear much less frequently. However, upon prolonged application of negative (i.e., physiological) voltages, the MMC channels appeared to be driven to (a) long-lasting closed state(s). In our pulse protocol, the pulses were too short for this phenomenon to be noticeable; it can however be observed with longer pulses.

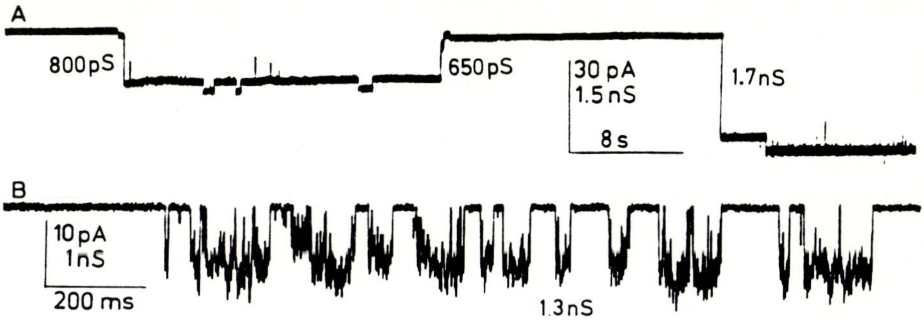

Fig. 4. *Examples of activity by VDAC in azolectin liposomes.* Medium: symmetrical 150 mM KCl. A) "Slow" behavior at -20 mV. B) Faster activity with a peak conductance of about 1.3 nS, at -10 mV.

In liposomes, usually one or more pores incorporated within a few minutes after seal formation.

In this experimental system, the VDAC at times exhibited a kinetic behavior similar to that observed with the planar bilayer, (Fig. 4A), but more often faster kinetics were observed, as exemplified in Fig. 4B. As expected, the current traces exhibited a variety of conductance levels. Notably, we often observed conductance changes close to 650 pS, the full-conductance size of VDAC, suggesting that complete channel closures took place (Fig. 4A). Higher conductance steps, presumably arising from cooperative gating (Fig. 4A) were also observed. 1.0-1.3 nS activity, the preferred sizes of the MMC, was sometimes encountered (Fig. 4B). Some of the most frequently encountered conductance variations found a correspondence with transitions often seen in mitoplasts: values of 530-550, 450-470, 310-350, 220-250 and 110-130 pS might be mentioned (see Petronilli et al., 1989).

When subjected to voltage pulses, for comparison purposes, VDAC-doped liposome patches responded in a manner reminescent of MMC behavior. Voltage pulses of the size normally used in experiments on the MMC (10-50 mV) had only little effect, or resulted in the disappearence of shallow substate levels at negative potentials (not shown). Higher voltages produced the characteristic pattern so often observed with mitoplast

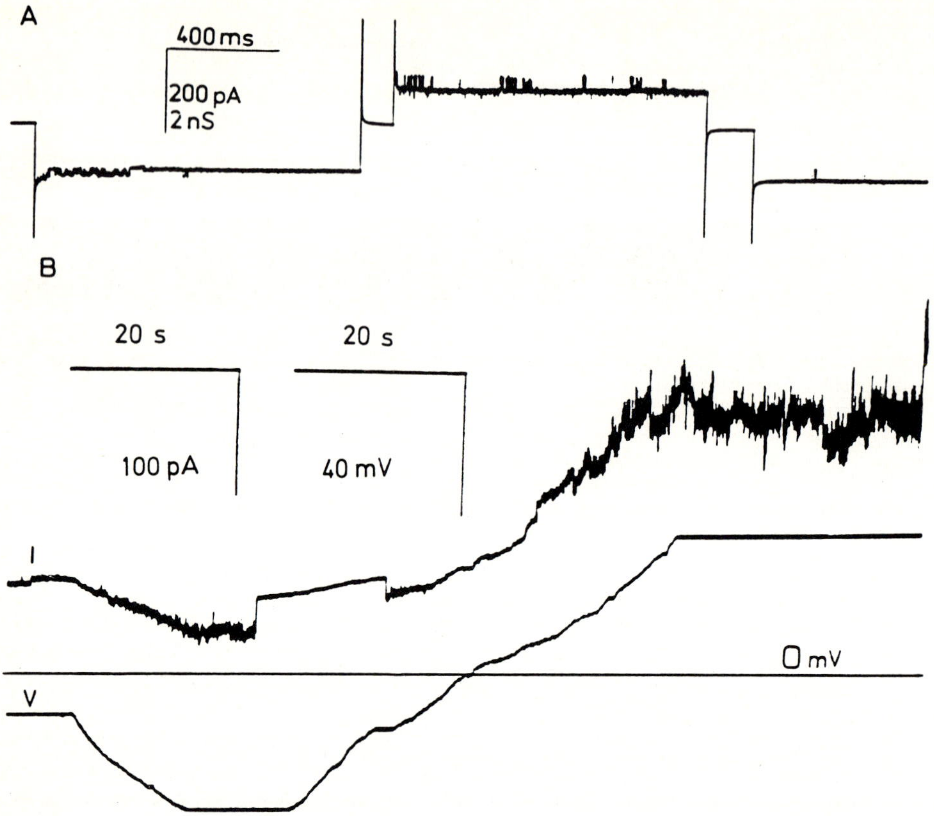

Fig. 5. *Voltage dependence of VDAC in liposomes.* Medium: symmetrical 0.15 M KCl. A) Behavior under voltage pulse protocol conditions. 1-sec. +/-100 mV pulses separated by 100-ms intervals at zero potential. B) An example of response to slow voltage changes.

membranes: at negative voltages high and stable conductance levels predominated, with occasional closures. At positive voltages, closed states were much in evidence (Fig. 5A).

Experiments involving slow voltage variations produced results analogous to the observations on mitoplast membranes in about 1/3 of the experiments. VDAC exhibited frequent transitions to and from substates at positive voltages and remained in a high conductance level, while open, at negative potentials. The latter favored, however, transitions to long-lasting closed state(s) (Fig. 5B). In the remaining experiments, no clear voltage dependence was observed. The slowness of the response of the channels to the voltage variations may have masked the dependence in some of these cases. Reports have appeared in the literature of an asymmetric voltage dependence of VDAC (Freitag et al., 1982; Moran et al. 1992) and of the conductance of patches of outer mitochondrial membrane (Tedeschi et al., 1987).

Insertion of VDAC into the mitoplast membrane

The insertion of VDAC into the mitoplast membrane was envisioned as a means to verify whether the high-conductance activity might be ascribed to the porin. The appearence of "unusual" channel activity in these patches would have indicated that the porin had inserted and that either it was intrinsically different from the high-conductance endogenous channels, or it could not express a similar behavior because of a failure to establish the required interactions with other membrane components. Instead, we have not been able to detect any qualitative difference in the electrical behavior of the patches when the pipette contained VDAC. The lack of alterations in the activity patterns under these experimental conditions may have two reasons: either VDAC does not insert into the mitoplast membrane, at variance with its behavior in the case of liposomes, or, once inserted, it behaves just like the endogenous channels. A choice between these two possibilities can only be provided by a statistical comparison of the activity of patches in the presence or absence of VDAC. This tedious work has begun, and the available results suggest that indeed the patches are more active if VDAC is present. However, a definite conclusion is not possible at the moment.

Selectivity

The MMC has been reported to be essentially unselective, with discrimination factors lower than 2 for several ions (Szabo' and Zoratti, 1992). The reversal potentials in those experiments were determined by plotting the average of amplitude determinations at various voltages, and fitting them

Table II. *K^+ vs. Cl^- Selectivity data for mitoplast patches and VDAC**.

System	KCl (pip) M	KCl (bath) M	Erev.[1] mV	P_K/P_{Cl}
VDAC	1.0	0.15	2.5	1.14
			4	1.24
			6	1.37
VDAC	1.0	0.15	4	1.24
			6.5	1.41
			10	1.70
VDAC	0.5	0.15	2	1.16
			7	1.67
MITOPL.	1.0	0.15	7	1.45
MITOPL.	1.0	0.15	1	1.05
MITOPL.	0.5	0.15	-13	0.37
MITOPL.	0.5	0.15	-4	0.75
			-6	0.64
MITOPL.	0.3	0.15	0	1.0

*Patches of VDAC-doped liposomes.
[1]Different reversal potentials in the same experiment refer to successive I/V scans.

by linear regression. We have now adopted a different experimental procedure, determining the reversal potentialsfrom computer-generated I/V curves obtained by continuously scanning the applied voltage. Table II presents the data obtained in this manner. Notice that the mitochondrial membranes may exhibited slightly cationic or slightly anionic selectivity. How much of this variability depends on the variable contribution to the currents of non-MMC channels remains to be determined. In our experiments the VDAC-doped azolectin patches were always slightly cation-selective.

The benzodiazepine receptor?

Dr.s S. Vicini and K. Kroeger, suggested to us in 1989 that we check the effects of benzodiazepines, in particular of those known to bind to the peripheral benzodiazepine receptor

(Anholt et al, 1986; McEnery et al., 1992), to patch-clamped mitoplasts. In 23 cases out of 30 experiments, addition of μM levels of the high-affinity benzodiazepine analogue Alpidem (Langer et al., 1990) resulted in the appearence of high, fast currents. Fig. 6 shows a representative record illustrating these currents. Currents were also elicited, although less reproducibly, by Alpidem in the 15-500 nM concentration range. No such effect was observed in control experiments on liposomes and on bacterial protoplasts. The "Alpidem currents" often increased dramatically in the course of a few seconds, causing the termination of the experiment before significative attempts at inhibition could be made. In a few experiments, the activity was much better behaved, allowing the tentative identification of conductance transitions as due to either VDAC or the MMC, on the basis of their size (Fig.6).

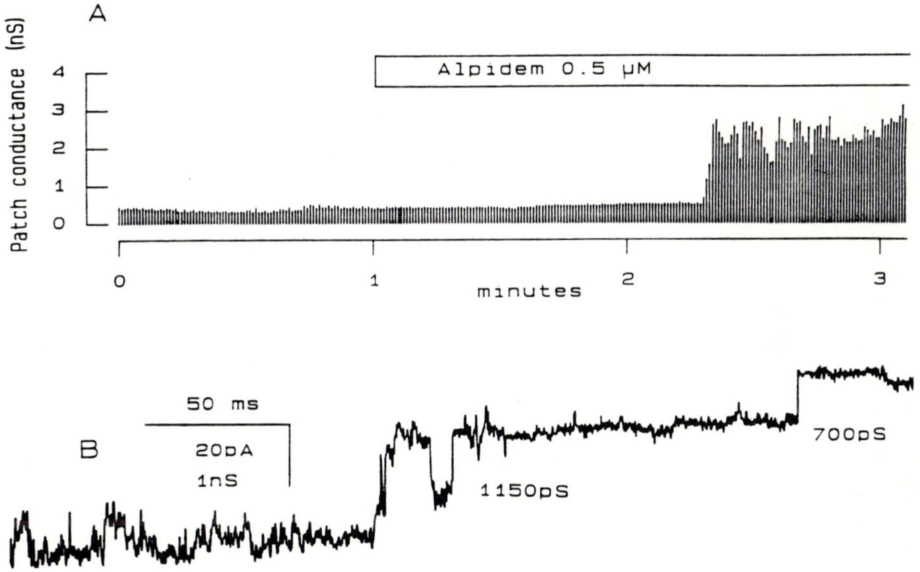

Fig. 6. *The effect of Alpidem on a silent mitoplast patch.* A) Plot of the total patch conductance, averaged over 0.82-sec. intervals, vs. time. The bar indicates perfusion of the patch-clamp chamber (capacity: approx. 1 ml) with 0.5 μM Alpidem. The conductance increase took place after about 2 ml had flown in and out. B) Current trace recorded after the conductance increase, showing conductance steps of sizes similar to those of the MMC and of VDAC. V: -20 mV.

Discussion

The observations summarized above eliminate most of the arguments against an identification of VDAC as the pore-forming component of the MMC/PTP. When reconstituted in liposomes and examined by patch clamp, the kinetic behavior, voltage dependence and selectivity of VDAC turn out to be compatible with those of the MMC in mitoplasts, despite the obvious and important differences between the two experimental systems. Both the MMC and VDAC exhibit substates, and the maximal conductance of the MMC is twice that of VDAC, suggesting that it is formed by two cooperating VDAC molecules. Indeed, estimates of the minimum size of the PTP in experiments monitoring the osmotic effects of solutes (Massari and Azzone, 1972) and of the size of the VDAC pore from ultrastructural studies and functional assays (Mannella et al., 1992) nearly coincide. VDAC can exhibit cooperative behavior in proteoliposomes, and the behavior of the MMC strongly suggests that it is formed by two approx. 650pS channels (Fig. 1; Szabo' et al., 1992).

The major remaining argument against this identification is that VDAC is generally thought to populate only the outer membrane, while the MMC clearly establishes a communication between the mitochondrial matrix and the environment. The solution to this difficulty may reside in the presence of contact sites: MMC-like conductances have been observed in contact site fractions (Moran et al., 1990). The adenine nucleotide carrier and the porin may help the formation of the contact sites themselves (or at least a class of them) through their interaction (Brdiczka, 1991). It is plausible that our mitoplasts retain contact sites and outer membrane fragments after osmotic shock. The presence of VDAC in mitoplasts prepared by this method has been reported (Schwaiger et al., 1987), while other studies have suggested that the formation of contact sites is favored by high Ca^{++} levels (vanVenetje et al., 1982; Brdiczka, 1991). Whether the induction of MMC activity by high Ca++ might be related to this latter phenomenon ought to be explored. Alternatively, the porin may actually be located in the inner membrane, where the high physiological potential would be expected to keep it in the closed state under normal conditions.

The voltage dependence we observed in both mitoplast and VDAC-doped liposome patches are at variance with literature reports of a symmetrical voltage dependence of VDAC in planar lipid bilayer experiments (Colombini, 1987) and our own results in that system (Fig. 3). Also, the fast kinetics observed in many experiments find little counterpart in planar bilayer experiments (but see, e.g., Fig. 3B). Various tentative explanations might be offered for these "discrepancies". One would invoke the different biophysical properties of the membranes used. Another is that the activity of the mitoplast patches and of many of the liposome patches might actually derive not from VDAC, but from the "peptide sensitive channel" described by the Paris group (Henry et al., elsewhere in this book), whose properties provide a closer match to our observations in patch-clamp work. This explanation would imply that our VDAC preparations contained an undetected contamination by this channel, which would have been observed preferentially because of inactivation of the VDAC in patch-clamp experiments (but not in lipid bilayer experiments).

While this interesting possibility needs to be considered, it seems to us appropriate at this point to propose as a working hypothesis that the PTP/MMC is composed of two cooperating VDAC molecules, plus presumably other components, including the ADC, having a regulatory role.

Acknowledgements

We are indebted to Dr.s Stefano Vicini and Karl Krueger for suggesting the use of Alpidem, to Prof. Daniel Wolff for setting up the planar lipid bilayer apparatus and to Prof. G.F. Azzone for support. This work was financed in part by a grant of the E.E.C. under the program "Science".

References

Anholt, R.R.H., Pedersen, E., DeSouza, B. and Snyder, S.H (1986) The Peripheral-type Benzodiazepine Receptor. Localization to the mitochondrial outer membrane. *J. Biol. Chem.* **261**, 576-583.

Antonenko,Y.N., Kinnally,K.W., Perini,S. and Tedeschi, H. (1991) Selective effects of inhibitors on inner mito chondrial membrane channels. *FEBS Lett.* **285**, 89-93.

Benz, R., Kottke, M. and Brdiczka, D. (1990) The cationi cally selective state of the mitochondrial outer membrane pore: a study with intact mitochondria and reconstituted mitochondrial porin. *Biochim. Biophys. Acta* **1022**, 311-318.

Bernardi, P., Vassanelli, S., Veronese, P., Colonna, R. Szabo', I. and Zoratti, M. (1992) Modulation of the Mitochondrial Permeability Transition Pore. Effect of Protons and divalent Cations. *J.Biol.Chem.* **267**, 2934-2939.

Bernardi, P. (1992) Modulation of the Mitochondrial Cyclo sporin A-sensitive Permeability Transition Pore by the Proton Electrochemical Gradient. Evidence that the Pore can be Opened by Membrane Depolarization. *J. Biol. Chem.* **267**, 8834-8839.

Berrier, C. (1992) Doctorate Thesis, Universite' Paris Sud, Orsay.

Brdiczka, D. (1991) Contact sites between mitochondrial envelope membranes. Structure and function in energy- and protein-transfer. *Biochim. Biophys. Acta* **1071**, 291-312.

Broekemeier, K., Dempsey, M.E. and Pfeiffer, D.R. (1989) Cyclosporin A is a potent inhibitor of the inner membrane permeability transition in liver mitochon dria. *J. Biol. Chem.* **264**, 7826-7830.

Broekemeier, K. M., Carpenter-Deyo, L., Reed, D. J. and Pfeiffer, D.R. (1992) Cyclosporin A protects hepato cytes subjected to high Ca^{2+} and oxidative stress. *FEBS Lett.* **304**, 192-194.

Colombini, M. (1989) Voltage Gating in the Mitochondrial Channel, VDAC. *J. Membr. Biol.* **111**, 103-111.

Crompton,M., Ellinger,H. and Costi,A. (1988) Inhibition by cyclosporin A of a Ca^{2+}-dependent pore activated by inorganic phosphate and oxidative stress in heart mitochondria. *Biochem. J.* **255**, 357-360.

Crompton, M., McGuiness, O. and Nazareth, W. (1992) The involvement of cyclosporin A binding proteins in regulating and uncoupling mitochondrial energy trans duction. *Biochim. Biophys. Acta* **1101**, 214-217.

De Pinto, V., Prezioso,G. and Palmieri,F. (1987) A simple and rapid method for the purification of the mito chondrial porin from mammalian tissues. *Biochim. Biophys. Acta* **905**, 499-502.

Fournier, N., Ducet, G. and Crevat, A. (1987) Action of cyclosporine on mitochondrial calcium fluxes. *J. Bioenerg. Biomembr.* **19**, 297-303.

Freitag, H., Neupert, W. and Benz, R. (1982) Purification

and Characterization of a Pore Protein of the Outer Mitochondrial Membrane from *Neurospora crassa*. *Eur. J. Biochem*. **123**, 629-636.

Gunter, T.E. and Pfeiffer, D.R. (1990) Mechanisms by which mitochondria transport calcium. *Am J. Physiol*. **258**, C755-C786.

Halestrap, A.P. and Davidson, A.M. (1990) Inhibition of Ca^{2+}-induced large-amplitude swelling of liver and heart mitochondria by cyclosporin is probably caused by the inhibitor binding to mitochondrial-matrix peptidyl-prolyl cis-trans isomerase and preventing it interacting with the adenine nucleotide transloca se. *Biochem J*. **268**, 153-160.

Hunter, D.R., Haworth, R.A. and Southard, J.H. (1976) Relationship between configuration, function and per meability in calcium-treated mitochondria. *J. Biol. Chem*. **251**, 5069-5077.

Hunter,D.R. and Haworth,R.A. (1979) The Ca^{2+}-induced membrane transition in mitochondria. III. Transitional Ca^{2+} release. *Arch. Biochem. Biophys*. **195**, 468-477.

Hunter, D.R. and Haworth, R.A. (1980) Allosteric inhibi tion of the Ca^{2+}-activated hydrophylic channel of the mitochondrial inner membrane by nucleotides. *J. Membr. Biol*. **54**, 231-236.

Kinnally, K.W., Campo, M.L. and Tedeschi, H. (1989) Mito chondrial channel activity studied by patch-clamping mitoplasts. *J. Bioenerg. Biomembr*. **21**, 497-506.

Langer, S.Z., Arbilla, S., Tan, S., Lloyd, G., George, P. Allen,J. and Wick, A.E. (1990) Selectivity for Omega Receptor Subtypes as a Strategy for the Development of Anxiolytic Drugs. *Pharmacopsychiatry* **23**, 103-107.

LeQuoc,K. and LeQuoc,D. (1988) Involvement of the ADP/ATP Carrier in Calcium-Induced Perturbations of the Mito chondrial Inner Membrane Permeability: Importance of the Orientation of the Nucleotide Binding Site. *Arch Biochem. Biophys*. **265**, 249-257.

Liu, M.Y. and Colombini,M. (1992) A Soluble Mitochondrial Protein Increases the Voltage Dependence of the Mito chondrial Channel, VDAC. *J. Bioenerg. Biomembr*. **24**, 41-46.

Mannella, C., Forte, M. and Colombini, M. (1992) Towards the Molecular Structure of the Mitochondrial Channel VDAC. *J. Bioenerg. Biomembr*. **24**, 7-19.

Massari, S. and Azzone, G.F. (1972) The equivalent pore radius of intact and damaged mitochondria and the mechanism of active shrinkage. *Biochim. Biophys.Acta* **283**, 23-29.

McEnery, M.W., Snowman, A.M., Trifiletti,R.R. and Snyder, S.H. (1992) Isolation of the mitochondrial benzodia zepine receptor: Association with the voltage-depen dent anion channel and the adenine nucleotide carrier. Proc.Natl.Acad.Sci. USA 89, 3170-3174.

Moran,O., Sandri,G., Panfili,E., Stuehmer,W. and Sorgato, M.C. (1990) Electrophysiological Characterization of Contact Sites in Brain Mitochondria. *J. Biol. Chem*. **265**, 908-913.

Moran, O., Sciancalepore, M., Sandri, G., Panfili, E., Bassi, R., Ballarin, C. and Sorgato, M.C. (1992) Ionic permeability of the mitochondrial outer membrane. *Eur. Biophys. J.* **20**, 311-319.

Novgorodov,S.A., Gudz,T., Kushnareva,Y.E., Zorov,D.B. and Kudrjashov, Y.B. (1990) Effect of the ADP/ATP antiporter conformational state on the suppression of the nonspecific permeability of the inner mitochondrial membrane by cyclosporine A. *FEBS Lett.* **277**, 123-126.

Novgorodov, S.A., Gudz,T.I., Kushnareva,Y.E., Eriksson,O. and Leikin, Y.N. (1991) Effects of the membrane potential upon the Ca^{2+}- and cumene hydroperoxide-induced permeabilization of the inner mitochondrial membrane. *FEBS Lett.* **295**, 77-80.

Novgorodov,S.A., Gudz,T.I., Milgrom,Y.M. and Brierley,G.P (1992) The Permeability Transition in Heart Mitochondria Is Regulated Synergistically by ADP and Cyclosporin A. *J. Biol. Chem.* **267**, 16274-16282.

Petronilli,V., Szabo',I. and Zoratti, M. (1989) The inner mitochondrial membrane contains ion-conducting channels similar to those found in bacteria. *FEBS Lett.* **259**, 137-143.

Schwaiger,M., Herzog,V. and Neupert, W. (1987) Characterization of Translocation Contact Sites Involved in the Import of Mitochondrial Proteins. *J. Cell Biol.* **105**, 235-246.

Sorgato, M.C., Keller, B.U. and Stuehmer,W. (1987) Patch-clamping of the inner mitochondrial membrane reveals a voltage-dependent channel. *Nature* **330**, 498-500.

Szabo',I. and Zoratti, M. (1991) The Giant Channel of the Inner Mitochondrial Membrane Is Inhibited by Cyclosporin A. *J. Biol. Chem.* **266**, 3376-3379.

Szabo', I. and Zoratti, M. (1992) The Mitochondrial Megachannel is the Permeability Transition Pore. *J. Bioenerg. Biomembr.* **24**, 111-117.

Szabo',I., Bernardi, P. and Zoratti, M. (1992) Modulation of the Mitochondrial Megachannel by Divalent Cations and Protons. *J. Biol. Chem.* **267**, 2940-2946.

Tedeschi, H., Mannella,C.A. and Bowman, C.L. (1987) Patch Clamping the Outer Mitochondrial Membrane. *J. Membr. Biol.* **97**, 21-29.

van Venetje, R. and Verkleij, A.J. (1982) Possible role of non-bilayer lipids in the structure of mitochondria. A freeze-fracture electron microscopy study. *Biochim. Biophys. Acta* **692**, 397-405.

Wunder, U.R. and Colombini, M. (1991) Patch Clamping VDAC in Liposomes Containing Whole Mitochondrial Membranes. *J. Membr. Biol.* **123**, 83-91.

MITOCHONDRIAL CHANNELS: AN INTEGRATED VIEW

Kathleen W. Kinnally and Henry Tedeschi
Department of Biological Sciences
State University of New York at Albany
1400 Washington Ave.
Albany, N.Y. 12222

Introduction

The presence of channels in the outer mitochondrial membrane has been recognized for some time. In particular the voltage dependent anion channel (VDAC[1] or mitochondrial porin) has been studied most frequently after reconstitution in planar bilayers (e.g. Schein et al., 1976; Colombini, 1979; Benz, 1985). In contrast early information on the inner membrane was primarily derived from permeability studies carried out with mitochondrial suspensions. The use of patch-clamping which allowed the study of channels directly in their native or near-native environment is a recent development. Correlations between these two kinds of studies are now underway. Although a good deal of new information is now available, its interpretation and evaluation in terms of nature and significance of channels to mitochondrial function is far from clear.

---

[1]The following abbreviations are used: **AAA**: Alkaline pH-induced anion selective activity. **ACA**: alkaline pH-induced cation selective activity. **ADT**: adenine nucleotide translocator. **BHT**: butylated hydroxitoluene. **DCCD**: dicyclohexylcarbodiimide. **IMAC**: inner membrane anion channel. **IMM**: inner mitochondrial membrane. **LCC**: low conductance channels. **mBzR**: benzodiazepine receptor. **MCC**: multiple conductance channel. **OMM**: outer mitochondrial membrane. **PTP**: permeability transition pore. **VDAC**: voltage dependent anionic channel. **VDCC**: voltage dependent cation channel.

NATO ASI Series, Vol. H 83
Molecular Biology of Mitochondrial Transport Systems
Edited by M. Forte and M. Colombini

In recent years, patch-clamping has permitted the direct demonstration of single channel activity in the outer (Henry et al., 1989; Wunder and Colombini, 1991, Moran et al. 1992) and inner mitochondrial membranes (for a review see Kinnally et al., 1992). The present review will examine recent results of patch-clamping experiments with some emphasis on the newer findings of our laboratory but with the intent of drawing connections between the various studies. The discussion, however, is far broader and we hope to raise several questions which we found intriguing with the hope that it will lead to new experimental approaches.

Outer membrane channels

While other channels may exist in the outer mitochondrial membrane (OMM), VDAC constitutes a large percentage of OMM protein and may control overall OMM permeability. VDAC have an average density of few thousand molecules per mitochondrion (for example calculated from data of DePinto et al., 1987 and Lindén et al., 1984 for mammalian mitochondria and Freitag et al, 1982 for Neurospora mitochondria). So, there should be several hundred VDAC in a patch with a diameter of 0.5 μm assuming a random distribution. We would expect a maximal patch resistance of about 1.7 GΩ if just one VDAC were open (VDAC conductance is ~650 pS in 150 mM KCl). Patch resistances as high as 10 GΩ were obtained with intact mitochondria (Moran et al., 1992) suggesting that no VDAC were open in these patches. Since even at lower resistances (e.g. 500 MΩ) most VDAC must have been closed, it must be concluded that in these mitochondria most VDAC were in an inactive state. Similarly, as another OMM channel, voltage dependent cation channel (VDCC), has about the same conductance as VDAC (Henry et al.,1989), they too were probably inactive in these cases.

The protein partially purified by Holden and Colombini (1988) and Liu and Colombini (1991,1992) may be a candidate for maintaining the VDAC inactive. The previous reports of much lower patch resistances in intact mitochondria (e.g. Tedeschi et al., 1987; Kinnally et al., 1987) suggest that the VDAC can be activated under some conditions. However, the presence of a modulator need not be the only mechanism for VDAC closure. The studies of Zimmerberg and Parsegian (1986, 1987) have shown that even moderate concentrations of polymers such as dextran close VDAC reconstituted in bilayers by an osmotic mechanism. In vivo, the cytoplasmic and intra-membrane space protein concentrations should be well in excess of these levels. In isolated mitochondria the intra-membrane space protein may also have effects through this mechanism. It may be of some significance that VDAC can be closed almost completely by voltage under at least some conditions (e.g. Ermishkin and Mirzabekov, 1990). The ease by which nanomolar levels of certain drugs block mitochondrial channels (Zorov et al., 1992a) suggests the presence of other regulatory mechanisms.

There have been several recent patch-clamp studies of the OMM. Wunder and Colombini (1991) report experiments in which OMM preparations from Neurospora were reconstituted into phosphatidylcholine and phosphatidic acid liposomes. They observed a conductance change of ~650 pS upon insertion or exit from the bilayer with predominant transitions of 450 pS exhibiting a voltage dependence (i.e. closing at both positive and negative voltages) characteristic of VDAC. Moran et al. (1992) observed in intact mitochondria several conductance levels ranging from less than 50 pS to about 300 pS and the same conductance levels were also detected when isolated OMM were incorporated into L-α-phosphatidylcholine liposomes. However, in contrast to those reported by Wunder and Colombini (1991) the voltage dependence was asymmetric. An additional conductance of 535 pS was also reported for the reconstituted OMM.

Dihanich et al. (1989) have reported a slightly cation selective activity of ~300 pS in VDAC-less mutants using planar bilayers. Similarly, Henry et al.(1989) have described a voltage dependent cation channel (VDCC) using tip-dip method with proteoliposomes derived from mitochondrial membranes. VDCC has two major conductance levels of 220 pS each in NaCl (which would be somewhat higher in KCl).

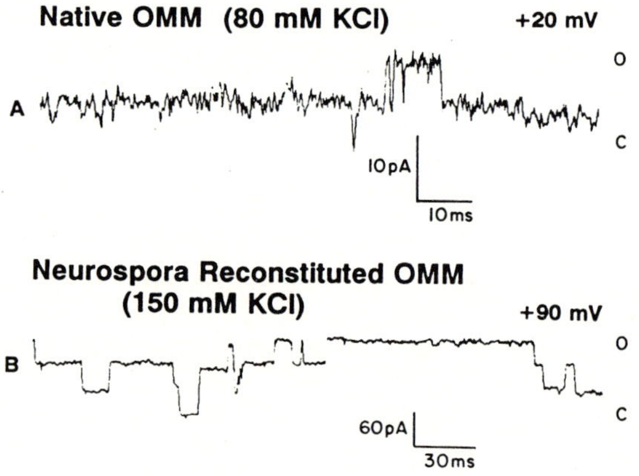

Figure 1 580 pS transitions recorded from whole mitochondria and reconstituted OMM: Current traces show ~580 pS transitions recorded from mouse liver mitochondria in **A** and fused Neurospora OMM-liposomes in **B**. **A:** Recording from a 2GΩ seal excised patch in symetrical 80 mM KCl 140 mM sucrose 5 mM HEPES pH 7.4. Transitions were 310 pS which, normalized to 150 mM KCl, is 580 pS. **B**: Recording from an excised patch containing 4 channels in symetrical 150 mM KCl 5 mM HEPES pH 7.4 where transitions correspond to 585 pS. Data was bandwidth limited to 1 kHz. Neurospora crassa OMM were the gift of C.Mannella and were prepared as in C.A.Mannella (1982). OMM were reconstituted by making giant liposomes with lecithin using the dehydration-rehydration method of Criado and Keller (1987) as previously reported (Costa et al., 1991).

We have obtained results similar to each report. As shown in Fig.1, we have observed transitions of ~580 pS (normalized from ~310 pS in 80 mM KCl medium to 150 mM KCl) in mouse liver mitochondria and have similar results in fused Neurospora OMM-liposomes. In most cases, our results resembled those of Moran et al. (1992) and Henry et al (1989) since we saw transitions of approximately 50, 100, 200 and 350 pS and at least some of these were sightly cation selective. Fig. 2 shows typical traces obtained with reconstituted OMM from kidney (A,B), liver (C) and Neurospora (D). Generally the voltage dependence was somewhat asymmetric. However, in some cases (approx. 10%) the results were more similar to those reported by Wunder and Colombini with a decrease in conductance at positive and negative voltages. The voltage dependence, however, was often slight suggesting some lability.

The transition sizes recorded were at times much greater than those reported by others and suggest that transitions can occur between the open and closed state bypassing the half-open state. The single channel characteristics of the VDAC-sized transitions of 650 pS shown in the current traces and legend of Figure 3 were voltage dependent. For example, the mean open time and burst length were long at 30 mV and much shorter at $\pm$ 80 mV (and -30 mV). Benz and Brdiczka (1992) have also reported a low sensitivity to voltage for VDAC (i.e. G/G_0 of 0.5 at $\pm$80 mV) in bilayer studies.

The existence of channels other than VDAC in the OMM has been demonstrated (e.g. Dihanich et al.,1989) and probably accounts for some of the heterogeneity observed. However, the details outlined above, i.e. that the very abundant VDAC are likely to be closed in the native membrane, suggest a high degree of modulation which may account for some of the differences in observed activities. We consider it likely that such modulation could provide a degree of plasticity not generally recognized.

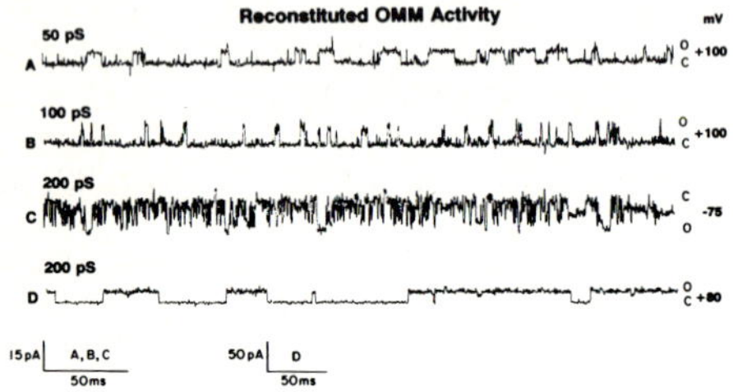

Figure 2 50-200 pS activity from reconstituted OMM: Current traces show a variety of activity in both conductance and single channel characteristics. Records from excised patches of fused outer mitochondrial membrane-liposome in 150 mM KCl 5 mM HEPES 1 mM EGTA 1.05 mM $CaCl_2$ pH 7.4. Data was bandwidth limited to 2 kHz (5 kHz sampling). The OMM source was mitochondria from kidney in A and B, liver in C and Neurospora in D. Other patch-clamp methods were previously described (e.g. Campo et al., 1992).

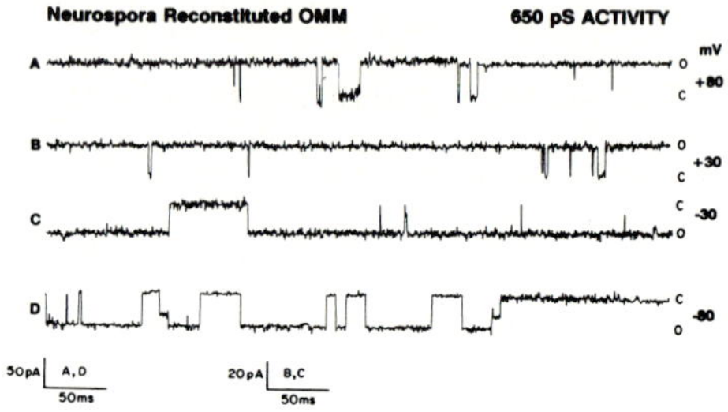

Figure 3 VDAC-sized transitions are voltage dependent: Current traces show 650 pS activity recorded from an excised patch of fused Neurospora outer mitochondrial membrane-liposome in 150 mM KCl 5 mM HEPES pH 7.4. Other details as in Fig. 2. The single channel charcteristics illustrate the assymetrical voltage dependence where the open state is maximized at 30 mV (nP_0 of 0.99) and minimized at -80 mV (nP_0 of 0.61):

	mV	80	30	-30	-80
mot[a]	ms	60	145	62	39
mct[b]	ms	1.5	0.9	1.4	20.9
bl[c]	s	0.8	2.3	1.0	0.1

[a]mean open time
[b]mean closed time
[c]burst length

Complexes involving VDAC

Studies of the interactions of VDAC with enzymes intimately involved with the cell's bioenergetics have suggested interplays with inner membrane components [e.g. adenine nucleotide translocator (ADT)] and introduced an entirely new perspective to the possible physiological significance of the mitochondrial channels. As originally described by Hackenbrock (1968), the inner and the outer membranes make contact at specific sites. A variety of experiments have suggested a special organization and role of these sites. Patch-clamping liposomes containing membrane preparations from brain mitochondria enriched in contact sites (Sandri et al., 1988) revealed multiple conductance channel activity including 475, 550, 660 and 970 pS (Moran et al., 1990) which are levels we ascribe to MCC activity. The association of MCC with contact sites suggests that outer and inner mitochondrial channels are in register at contact sites to provide patent channels between the medium and the patching pipette (Kinnally et al., 1992).

Experiments on the biochemical properties of mitochondrial components support the concept that the contact sites are organized in complexes. Several studies indicate an interaction between VDAC and other components: hexokinase (Nakashima et al., 1986, Lindén et al., 1982), creatine kinase and Ca^{2+} (see Benz and Brdiczka, 1992), glutathione transferase (Krause et al, 1986) and glycerol kinase (Ostlund et al, 1983). The association of hexokinase (Gots and Bessman, 1974, Arora, et al. 1988; BeltrandelRio and Wilson, 1991) or creatine kinase (Yang et al., 1977; Gellerich and Saks, 1982; Erickson-Viitanen, 1982) with the OMM favors the use of intra-mitochondrial synthesized ATP. These observations have led to the suggestion of a functional complex of VDAC and ADT. In the appropriate tissues, hexokinase (e.g. McEnery, 1992) or creatine kinase

(Schlegel et al. 1990) are also thought to form part of this complex. The effect of a synthetic polyanion (for review see Benz and Brdiczka, 1992) or an endogenous protein regulator (Liu and Colombini, 1992) on membrane function in intact mitochondria supports these views as both affect mitochondrial functions associated with ATP utilization. However, the fact that both polyanions and modulators are effective when placed on the outside of the mitochondrion remains a puzzle.

Other techniques support an organization of contact sites. An association of contact sites with hexokinase (Brdiczka et al., 1986, Kottke et al., 1988), creatine kinase (Kottke et al., 1988) and VDAC clusters (Brdiczka et al., 1986) has been claimed from electron microscopy studies using immunogold-labelling techniques (Bridcza et al. 1986; Kottke et al., 1988) or cytochemical techniques (Biermans et al, 1989). In agreement with the proposed models, a functioning mitochondrial benzodiazepine receptor (mBzR) complex recently been isolated. It consisted of an 18 kDa protein as well as ADT and VDAC, the latter two identified by Western blot analysis (McEnery et al., 1992). These findings suggest the possibility of examining the relationship between outer and inner mitochondrial membrane channels with the mBzR through an examination of the sensitivity of the channel activity to benzodiazepines. In accordance with these expectations these drugs were found to block inner mitochondrial membrane channels (Zorov et al. 1992a).

Inner membrane channels

The basic properties of the four inner mitochondrial membrane channels are summarized in Table 1. This display includes MCC (multiple conductance channel) activity, ~100 pS voltage dependent activity and three low conductance channel (LCC) activities. It is interesting to note two points. Except for the ATP-sensitive K^+ channel, none of the activities are highly selective. Secondly, if the Ca^{2+} is maintained at or below physiological levels throughout preparation and the pH is close to neutral, no channel activity is recorded (Kinnally et al., 1992). This finding suggests a high degree of regulation of these channels.

TABLE 1

Class	Conductance	Selectivity	Effectors (+/-)[a]
Multiple Conductance (MCC)	~40- >1000	sl. cation (lower substates)	voltage, Ca^{2+} (+), Mg^{2+}(-), ADP(-)
~100 pS	50-140	sl. anion	voltage, Ca^{2+}(-)
Low Conductance (LCC)			
K^+ channel	15	K^+	ATP(-)
ACA	15	sl. cation	pH, Mg^{2+}(-)
AAA	45	sl. anion	pH, Mg^{2+}(-)

[a] (+/-): + activating, - inhibiting

<u>The ~100 pS channel</u>

Studies of the ~100 pS channel activity first described by Sorgato et al. (1987) have recently appeared (see separate section below for inhibitor studies). Klitsch and Siemen (1991) report "whole mitoplast" recordings of the ~100 pS channel in brown adipocytes which indicate an inhibition by purine nucleotides. This configuration was supposedly obtained by exposing the patches to high voltage. Evidence for this configuration is a reversal of polarity seen in IV curves, a slight increase in capacitance and a decrease in resistance after high voltage treatment. It should be noted however that we have observed the activation of MCC by voltage in excised patches. In most of these MCC was found to have a polarity opposite to the 100 pS channel (e.g. Fig. 1C of Kinnally et al., 1989). Furthermore, Inoue et al. (1991) claim an insensitivity of the ~100 pS channel to ATP and Szabó and Zoratti (1992) report an inhibition of MCC by ADP although in both studies the adenine nucleotides were introduced on the matrix side of the membrane.

We have previously proposed (a) a ~140 pS state of the ~100 pS channel (Fig. 7 of Kinnally et al, 1992) based on the higher conductance in the presence of amiodarone (Antonenko et al, 1991a) (ascribed to inhibition below the "apparent closed" state) and (b) a ~50 pS substate (Fig. 6 of Campo et al., 1992) based on observed transitions of an activity of predominantly ~100 pS. Similarly, Klitsch and Siemen (1991) observed infrequent channel openings 1/3, 1/2 and 2/3 of the opening of 108 pS. We also proposed that the 50% conductance level is a substate. As shown in Fig. 4, we have observed the higher conductance level and the 50 pS channel activity occurring without ~100 pS activity. These substates have the same voltage dependence as the ~100 pS activity. In addition the 50 pS conductance level has similar single channel characteristics and the same slight anion-selectivity as the ~100 pS channel (see Fig.5). Upon imposing a five-fold KCl gradient, we found the reversal potential for the 50 pS activity to be -22 mV, the same as that reported by Sorgato et al.(1987) for the ~100 pS channel. In addition, bath substitution of Cl^- by acetate$^-$ revealed reversal potential shifts consistent with higher Cl^- than acetate permeability. Also consistent with the anion selectivity was the finding that cation substitution of choline$^+$ for K^+ had no effect on the reversal potential. The increase in conductance in the acetate (also in choline chloride) is puzzling, at this time we have no explanation for it.

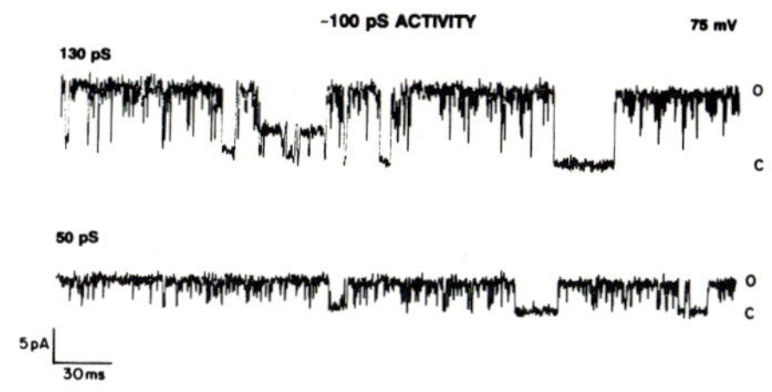

Figure 4 Substates of the ~100 pS activity: Current trace recorded from excised patch (rat heart mitoplast) at 75 mV shows 130 pS activity with a substate of 47 pS in the upper trace and in a different patch, 50 pS activity in the lower trace. Medium was 150 mM KCl 5 mM HEPES 0.75 mM $CaCl_2$ pH 7.4. Other details as in Fig. 3. At 75 mV, the mean open time distribution was fit by a single exponential with a time constant of 3.0 ms for the 50 pS activity.

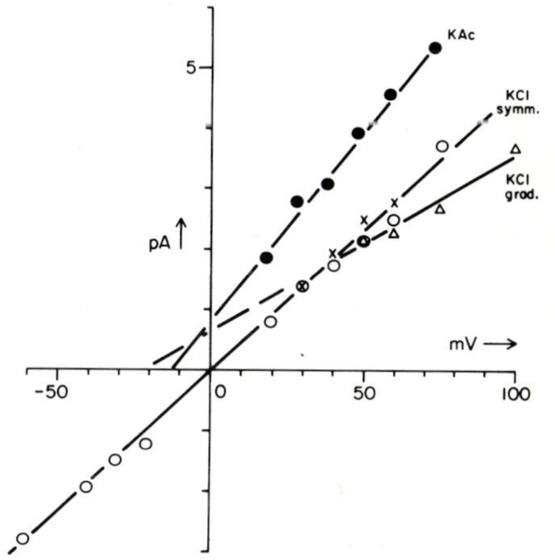

Figure 5 Selectivity of the 50 pS activity: Single channel amplitudes determined from total amplitude diagrams (bin width of 0.1 pA) at different voltages were obtained under different ionic conditions from an excised patch from rat heart mitoplast. The pipette contained 150 mM KCl, 5 mM HEPES, 1 mM EGTA, 0.75 mM $CaCl_2$ pH 7.4 and substitutions were made in the bath medium: 150 mM KCl (○), 120 mM KAcetate, 30 mM KCl (●), 30 mM KCl 240 mM sucrose (▵). Other details as in Fig. 3. Choline chloride substitution resulted in no change in reversal potential, data not shown. Bath substitutions were completely reversible as the X and O were the control (symmetrical 150 mM KCl) after and before the application of ion gradients.

The channel activity induced by alkaline pH

During our previous study of the effect of alkaline pH we observed the activation of a ~15 pS channel. On occasions we also observed a larger channel (e.g. see fig, 1 of ref. Antonenko et al, 1991b). Subsequent studies revealed the activation of a ~45 pS channel (Antonenko, Kinnally and Tedeschi, unpublished results). Sample current traces of these two alkaline pH-induced activities are shown in Fig.6. A study of the ion selectivity of the two alkaline pH-induced channels using asymmetric ion substitutions and reversal potential techniques (e.g. Anderson et al., 1986) was also carried out. The ~15 pS channel was found to be cation-selective and to be insensitive to the presence of anions (we refer to these as the alkaline pH-induced cation selective activity, ACA). Conversely the ~45 pS was found to be anion-selective (we refer to these as the alkaline pH-induced anion selective activity, AAA). However, the selectivity among the ions of the same charge was very slight.

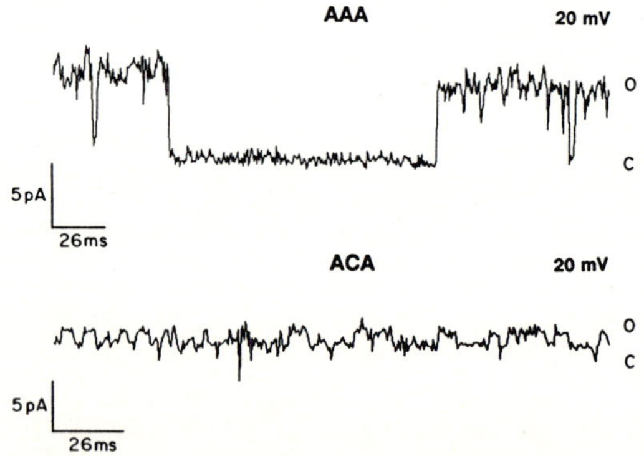

Figure 6 pH-dependent activities: Current traces show alkaline pH activated anion activity (AAA) and cation activity (ACA) of 45 and 15 pS respectively. Data was recorded from excised patches from mouse liver mitoplasts in 150 mM KCl, 5 mM HEPES, 10 μM $CaCl_2$ pH 8.3. Other details as in Fig.3.

A permeability transition that can be induced in mitochondrial suspensions by alkaline pH and by lowering matrix Mg^{2+} (for review see Garlid and Beavis, 1986; Beavis, 1992) has been ascribed to the activation of a specific channel referred to as inner membrane anion channel (IMAC). Similarly, a so called K^+-uniporter can be activated in mitochondrial suspensions by depletion of Mg^{2+} (e.g. Bernardi et al., 1989; Nicolli et al., 1991). It appears to be relatively unselective with regard to Rb^+, Na^+ and Li^+ (Nicolli et al. 1991) and might also correspond to the activation of a channel.

In view of the observations that alkaline pH and Mg^{2+} depletion induce an anion-selective permeability transition, it would be tempting to postulate that the 45 pS channel activity corresponds to IMAC. Our previous study (Antonenko et al. 1991b) examined the pharmacology and dependence on Mg^{2+}, or pH of the currents induced by alkaline conditions and found them to parallel closely those of IMAC (e.g. Beavis, 1989). However, a strong argument against considering IMAC and AAA reflections of the same activity comes from selectivity data. The ratio of permeabilities to Cl^- over glucuronate was reported to be higher than 400 for IMAC (Beavis, 1992) whereas Antonenko et al. found it to be approximately 3 for AAA (Antonenko, Kinnally and Tedeschi, unpublished results). Similarly the ~15 pS channel might correspond to the K^+ uniporter. This view would be supported by the fact that the two are relatively unselective in relation to cations and inhibited by Mg^{2+}. However, it is our feeling that present information is not sufficient at this time for a firm conclusion.

Multiconductance channel activity (MCC)

The MCC activity first described in 1988 (Kinnally et al., 1988) exhibits a variety of conductance states (e.g. see Kinnally et al., 1989, Petronilli et al., 1989; Zorov et al., 1992b). The channel can be induced by Ca^{2+} (e.g. Kinnally et al., 1991; Szabó et al., 1992; Szabó and Zoratti, 1992) or voltage of usually >±60 mV (Zorov et al, 1992b). Most frequently the activity is voltage dependent with opening in the negative and closing to lower conductance levels in the positive range. During voltage activation or the introduction of inhibitors such as amiodarone (Zorov et al, 1992b) or cyclosporine A (Zorov et al. 1992c) a stepping through the various conductance levels is observed. MCC can remain in the open or closed state for prolonged periods of time (up to many minutes). Moran et al (1990) demonstrated activity in preparations enriched in contact sites (Sandri et al., 1988) that we consider to correspond to MCC.

A permeability transition pore (PTP) has been postulated to account for the permeability transition induced in mitochondrial suspensions by elevated Ca^{2+} levels (for review see Gunter and Pfeiffer, 1990). Studies of MCC (Szabó and Zoratti, 1991, 1992, Szabó et al, 1992, called the mitochondrial megachannel, MMC, in their reports) and parallel experiments of PTP using mitochondrial suspensions (Bernardi et al., 1992) have presented convincing evidence that the two are equivalent. The studies include the dependence on divalent cations, ADP, pH and the effect of drugs.

Are there other channels?

There is considerable evidence that mitochondrial translocases can act as channels in reconstituted systems (e.g. Mironova et al.,1981, 1982, Costa et al., 1991, Paliwal et al., 1992). Therefore, the possibility that these proteins could function as channels as previously proposed (Tedeschi, 1981) has to be considered.

Pharmacology of inner membrane channels

There are several factors that emerge from the sensitivity of inner mitochondrial membrane (IMM) channels to chemicals. One of these is an embarrassment of riches as many are effective. Evidence comes from the less direct permeability studies using mitochondrial suspensions as well as from patch-clamp studies.

In mitochondrial suspensions a broad range of drugs have been found to inhibit the inner membrane anion channel (IMAC). These include cardiac drugs (propranolol, amiodarone), quinine, local anesthetics (dibucaine, dupivacaine) as well anxyolitic and muscle relaxing drugs (Ro5-4864,clonazepam) (for a review see Beavis, 1992).

Experiments in our laboratory using patch-clamp techniques have made similar observations. We found inhibition of MCC and the ~100 pS channel activity by amiodarone, propranolol (see Kinnally et al., 1992), antimycin A (Campo et al., 1992) and dibucaine (Holmuhamedov, Kinnally and Tedeschi, 1992, unpublished results). As shown in Fig. 7, aluminum can be added to this list as it blocks both MCC activity and the 100 pS activity (Zorov et al., 1991). The amphiphilic drugs (amiodarone,quinine, propranolol) also block the alkaline pH-induced conductance. We have recently reported the inhibition of MCC and the ~100 pS channel in cardiac mitochondria by high affinity ligands of the mBzR (Zorov et al., 1992a).

One of the unique features of the effect of these inhibitors on inner mitochondrial membrane channels is the activation of channels by high concentrations of some drugs. We have found this to be the case for antimycin A (Campo et al., 1992) as shown in Fig.8 and we have now observed it with protoporphyrin IX, a high affinity ligand of mBzR (Kinnally and Tedeschi, unpublished results). While the effects of amiodarone are not readily reversible, they are consistent with an activation. Recently, Gudz et al. (1992), reported an inhibition by low and activation by high concentrations of amiodarone, DCCD and BHT of the permeability transition pore. Although other interpretations are possible, the activation phenomenon can be explained most simply by assuming the presence of additional receptor sites of lower affinity than the inhibitor sites as illustrated in Fig. 9. At high concentrations the agent can allosterically negate the inhibitory effect. The addition of activating concentrations of a ligand were found to be blocked by inhibitory concentrations of a second higher affinity ligand (Ro5-4864 blocked protoporphyrin IX activation, Kinnally, unpublished results).

The sensitivity of various mitochondrial channels to the same inhibitors raises an important question. Are they all manifestations of the same channel protein? For example, could the ~100 pS channel be one of the substates of MCC? We have shown that the voltage-activation of MCC or the inhibition of MCC by amiodarone (Zorov et al., 1992b) and cyclosporin (Zorov et al., 1992c) proceed in steps which span the range of conductance from 45 (and perhaps less) to approximately ~1.5 nS. At least some of these activities differ in ion-selectivity (see Kinnally et al, 1989). However, differing ion selectivity cannot be used to answer this question since, at least in VDAC, substates have been shown to have different selectivity (Ludwig et al., 1989; Ermishkin et al., 1990). Therefore the question remains unanswered at this time.

The idea that VDAC may be part of other channel complexes and therefore involved in other activities is supported by the report of MCC-like activity (Moran et al., 1990) in membranes enriched with contact sites and the claim that VDAC cluster at contact sites (Brdiczka et al., 1986). It has been suggested that VDAC can play a vital role in metabolism because it can selectively block the passage of ADP and ATP in its half-closed state (e.g. Ludwig et al., 1989). A fine regulative role of VDAC may require connection in series to inner mitochondrial channels. A variety of data do suggest an association of VDAC with inner membrane components as presented above. The next section examines the possible role of contact sites.

Another question (in part a different face of the same questions discussed above) is no less pertinent. Why is this plethora of unrelated drugs affective in inhibiting supposedly distinct mitochondrial channels? The effect of the compounds may be directly on the channel proteins at highly homologous binding domains. However, several indirect effects are possible. (a) Since these compounds are hydrophobic they could exert their effect by altering the properties of the membrane. Drug effects on lipid components have been documented for both artificial and plasma membranes. Nevertheless, it is unlikely these compounds act through a such a mechanism since their effects on membrane lipids are frequently in opposite directions [e.g. reportedly quinidine decreases lipid mobility (Needham et al., 1987), propranolol increases it (Dash and Rao, 1990)]. Furthermore, in some rare cases the specificity of the drugs can be exquisite even when the drugs are closely related structurally, e.g. we found the benzodiazepine Ro 5-4864 which binds to the peripheral receptor with high affinity is very effective whereas clonazepam which has a lower affinity is not (Zorov et al., 1992a). (b) The channels may be capable of responding to different drugs because of shared binding proteins each specific for one ligand. These proteins would be free to move in the plane of the membrane. Drug binding may allosterically alter their interaction with the channels.

(c) A single receptor could respond to a variety of drugs and be capable of binding to several target channels. The mBzR can interact with a broad spectrum of drugs apparently structurally unrelated. These compounds include antimycin A, propranolol, protoporphyrin IX and Ro5-4864 (e.g. see Hirsch et al, 1988b, Verma and Snyder, 1988), all of which can block channel activity (see above).

(d) Open channels could be restricted to contact sites. And, the assembly and disassembly of contact sites have been claimed to be dynamic (see above). Such a model could allow the coordinate regulation of channel activity. Results obtained with mitoplasts need not be excluded from this interpretation since outer membrane pieces generally persist even after removal of the outer membrane (e.g. see Sorgato et al., 1987). The next section will examine in some detail this alternative.

Channels and contact sites: possible models

A functional channel might require the presence of inner and outer membrane channels in series which would require the apposition of inner and outer membrane for patent ion pathways. Hackenbrock (1968) calculates the presence of about 37 contact sites sq.μ^{-1} in mitochondria in state 4. Assuming a random distribution, this would correspond to ~7 contact sites for a patch with a diameter of 0.5 µm. It would therefore appear that there are always likely to be contact sites in a mitoplast patch provided that the contact sites remain attached during the preparatory procedures. An association between metabolic state and number or extent of contact between inner and outer membrane has been suggested by electron microscope data (Knoll and Brdiczka, 1983; Brdiczka et al., 1986).

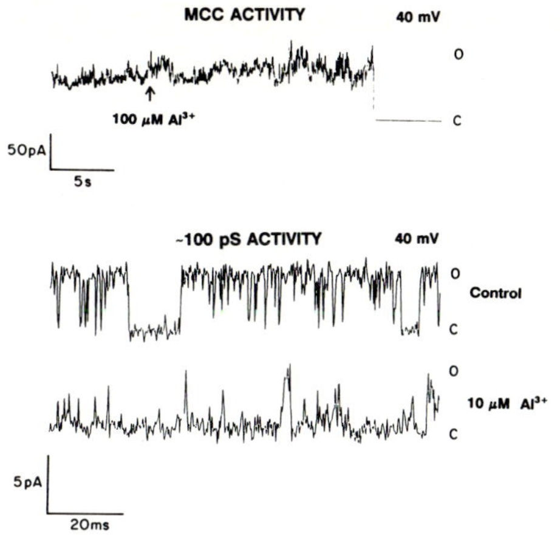

Figure 7 Aluminum inhibition of inner membrane channel activity: Current trace shows kinetics of MCC inhibition by aluminum at +40 mV (bandwidth limited to 50 Hz). $AlSO_4$ in medium was added as indicated and resulted in closure of ~1.4 nS (57 pA) to the fully closed state (0.3 pA) which was not reversible with perfusion with 10 mL of Al-free medium (150 mM KCl, 5 mM HEPES, 1 mM EGTA, 0.75 mM $CaCl_2$ pH 7.4. Current traces before (control) and 37 seconds after the addition of 10 μM $AlSO_4$ show inhibition of the 110 pS activity. Transitions were 5.2 pA and the voltage was 40 mV. The excised patches were from mouse liver mitoplasts. Other details as in Fig.3.

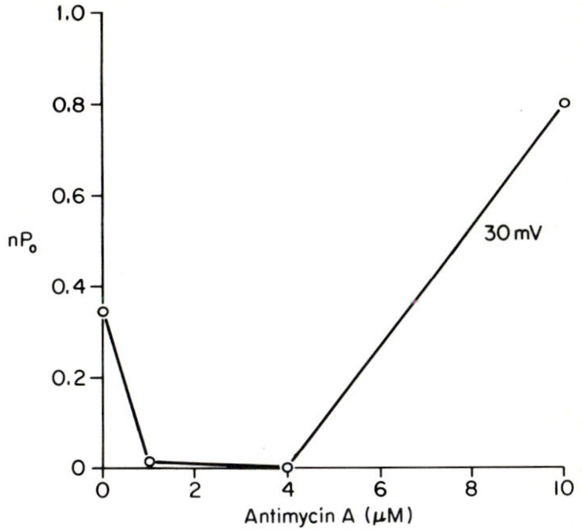

Figure 8 Antimycin A inhibition and activation of MCC: The open probaility calculated from amplitude diagrams shows the inhibition at 1-4 μM and activation at 10 μM. Data obtained on this excised mouse liver mitoplast was band width limited to 2 kHz and the medium was 150 mM KCl, 5 mM HEPES, 1 mM EGTA, 0.75 mM $CaCl_2$ pH 7.4. Taken from Campo et al. (1992) with permission.

A good deal of the data could be explained if contact sites were regarded as dynamic complexes of proteins, formed and disassembled depending on metabolic or other conditions (e.g. Ca^{2+}, presence of certain inhibitors, etc.). The presence of contact sites in mitoplasts has been shown by EM studies. The dependence of contact sites integrity and the observation of MCC activity on Ca^{2+} level has been well documented. We have shown that the higher conductance levels of MCC can be reached gradually from lower conductance levels (Zorov et al., 1992c), which could correspond to an assembly from subunits, from increasing effective pore diameter or from an increasing extent of interaction of inner and outer membrane to form patent channels. A graphical representation of the model is shown in Fig. 10. The model assumes that open channels can only form at contact sites. The "stoppers" in the figure represent the assumption that the channels are closed at other sites.

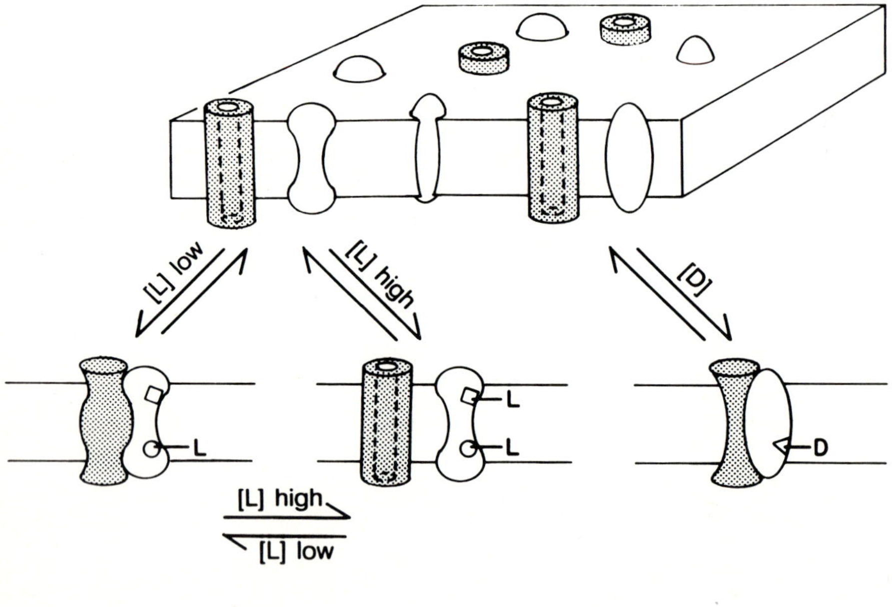

Figure 9 Hypothetical model of generic channels in a membrane to illustrate the inhibition by low ligand (L) or drug (D) levels by binding to high affinity sites (▵,○) and the activation by high ligand levels due to binding of ligands to a lower affinity site (□).

We are basically extending the model of others which propose functional complexes at contact sites (see above) with the addition of IMM channels. Activation (e.g. Ca^{2+} activation of MCC) would involve the formalization of a contact site (apposition of inner and outer membrane channels to form patent passages). Different conditions could favor some substates of MCC with different properties (e.g. ion selectivity). The presence of channels in a single complex and its dependence on the contact sites would permit inhibition by unrelated drugs and factors. The involvement of the mBzR in MCC, ~100 pS activity as well as IMAC implicate the involvement of inner and outer membrane components.

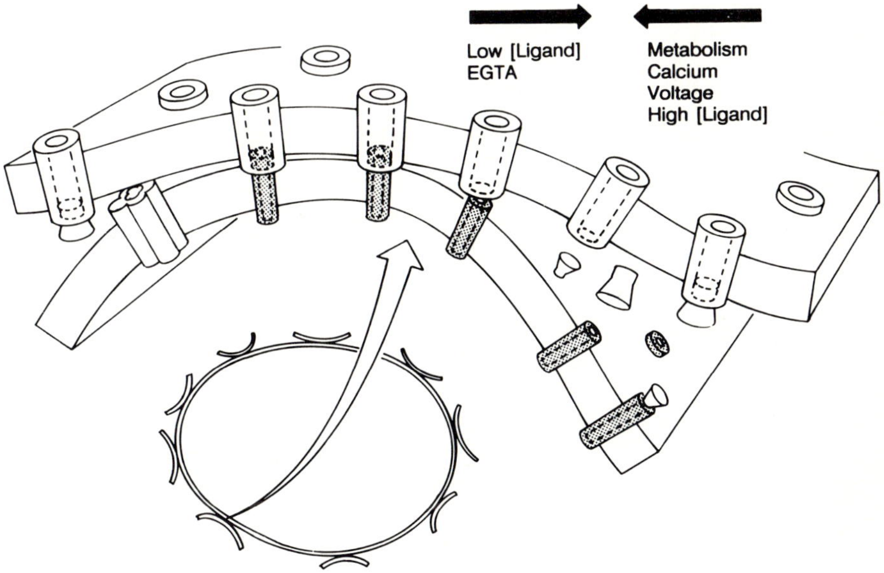

Figure 10 Surrealistic model involving the interaction of the inner and outer mitochondrial membrane channels to provide a patent ion pathway. Regulation of the apposition of the two membranes could control activity. Channels outside of the contact site are thought to generally be inactive as indicated by "corks".

A variety of observations are in harmony with this model. Contact site assembly is increased by Ca^{2+}. Ca^{2+}-binding and Ca^{2+}-translocase appear to be involved with contact sites (Kottke et al., 1988). In addition, MCC activity is probably located at contact sites (Moran et al. 1990) and VDAC are thought to be clustered there.

The activation and inactivation could involve separate target receptors each a component of the complex (e.g VDAC, ADT, when appropriate hexokinase or phosphocreatine kinase, etc.). Control of the process could be indirect since Brdiczka et al. (1986) claim a correlation between metabolic states and the presence and size of contact sites. Early EM observations with freeze fracture (e.g. Hackenbrock, 1972) have claimed a clustering of macromolecules in the inner and the outer membrane brought about by succinate and a dispersion brought about by antimycin A. We have shown that antimycin A blocks both channels studied. Amytal that also blocks the electron transport chain and is thought to decrease contact sites (Biermans et al., 1989). In addition, drugs which block channel activity, such as ligands of mBzR (Vorobjev and Zorov, 1983; Hirsch et al, 1988, Moreno-Sánchez et al., 1991) and amiodarone (Fromenty et al., 1990) have been shown to block electron transport.

Formation of contact sites need not be the only factor and activation of individual kinds of channels may require special signals (low Mg^{2+}, alkaline pH, etc.).

Conclusions

Channels of the inner and outer mitochondrial membranes are for the most part quiescent and must be induced to be observed with patch-clamp techniques. Their activity is subject to fine regulation by a variety of physiological factors, e.g. voltage, Ca^{2+}, Mg^{2+}, pH and nucleotides. A variety of structurally unrelated drugs have similar effects on different channel activities suggesting the involvement of a common intermediate or an interaction of inner and outer membrane channels. In addition, the observation of activation by high concentrations of some inhibitors is novel and significant. While we have proposed general models to account for some of these observations here and in earlier papers, we are still far from understanding the functioning of either outer or inner mitochondrial membrane channels. A few questions have been answered, but more importantly we now can pose many more in a way which will permit experimental analysis.

Acknowledgements

We thank Carmen A. Mannella for his gift of Neurospora outer mitochondrial membranes and discussion. We thank James DeBritz and Nicely Guzman for technical assistance. This work was supported in part by NSF grant MCB9117658.

References

Andersen OS, Green WN, Urban BW (1986) Ion conduction through sodium channels in planar lipid bilayers. In: Ion Channel Reconstitution. Miller C, (ed) p. 385-404 (see p. 392-398)

Antonenko YuN, Kinnally KW, Perini S, Tedeschi H (1991a) Selective effect of inhibitors on inner mitochondrial membrane channels. FEBS Lett 285:89-93

Antonenko YuN, Kinnally KW, Tedeschi H. (1991b) Identification of anion and cation pathways in the inner mitochondrial membrane by patch clamping of mouse liver mitoplasts. J Membr Biol 124:151-158

Arora KK, Pedersen PL (1988) Functional significance of mitochondrial bound hexokinase in tumor cell metabolism. Evidence for preferential phosphorylation of glucose by intramitochondrially generated ATP. J Biol Chem 263:17422-17428

Barbour RL, Ribaudo J, Chan SHP (1984) Effect of creatine kinase activity on mitochondrila ADP/ATP transport. J Biol Chem 259:8246-8251

Beavis AD (1989) On the inhibition of the mitochondrial inner membrane anion uniporter by cationic amphiphiles and other drugs. J Biol Chem 264:1508-1518

Beavis AD (1992) Properties of the inner membrane anion channel in intact mitochondria. J Bioenerg Biomemb 24:77-90

BelrandelRio H, Wilson JE (1991) Hexokinase of rat brain mitochondria: relative importance of adenylate kinase and oxidative phosphorylation as sources of substrate ATP, and interaction with intramitochondrial compartments of ATP and ADP. Arch Biochem Biophys 286:183-194

Benz R (1985) Porin from bacterial and mitochondrial outer membrane. CRC Crit Rev Biochem 19:145-190

Benz R, Brdiczka D (1992) The cation-selective substate of the mitochondrial membrane pore: single channel conductance and influence on intermembrane and peripheral kinases. J Bioenerg Biomemb 24:33-39

Benz R, Kottke M, Brdiczka D (1990) The cationically selective state of the mitochondrial outer mebrane pore: a study with intact mitochondria and reconstituted mitochondrial porin. Biochim Biophys Acta 1022:311-318

Bernardi P, Angrilli A, Ambrosin V, Azzone GF (1989) Activation of latent K^+ uniport in mitochondria treated with the ionophore A23187. J Biol Chem 264:18902-18906

Bernardi P, Vassanelli S, Veronese P, Colonna R, Szabó I, Zoratti M (1992) Modulation of the mitochondrial transition pore: effect of protons and divalent cations. J Biol Chem 267:2934-2939

Biermans W, De Bie M, Nijs B, Jacob W (1989) Ultrastructural localisation of creatine kinase activity in contact sites between inner and outer mitochondrial membranes of rat myocardium. Biochim Biophys Acta 974:74-80

Brdiczka D, Knoll G, Riesinger I, Wieler U, Klug G, Benz R, Krause J (1986) Microcompartmentation at the mitochondrial surface: its function in metabolic regulation. In: Myocardial and Skeletal Muscle Bionergetics. Brautbar N (ed) Plenum Press New York and London pp. 55-69

Campo ML, Kinnally KW, Tedeschi H (1992) The effect of antimycin A on mouse liver inner mitochondrila membrane channel activity. J Biol Chem 267:8123-8127

Colombini M (1979) A candidate for the permeability pathway of the outer mitochondrial membrane. Nature (London) 279:643-645

Colombini M, Yeung CL, Tung J, König T (1987) The mitochondrial outer membrane channel VDAC, is regulated by a synthetic polyanion. Biochim Biophys Acta 905:279-286

Costa G, Kinnally KW, Diwan JJ (1991) Patch-clamp analysis of a partially purified ion channel from rat liver mitochondria. Biochem Biophys Res Comm 175:305-310

Criado M, Keller BU (1987) A membrane fusion strategy for single-channel recordings of membranes ussually non-accessible to patch-clamp pipette electrodes. FEBS Lett 224:172-176

Dash D, Rao GR (1990) Characterization of the effects of propranolol on the physical state of platelet membrane. Arch Biochem Biophys 276:343-347

De Pinto V, Ludwig O, Krause J, Benz R, Palmieri F (1987) Porin pores of mitochondrial outer membranes from high and low eukaryotic cells: biochemical and biophysical characterization. Biochim Biophys Acta 894:109-119

Dihanich M, Schmid A, Oppliger W, Benz R (1989) Identification of a pore in the mitochondrial outer membrane of a porin-deficient yeast mutant. Eur J Biochem 181:703-708

Erickson-Viitanen S, Viitanen P, Geiger PJ, Yang WCT, Bessman SP (1982) Compartmentation of mitochondrial creatine phosphokinase. I. Direct demonstration of compartmentation with the use of labeled precursors. J Biol Chem 257:14395-1404

Ermishkin LN, Mirzabekov TA (1990) Redistribution of the electric field within the pore contributes to the voltage dependence of mitochondrial porin channel. Biochim Biophys Acta 1021:161-168

Freitag H, Neupert W, Benz R (1982) Purification and characterisation of a pore protein of the outer membrane from Neurospora crassa. Eur J Biochem 123:619-636

Fromenty B, Fisch C, Berson A, Letteron P, Larrey D, Pessayre D (1990) Dual effect of amiodarone on mitochondrial respiration. Initial protonophoric uncoupling effect followed by inhibition of the respiratory chain at the levels of complex I and II. J Pharmacol Exp Ther 255:1377-1384

Garlid KD, Beavis AD (1986) Evidence for the existence of an inner membrane anion channel in mitochondria. Biochim Biophys Acta 853:187-204

Gellerich F, Saks VA (1982) Control of heart mitochondrial oxygen consumption by creatine kinase: the importance of enzyme localization. Biochem Biophys Res Comm 105:1473-1481

Gots RE, Bessman SP (1974) The functional compartmentation of mitochondrial hexokinase. Arch Biochem Biophys 163:7-14

Gudz TI, Novgorodov SA, Pfieffer DR (1992) Structurally and pharmacologically dissimilar compounds with common effects on the mitochondrial permeability transition. The Seventh European Bionergetics Conference. Short communications 7:125

Gunter TE, Pfeiffer DR (1990) Mechanisms by which mitochondria transport calcium. Amer J Physiol, Cell Physiol 27:C755-786

Hackenbrock CR (1968) Chemical and physical fixation of isolated mitochondria in low and high energy states. Proc Natl Acad Sci USA 61:598-605

Hackenbrock CR (1972) States of activity and structure in mitochondrial membranes. Ann NY Acad Scie 192:492-504

Henry JP, Chich J-F, Goldschmidt D, Thieffry M (1989) Blockade of a mitochondrial cationic channel by an adressing peptide: an electrophysiological study. J Membr Biol 112:139-147

Hirsch JD, Beyer CF, Malkowitz L, Beer B, Blume AJ (1988a) Mitochondrial benzodizepine receptors mediate inhibition of mitochondrial respiratory control. Molecular Pharm 34:157-163

Hirsch JD, Beyer CF, Malkowitz L, Loullis CC, Blume AJ (1988b) Characterization of ligand binding to mitochondrial benzodiazepine receptors. Molec Pharm 34:164-172

Holden MJ, Colombini M, (1988) The mitochondrial outer membrane channel, VDAC, is modulated by a soluble protein. FEBS Lett 241:105-109

Inoue I, Nagase H, Kishi K, Higuti T (1991) ATP-sensitive K^+ channel in the mitochondrial inner membrane. Nature 352:244-247

Kinnally KW, Tedeschi H, Mannella CA (1987) Evidence for a novel voltage activated channel in the outer mitochondrial membrane. FEBS Lett 226:83-87

Kinnally KW, Campo ML, Tedeschi, H. (1988) Evidence for multiple channels in mitochondrial inner membranes: results from patch-clamping. Abstracts, 14th International Congress of Biochemistry, p.168

Kinnally KW, Campo ML, Tedeschi H (1989) Mitochondrial channel activity studied by patch clamping mitoplasts. J Bioenerg Biomemb 21:497-506

Kinnally KW, Zorov DB, Antonenko YuN, Perini S (1991) Calcium modulation of inner mitochondrial channel activity. Biochem Biophys Res Comm 176:1183-1188

Kinnally KW, Antonenko YuN, Zorov DB (1992) Modulation of inner mitochondrial membrane channel activity. J Bioenerg 24:99-110

Klitsch T, Siemen D (1991) Inner mitochondrial membrane anion channel is present in brown adipocytes but is not identical with the uncoupling protein. J Membr Biol 122:69-75

Knoll G, Brdiczka D (1983) Changes in freeeze-fractured mitochondrial membranes correlated to their energy state. Dynamic interactions of the boundary membranes. Biochim Biophys Acta 733:102-110

Kottke M, Adam V, Riesinger I, Bremm G, Bosch W, Brdiczka D, Sandri G, Panfili E (1988) Mitochondrial boundary membrane contact sites in brain: points of hexokinase and creatine kinase location, and control of Ca^{2+} transport. Biochim Biophys Acta 935:87-102

Krause J, Hay R, Kowollik C, Brdiczka D (1986) Crosslinking analysis of yeast mitochondria outermembrane. Biochim Biophys Acta 860:690-698

Lindén M, Gellerfors P, Nelson BD (1982) Pore protein and the hexokinase-binding protein of rat liver mitochondria are identical. FEBS Lett 141:189-192

Lindén M, Andersson G, Gellefors P, Nelson BD (1984) Subcellular distribution of rat liver porin. Biochim Biophys Acta 770:93-96

Liu MY, Colombini M (1991) Voltage gating of the mitochondrial outer membrane channel VDAC is regulated by a very conserved protein. Am J Physiol 260:C371-C374

Liu MY, Colombini M (1992) Regulation of mitochondrial respiration by controlling the permeability of the outer membrane through the mitochondrial channel, VDAC. Biochim Biophys Acta 1098:255-260

Ludwig O, Benz R, Schulz JE (1989) Porin of Paramecium mitochondria. Isolation, characterization and ion-selectivity of the closed state. Biochim Biophys Acta 978:319-327

Mannella CA (1982) Structure of the outer mitochondrial membrane: Oredered arrays of pore-like subunits in outer mitochondrial membrane fractions from Neurospora crassa. J Cell Biology 94:680-687

McEnery MW (1992) The mitochondrial benzodiazepine receptor: evidence for association with voltage-dependent anion channel (VDAC). J Bioenerg Biomembr 24:63-70

McEnery MW, Snowman AM, Trifiletti RR, Snyder SH (1992) Isolation of the mitochondrial benzodiazepine receptor: association with the voltage dependent anion channel and the adenine nuclotide carrier. Proc Natl Acad Sci USA 89:3170-3174

Mironova GD, Fedotcheva NI, Makarov PR, Pronevich LA, Mironova GP (1981) A protein of the bovine heart mitochondria inducing channel potassium conductivity in bilayer lipid membranes. Biofizika 26:451-457

Mironova GD, Sirota TV, Pronevich LA, Trofimenko N, Mironov GP, Gregorjev PA, Kondrashova MN (1982) Isolation and properties of Ca^{2+}-transporting glycoprotein and peptide from beef heart mitochondria. J Bioenerg Biomemb 14:213-225

Moran O, Sandri G, Panfili E, Stühmer W, Sorgato MC (1990) Electrophysiological characterization of contact sites in brain mitochondria. J Biol Chem 265:908-913

Moran O, Sciancalepore M, Sandri G, Panfili E, Bassi R, Ballarin C, Sorgato MC (1992) Ionic permeability of the mitochondrial outer membrane. Eur Biophys J 21:311-319

Moreadith RW, Jacobus WE (1982) Creatine kinase of heart mitochondria. Functional coupling of ADP transfer to the adenine nucleotide translocase. J Biol Chem 257:899-905

Moreno-Sánchez R, Hogue BA, Bravo C, Newman AH, Basile AS, Chiang PK (1991) Inhibition of substrate oxidation in mitochondria by peripheral-type benzodiazepine receptor ligand AHN 086. Biochem Pharm 41:1479-1484

Nakashima RA, Mangan PS, Colombini M, Pedersen PL (1986) Hexokinase receptor complex in hepatoma mitochondria: evidence from N, N'-Dicyclohexylcarbodiimide-labeling. Studies for the involvement of VDAC. Biochem 25:1015-1021

Needham L, Dodd NJF, Houslay MD (1987) Quinidine and mellitin both decrease the fluidity of liver plasma membranes and both inhibit hormone-stimulated adenylate cyclase activity. Biochim Biophys Acta 899:44-50

Nicolli A, Redetti A, Bernardi P (1991) The K^+ conductance of the inner mitochondrial membrane. J Biol Chem 266:9465-9470

Ostlund AK, Gohring U, Krause J, Brdiczka D (1983) The binding of glycerol kinase to the outer membrane of rat liver mitochondria: its importance in metabolic regulation. Biochem Med 30:231-245

Paliwal R, Costa G, Diwan J (1992) Purification and patch clamp analysis of a 40-pS channel from rat liver mitochondria. Biochem 31:2223-2229

Petronilli V, Szabó I, Zoratti M (1989) The inner mitochondrial membrane contains ion-conducting channels similar to those found in bacteria. FEBS Lett 259:137-143

Saks VA, Kupriyanov VV, Elizarove GV (1980) Studies of energy transport in heart cells. The importance of creatine kinase localization for the coupling of mitochondrial phosphoryl creatine production to oxidative phosphorylation. J Biol Chem 255:755-763

Saks VA, Khychua ZA, Kuznetsov AV (1987) Specific inhibition of ATP-ADP translocase in cardiac mitoplasts by antibodies against creatine kinase. Biochim Biophys Acta 891:138-144

Sandri G, Siagri M, Panfili E (1988) Influence of Ca^{2+} on the isolation from rat-brain mitochondria of a fraction enriched in boundary membrane contact sites. Cell Calcium 9:159-165

Schein SJ, Colombini M, Finkelstein A (1976) Reconstitution in planar lipid bilayers of voltage-dependent anion-selective channel obtained from Paramecium mitochondria. J Membr Biol 30:99-120

Schlegel J, Wyss M, Eppenberger HM, Walliman T (1990) Functional studies with the octameric and dimeric form of mitochondrial creatine kinase. J Biol Chem 265:9221-9227

Sorgato MC, Keller BU, Stühmer W (1987) Patch-Clamping of the inner mitochondrial membrane reveals a voltage-dependent ion channel. Nature (London) 330:498-500

Szabó I, Zoratti M (1991) The giant channel of the inner mitochondrial membrane is inhibited by cyclosporin A. J Biol Chem 266:3376-3379

Szabó I, Zoratti M (1992) The mitochondrial megachannel is the permeability transition pore. J Bioenerg Biomemb 24:111-117

Szabó I, Bernardi P, Zoratti M (1992) Modulation of the mitochondrial megachannel by divalent cations and protons. J Biol Chem 267:2940-1946

Tedeschi H (1981) The transport of cations in mitochondria. Biochim Biophys Acta 639:157-196

Tedeschi H, Mannella CA, Bowman CL (1987) Patch clamping the outer mitochondrial membrane. J Membr Biol 97:21-29

Varobjev IA, Zorov DB (1983) Diazepam inhibits cell respiration and induce fragmentation of mitochondrial reticulum. FEBS Lett 163:311-314

Verma A, Snyder SH (1988) Characterization of porphyrin interactions with peripheral type benzodiazepine receptors. Molec Pharm 34:800-805

Weiler U, Riesinger I, Knoll G, Brdiczka D (1985) The regulation of mitochondrial-bound hexokinase in the liver. Biochem Med 33:223-235

Wunder UR, Colombini M (1991) Patch clamping VDAC in liposomes containing whole mitochondrial membranes. J Membr Biol 123:83-91

Yang WCT, Geiger PJ, Bessman SP, Borrebaek B (1977) Formation of creatine phosphate and ^{32}P-labelled ATP by isolated rabbit heart mitochondria. Biochem Biophys Res Comm 76:882-887

Zimmerberg J, Parsegian VA (1986) Polymer inaccessible volume changes during opening and closing of a voltage-dependent ionic channel. Nature (London) 323:36-39

Zimmerberg J, Parsegian VA (1987) Water movement during channel opening and closing. J Bioenerg Biomemb 19:351-358

Zorov DB, Antonenko YuN, Kinnally KW, McEnery MW, Tedeschi H (1992a) Effect of peripheral benzodiazepene receptor ligands on mitochondrial channel activity. The Seventh European Bionergetics Conference. Short communications 7:125

Zorov DB, Kinnally KW, Perini S, Tedeschi H (1992b) Multiple conductance levels in rat heart inner mitochondrial membranes studied by patch clamping. Biochim Biophys Acta 1105:263-270

Zorov DB, Kinnally KW, Tedeschi H (1992c) Voltage activation of the heart inner mitochondrial membrane channels. J Bioenerg Biomemb 24:119-124

Zorov DB, Kinnally KW, Perini S, Tedeschi H (1991) Aluminum inhibition of multiconductance channel activity in the inner mitochondrial membrane. American Society for Cell Biology, 31st Annual meeting J Cell Biol 111:60a

PURIFICATION AND PATCH CLAMP ANALYSIS OF TWO MITOCHONDRIAL CHANNELS

Joyce J. Diwan and Germán Costa
Biology Department and Center for Biophysics
Rensselaer Polytechnic Institute
Troy, NY 12180-3590
U.S.A.

Patch clamping has revealed the presence in mitochondrial membranes of several channel types, of conductance ranging from 6 pS to 1.3 nS in the presence of 150 mM KCl or NaCl. A voltage-gated channel of about 107 pS conductance has been found in patches of mitoplasts and artificial membranes incorporating mitochondrial membrane fragments or fractions (Sorgato et al., 1987 & 1989; Kinnally et al. 1989; Moran et al., 1990; Klitsch & Siemen, 1991; Inoue et al, 1991). An ATP-sensitive K^+ channel of 10 pS conductance in the presence of 100 mM KCl has been seen in patches of mitochondrial inner membrane (Inoue et al., 1991). A 1.3 nS mega-channel, which is opened by mM levels of Ca^{++} and blocked by cyclosporin A (Petronilli et al., 1989; Szabó & Zoratti, 1991 & 1992), appears to mediate the permeability transition associated with Ca^{++} loading (Broekemeier et al., 1989). An apparently equivalent multiconductance channel (MCC) exhibiting transitions ranging from 30 pS to 5 nS is found to be activated by Ca^{++} or high voltage (Kinnally et al., 1991, 1992; Antonenko et al., 1991; Zorov et al., 1992). A water-soluble 60 kDa protein extracted from mitochondria in ethanol was earlier reported to induce variable conductance increments in black lipid membranes (Mironova, 1981).

The outer mitochondrial membrane channel VDAC exhibits characteristic conductance increments (about 500 pS with 100 mM KCl) and voltage-dependence in planar bilayer membranes (Colombini, 1989; Bowen et al., 1985). More varied conductance increments, ranging from less than 50 pS to 530 pS, have been observed in patch clamp studies of intact mitochondria and of liposomes incorporating outer mitochondrial membrane fragments (Moran et al., 1992, Wunder & Colombini, 1991). An apparently distinct outer membrane channel, detectable in mitochondria of a yeast deficient in VDAC, exhibits conductance increments of about 220 pS and is postulated to have a role in protein import (Thieffry et al., 1988; Fèvre et al., 1990; Chich et al., 1991).

A protein fraction has been obtained from detergent-solubilized rat liver mitochondrial membranes by affinity chromatography on immobilized quinine (Diwan et al., 1987, 1988, 1990). The quinine-affinity column eluate has been reconstituted into phospholipid (asolectin) vesicles by detergent dialysis. Using electrode recordings and fluorescent probe techniques, it was determined that vesicles reconstituted with the quinine-affinity column eluate were more permeable to K^+ and to the K^+ analog Tl^+ than vesicles prepared without protein (Diwan et al.,

NATO ASI Series, Vol. H 83
Molecular Biology of Mitochondrial Transport Systems
Edited by M. Forte and M. Colombini

1988 and 1990). Patch clamping has revealed the presence of two types of channels in membranes reconstituted with the quinine-affinity column eluate or fractions derived from this eluate. Conductance increments of about 40 pS and 130 pS have been observed in patches of vesicle membrane bathed symmetrically in buffered medium containing 150 mM KCl and 100 μM $CaCl_2$ (Costa et al., 1991; Paliwal et al., 1992).

METHODS

As described (Diwan et al., 1988 & 1990), epoxy-activated sepharose 6B (Pharmacia) was reacted with the OH group of quinine at alkaline pH. Submitochondrial particles prepared from 16 rat livers by standard procedures (Pedersen et al., 1978) and solubilized in 2% Triton X-100, 50 mM NaCl, 2 mM HEPES, and 10 mM EDTA (pH 7.5) were applied to two 4-5 ml columns of the Sepharose with covalently bound quinine. Each column was washed with a similar solution but of pH 6.6 with 100 mM NaCl. Protein was eluted with medium containing 1% Na cholate, 400 mM KCl, and 2 mM Hepes (pH 6.4). Media included 0.5 μg/ml leupeptin and 0.7 μg/ml pepstatin. The eluate was concentrated to 2-3 ml by ultrafiltration in an Amicon stirred cell with YM-30 membrane, and then either reconstituted directly or diluted to 40 ml with medium containing 1% octyl glucoside, 5% glycerol, and 2% Bio-Lyte ampholytes (Bio-Rad, pH 5-8) for isoelectric focusing in a Bio-Rad Rotofor cell (12 W constant power).

SDS PAGE was carried out by the method of Laemmli (1970). Gels were scanned with a Hoefer GS300 densitometer, and the results analyzed with the Hoefer GS370 data system.

For reconstitution, protein fractions were combined with octyl glucoside-solubilized asolectin (purified by the method of Kagawa & Racker, 1971) at a final concentration of 15 mg/ml asolectin and 200 mM octyl glucoside, and dialyzed at 4 ^{o}C against 2 liters of 25 mM KCl with 5 mM Tris P_i, pH 7.5. During 64 hours of dialysis, the medium was changed 4 times, with 8 g BioBeads (SM-2, 20-50 mesh, Bio-Rad) added in a dialysis sac each time after the initial 2 changes of medium. Control vesicles were prepared the same way, but with cholate elution medium or the Rotofor medium minus ampholytes substituted for protein fractions. Each vesicle suspension was passed through a 2 ml column of washed BioBeads (100-200 mesh), prior to sedimentation at 50,000 RPM X 1 hour 25 min in a Beckman SW60Ti rotor.

Prior to patch clamping (Neher, 1992; Sakmann, 1992), vesicles were enlarged by the dehydration/rehydration procedure of Criado and Keller (1987). The protein/lipid ratio of some samples was decreased by mixing in control vesicles before dehydration/rehydration. Pipets of 10-40 MΩ resistance before patching were prepared from borosilicate glass (1 mm with filament, World Precision Instr.) with a Sutter PC-84 pipet puller. A GΩ seal was obtained by bringing a patch pipet into contact with a vesicle adhering to a plastic culture dish or glass coverslip. Patches were excised by lifting the pipet away from the vesicle, or by brief transient removal of the pipet from the bath. Unless stated otherwise, excised patches were symmetrically bathed in standard patch medium consisting of 150 mM KCl, 5 mM HEPES, & 0.1 mM $CaCl_2$, pH 7.4. An Axon Axopatch 200 amplifier was used, with filtering at a bandwidth of 2 kHz. Other apparatus included a Nikon TMS phase contrast microscope, Newport vibration isolation workstation, Narashige MO-303 micromanipulator, & Tektronix 2201 digital storage oscilloscope. Current recordings via a Medical Systems PCM-2 digitizer were stored on VHS tape. Data were analyzed using Axon's pClamp software with a BitWise 80386/SX computer. Single channel currents presented (e.g. in Fig. 4) were estimated manually in the Demoevent subroutine of the Fetchan program of pClamp. Positive voltages reported refer to the interior of the pipet, and correspond to negative bath potentials.

RESULTS

The protein composition of the quinine-affinity column eluate, as resolved on an SDS polyacrylamide gel, is shown in Figure 1. Protein bands of molecular weight 96-99 kDa, 76-78 kDa, 56-58 kDa, 52-53 kDa, and 31-32 kDa are consistently seen. Other relatively minor protein bands are present in some preparations.

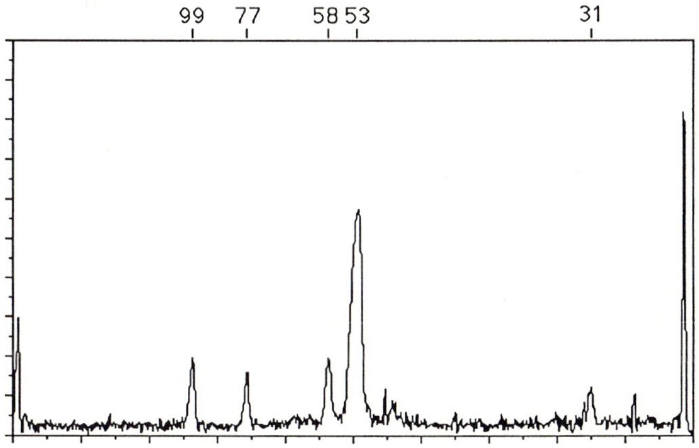

Fig. 1. Scan of an SDS polyacrylamide gel lane loaded with the quinine-affinity column eluate. The calculated kDa molecular weight is shown for each band.

Figure 2 shows a recording from a patch of membrane reconstituted with the quinine-affinity column eluate. In the recording shown, with the voltage clamped at -40 mV the current transitions were estimated to be about 2.6 pA, corresponding to a single channel conductance of 64 pA. This is one of the highest values observed for the smaller of the two channels in the quinine-affinity column eluate. The single channel conductance among 26 patches was found to average 43 ± 10 pS (mean ± SD). The conductance of the open channel has been found to not vary systematically with voltage. No evidence of voltage-dependent gating of the channel of about 40 pS conductance has been obtained. Conductance transitions have been seen at positive and negative potentials. Kinetic parameters, such as open and closed times, have been found to be variable. The analysis has been complicated by a tendency of the reconstituted channels to exhibit silent periods. It is not clear whether the sporadic activity may correspond to some as yet not understood inactivation process, or whether the silent periods may result from lateral diffusion of the channel into and out of the region of membrane across the pipet tip.

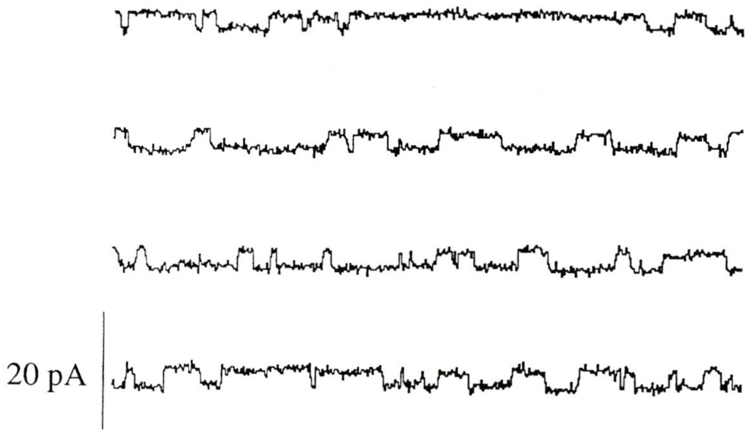

Fig. 2. Continuous recording of current vs time (102 ms/trace) at -40 mV from a patch of membrane reconstituted with the quinine-affinity column eluate, showing transitions estimated to average 64 pS.

Figure 3 shows activity of the larger of the two channels seen in patches of membrane reconstituted with the quinine-affinity column eluate. For the particular recording shown, with voltage clamped at +40 mV the conductance increments calculated from the current transitions were found to be 141 pS. Among 27 patches studied under these standard conditions, the average single channel conductance was found to be 126 ± 18 pS.

Fig. 3 also illustrates that, while some patches are observed to fluctuate between only two current levels, patches exhibiting activity of the channel averaging about 130 pS usually fluctuate among three or more current levels, suggesting the presence of two or more channels. This is true even for preparations yielding many patches apparently devoid of activity. Such findings suggest a tendency for the 130 pS channel to aggregate during reconstitution or vesicle fusion, causing some patches to contain multiple channels, while many patches lack channels.

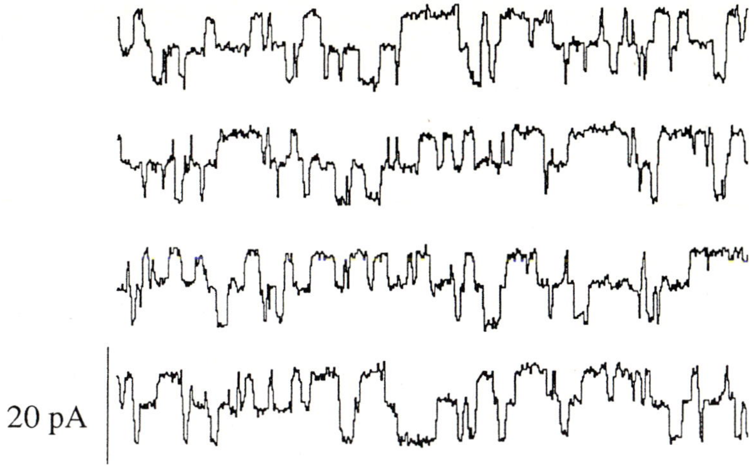

Fig. 3. Continuous recording of current vs time (102 ms/trace) at +40 mV from a patch of membrane reconstituted with the quinine-affinity column eluate, showing transitions estimated to average 141 pS.

In preliminary studies, the channels of about 130 - 140 pS were found to open only at positive bath potentials (or negative pipet potentials), suggesting voltage-dependent gating (Costa et al., 1991). Since those early studies were carried out, some patches have been observed which exhibited the opposite voltage dependence, while other patches have exhibited the transitions of about 130 pS at both negative and positive voltages. A possible interpretation is that the channel proteins may vary in their orientation across the vesicle membrane after reconstitution and vesicle fusion. A patch containing two voltage-gated channels of opposite orientation might exhibit channel openings at positive as well as negative voltages.

The quinine-affinity column eluate has been subfractionated by isoelectric focusing in a Rotofor Cell (Paliwal et al., 1992). The Rotofor procedure yields 20 fractions of varied pH. Table I summarizes voltage clamp recordings under standard conditions from excised patches of membrane reconstituted with selected Rotofor fractions. Conditions, including the ratio of protein/lipid in vesicle membranes, resulted in an apparent lack of activity in many patches. Active patches of membrane reconstituted with pooled Rotofor fractions of pH ranging from about 6.6 to 7.9 (designated the "high pH" fraction), exhibited mainly the conductance transitions of about 40 pS, although other activities were occasionally seen. In contrast, the conductance transitions of about 130 pS were observed more often in membranes reconstituted with a "low pH" pooled sample of Rotofor fractions of pH about 5.1 to 5.8.

Channel Activity	Rotofor Fraction Reconstituted High pH	Rotofor Fraction Reconstituted Low pH
No Observed Activity	19	24
Transitions approx. 30 - 65 pS	22	8
Transitions approx. 100-160 pS	5	12
Apparent transitions ≥ 200 pS	5	2
Too noisy for analysis	3	2
Total number of patches	49	46

Table I. Channel activity detected in membranes reconstituted with Rotofor Fractions. Pooled fractions designated High pH ranged in pH from about 6.6 - 7.9, while those designated Low pH ranged from about 5.1 - 5.8. Patches were symmetrically bathed in standard patch medium. Current was recorded from each patch at several voltages usually ranging from -80 to +80 mV.

The high pH (6.6 - 7.9) Rotofor fractions usually appear on Coomassie blue stained SDS polyacrylamide gels to contain exclusively or predominantly protein of 57 kDa (Paliwal et al., 1992). The minor band at 77 kDa usually distributes at somewhat lower pH, but has been seen to extend into fractions of pH 6.6 and higher in a few preparations. In recent experiments, in which the protease inhibitors leupeptin and pepstatin were again added after dilution of the samples prior to Rotofor purification, some contamination of the high pH Rotofor fraction with 53 kDa protein was also found. However activity of the 40 pS channel has been observed in membranes reconstituted with samples that, by the criterion of Coomassie blue staining, appeared to contain only protein of 57 kDa. The low pH (5.1 - 5.8) Rotofor fraction tends to contain predominantly protein of 53 kDa, but is usually found to be contaminated with one or more of the other proteins of the quinine-affinity column eluate, including 57 kDa protein which broadly distributes along the pH gradient (Paliwal et al., 1992).

Recent preliminary N-terminal sequence analysis has indicated that the 57 kDa band blotted from a 1D SDS polyacrylamide gel loaded with a sample of quinine-affinity column eluate contains a soluble matrix enzyme, which could be considered unlikely to be an ion channel (R. Murphy, C. Mannella, & J. Diwan, unpublished data). It is still unclear whether this is the only 57 kDa protein in the quinine-affinity column eluate. Two dimensional O'Farrell (1975) gels loaded with quinine-affinity column eluate consistently show at least three bands at 57 kDa (e.g., see Fig. 3 of Diwan et al., 1990). Questions of whether these bands constitute different polypeptides, or how such polypeptides might distribute along the pH gradient under the conditions of the Rotofor fractionation, remain to be addressed in future studies.

As reported (Paliwal et al., 1992), the 40 pS channel does not appear to discriminate much between K^+ and Na^+. The conductance was found to be only slightly less (mean of 38 ±

9 pS, 11 patches) when Na^+ was substituted for K^+ in the medium bathing the patch. As another test of channel selectivity, recordings were carried out with gradients of KCl across excised patches of vesicle membrane reconstituted with the high pH Rotofor fraction. In the experiment of Fig. 4, the patch was initially bathed symmetrically in standard patch medium but with 100 mM KCl. The patch clamp amplifier was driven by the microcomputer to clamp the voltage sequentially at a series of values, using the Clampex routine of the pClamp software. KCl in the bath was then increased by perfusion to 400 mM, and current transitions again recorded with the voltage clamped at various values. The essentially linear plots of single channel current against voltage in Fig. 4 indicate that the open channel is ohmic. The positive reversal potential with the higher KCl concentration in the bath indicates selectivity for the cation K^+ over the anion Cl^-. For the experiment shown, the permeability ratio calculated (Hille, 1992) from the estimated reversal potential of 28 mV with a 4-fold gradient of KCl (ignoring K^+ added as KOH in titrating the 5 mM Hepes buffer) is about 10. We have consistently observed cation selectivity; however for some experiments current transitions could not be recorded over a full range of voltages because of the tendency of the channel to exhibit apparently random silent periods (see above). Thus more studies are needed to determine precisely the selectivity ratio.

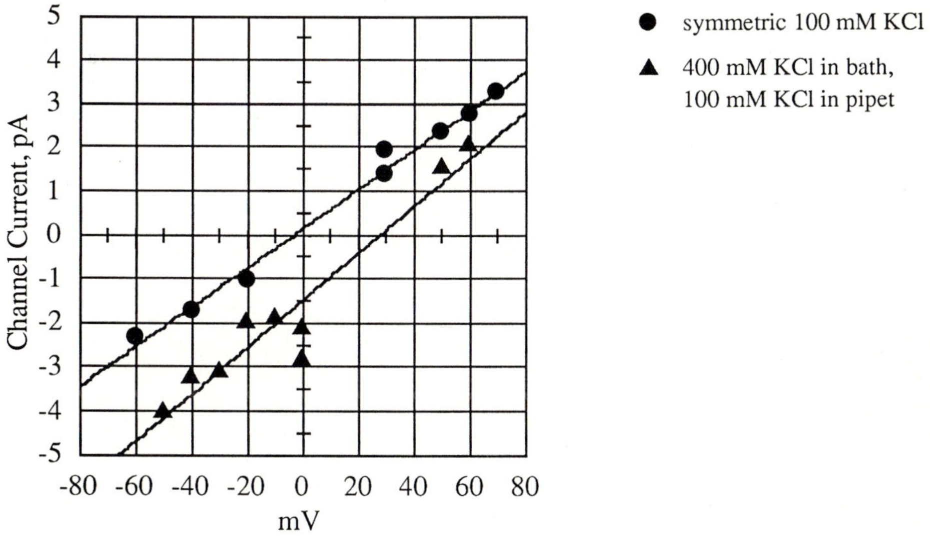

Fig. 4. Cation/anion selectivity. Current was recorded at various voltages with the patch bathed symmetrically in standard patch medium but containing 100 mM KCl, and again after perfusion of the bath with medium containing 400 mM KCl. Single channel currents in pA are plotted as a function of voltage in mV. The lines were calculated by the method of least squares.

SUMMARY AND DISCUSSION

Patch clamp analysis indicates that a fraction isolated from submitochondrial particles by affinity chromatography on immobilized quinine contains two channels, of approximately 40 pS and 130 pS conductance in the presence of 150 mM KCl. These two channel types have been independently observed in individual patches of membrane reconstituted with the quinine-affinity column eluate. The affinity-purified fraction has been subfractionated by isoelectric focusing, yielding some separation of the 40 pS and 130 pS channels, as assayed by reconstitution and patch clamping. However overlap of activities and protein constituents between fractions is seen. Results to date leave uncertain the assignment of particular proteins of the quinine-affinity column eluate to individual channel species.

Studies of selectivity have indicated that the 40 pS channel does not discriminate significantly between K^+ and Na^+. However the reversal potential observed when there is a transmembrane gradient of KCl indicates that the channel is selective for the cation (K^+) over the anion (Cl^-). Further studies, including recordings in the presence of additional ionic species, are needed to fully define the selectivity of the isolated reconstituted channels.

The 40 pS channel may correspond to one of the transport pathways earlier defined as monovalent cation uniporters on the basis of measurements of osmotic swelling of whole mitochondria (Brierley, 1983). At least one of the putative uniporters was found to discriminate little between K^+ and Na^+ (Brierley et al., 1978).

Conductance increments of about 40 pS in patches of inner mitochondrial membrane have been attributed to the multiconductance channel, MCC (Kinnally et al., 1992). The possibility should be considered that the affinity-purified fraction may include MCC. Attributing the 40 pS transitions to MCC would require assuming that experimental conditions and/or loss of a constituent during purification have prevented MCC from reaching its higher conductance states. As indicated in Table I, large conductance increments have been seen occasionally; but most patches containing the 40 pS channel have not exhibited the large transitions typical of MCC, even at the relatively high voltages reported to induce MCC activity.

Conductance transitions of varied magnitudes, including some less than 50 pS, have been detected in patches of outer mitochondrial membrane (Moran et al., 1992). Thus consideration also should be given to the possibility that the affinity-purified fractions might contain VDAC or some other outer membrane channel.

The channel of approximately 130 pS conductance resembles in some respects the voltage-gated 107 pS channel first detected in patches of mitoplast membrane by Sorgato et al. (1987). Conditions of purification and reconstitution may have led to the observed differences in magnitude and variability of conductance increments. Further study is needed before a judgment can be made regarding the identity of these channels.

Acknowledgment: This research was supported by USPHS Grant GM-20726. Dr. Ranjana Paliwal participated in some of the experiments summarized in Table I.

References:

Antonenko, Y. N., Kinnally, K. W., Perini, S., and Tedeschi, H. (1991) "Selective effect of inhibitors on inner mitochondrial membrane channels," FEBS Lett. 285, 89-93.

Bowen, K. A., Tam. K., & Colombini, M.(1985) "Evidence for titratable gating charges controlling the voltage dependence of the outer mitochondrial membrane channel, VDAC," J. Membr. Biol. 86, 51-59.

Brierley, G. P. (1983) "Monovalent cation transport by heart mitochondria," in: Pathobiology of Cell Membranes (Trump, B. and Arstila, A.V. eds) vol. III, pp. 23-61, Academic Press, New York.

Brierley, G. P., Jurkowitz, M., and Jung, D. W. (1978) Osmotic swelling of heart mitochondria in acetate and chloride salts. Arch. Biochem. Biophys. 190:181-192.

Broekemeier, K. M., Dempsey, M. E., and Pfeiffer, D. R. (1989) "Cyclosporin A is a potent inhibitor of the inner membrane permeability transition in liver mitochondria," J. Biol. Chem. 264, 7826-7830.

Chich, J.F., Goldschmidt, D., Thieffry, M. & Henry, J.P. (1991) "A peptide-sensitive channel of large conductance is localized on mitochondrial outer membrane," Eur. J. Biochem. 196, 29-35.

Colombini, M. (1989) "Voltage gating in the mitochondrial channel, VDAC," J. Membr. Biol. 111, 103-111.

Costa, G., Kinnally, K. W., and Diwan, J. J. (1991) "Patch clamp analysis of a partially purified ion channel from rat liver mitochondria," Biochem. Biophys. Res. Comm.175, 305-310.

Criado, M., and Keller, B. U. (1987) "A membrane fusion strategy for single-channel recordings of membranes usually non-accessible to patch-clamp pipette electrodes," FEBS Lett. 224, 172-176.

Diwan, J. J., Haley, T., and Sanadi, D. R. (1988) "Reconstitution of transmembrane K^+ transport with a 53 kilodalton mitochondrial protein," Biochem. Biophys. Res. Comm. 153, 224-230.

Diwan, J. J., Moore, C., Haley, T., Herbrandson, H., & Sanadi, D. R. (1987) "Interaction of quinine with mitochondrial K^+ transport mechanisms," In: Advances in Membrane Biochemistry and Bioenergetics (C. H. Kim, H. Tedeschi, J. J. Diwan, & J. C. Salerno, eds) pp. 401-408, Plenum, NY.

Diwan, J. J., Paliwal, R., Kaftan, E., and Bawa, R (1990) "A mitochondrial protein fraction catalyzing transport of the K^+ analog Tl^+," FEBS Lett. 273, 215-218.

Fèvre, F., Chich, J.-F., Lauquin, G. J. M., Henry, J.-P., & Thieffry, M. (1990) "Comparison of mitochondrial cationic channels in wild-type and porin-deficient mutant yeast," FEBS Lett. 262, 201-204.

Hille, B. (1992) Ionic Channels of Excitable Membranes, 2nd Edition, Sinauer, Sunderland, MA.

Inoue, I., Nagase, H., Kishi, K., and Higuti, T. (1991) "ATP-sensitive K^+ channel in the mitochondrial inner membrane," Nature 352, 244-247.

Kagawa, Y., and Racker, E. (1971) "Partial resolution of the enzymes catalyzing oxidative phosphorylation," J. Biol. Chem. 246, 5477-5487.

Kinnally, K. W., Antonenko, Y. N., and Zorov, D. (1992) "Modulation of inner mitochondrial membrane channel activity," J. Bioenerget. Biomembr. 24, 99-110.

Kinnally, K. W., Campo, M. L., and Tedeschi, H. (1989) "Mitochondrial channel activity studied by patch-clamping mitoplasts," J. Bioenerget. Biomembr. 21, 497-506.

Kinnally, K. W., Zorov, D., Antonenko, Y., and Perini, S. (1991) "Calcium modulation of mitochondrial inner membrane channel activity," Biochem. Biophys. Res. Comm. 176, 1183-1188.

Klitsch, T., and Siemen, D. (1991) "Inner mitochondrial membrane anion channel is present in brown adipocytes but is not identical with the uncoupling protein," J. Membr. Biol. 122, 69-75.

Laemmli, U. K. (1970) "Cleavage of structural proteins during the assembly of the head of bacteriophage T4," Nature 227, 680-685.

Mironova, G. D., Fedotcheva, N. I., Makarov, P. R., Pronevich, L. A., and Mironov, G. P. (1981) "Protein from beef heart mitochondria inducing the channel potassium conductivity of bilayer lipid membranes," Biofizica 26, 458-465.

Moran, O., Sandri, G., Panfili, E., Stühmer, W., and Sorgato, M. C. (1990) "Electrophysiological characterization of contact sites in brain mitochondria," J. Biol. Chem. 265, 908-913.

Moran, O., Sciancalepore, M., Sandri, G., Panfili, E., Bassi, R., Ballarin, C., and Sorgato, M. C. (1992) "Ionic permeability of the mitochondrial outer membrane," Eur. Biophys. J. 20, 311-319.

Neher, E. (1992) "Ion channels for communication between and within cells," Science 256, 498-502. (Nobel Lecture)

O'Farrell, P. H. (1975) "High resolution two-dimensional electrophoresis of proteins," J. Biol. Chem. 250, 4007-4021.

Paliwal, R., Costa, G., and Diwan, J. J. (1992) Purification and patch clamp analysis of a 40-pS channel from rat liver mitochondria," Biochemistry 31, 2223-2229.

Pedersen, P. L., Greenawalt, J. W., Reynafarje, B., Hullihen, J., Decker, G. L., Soper, J. W. and Bustamente, E. (1978) "Preparation and characterization of mitochondria and submitochondrial particles of rat liver and liver-derived tissues," Methods in Cell Biology 20, 411-481.

Petronilli, V., Szabò, I., and Zoratti, M. (1989) "The inner mitochondrial membrane contains ion-conducting channels similar to those found in bacteria," FEBS Lett. 259, 137-143.

Sakmann, B. (1992) "Elementary steps in synaptic transmission revealed by currents through single ion channels," Science 256 503-512. (Nobel Lecture)

Sorgato, M. C., Keller, B. U., and Stühmer, W. (1987) "Patch-clamping of the inner mitochondrial membrane reveals a voltage-dependent ion channel," Nature 330, 498-500.

Sorgato, M.C., Moran, O., DePinto, V., Keller, B. U., & Stuehmer, W. (1989) "Further investigation on the high-conductance ion channel of the inner membrane of mitochondria," J. Bioenerget. Biomembr. 21, 485-496.

Szabó, I., and Zoratti, M. (1991) "The giant channel of the inner mitochondrial membrane is inhibited by cyclosporin A," J. Biol. Chem. 266, 3376-3379.

Szabó, I. and Zoratti, M. (1992) "The mitochondrial megachannel is the permeability transition pore," J. Bioenerget. Biomembr. 24, 111-117.

Thieffry, M., Chich, J.-F., Goldschmidt, D., and Henry, J.-P. (1988) "Incorporation in lipid bilayers of a large conductance cationic channel from mitochondrial membranes," EMBO J. 7, 1449-1454.

Wunder, U. R. and Colombini, M. (1991) "Patch clamping VDAC in liposomes containing whole mitochondrial membranes," J. Membr. Biol. 123, 83-91.

Zorov, D. B., Kinnally, K. W. and Tedeschi, H. (1992) "Voltage activation of the heart inner mitochondrial membrane channels," J. Bioenerget. Biomembr. 24, 119-124.

THE MITOCHONDRIAL OUTER MEMBRANE CONTAINS AT LEAST TWO DISTINCT CHANNELS

Thieffry*, F. Fèvre*, M. Pelleschi+ and J.P. Henry+
*Laboratoire de Neurobiologie Cellulaire et Moléculaire, CNRS, 91198 Gif-sur-Yvette Cedex
4,+Service de Neurobiologie Physico-Chimique, URA-CNRS 1112
Institut de Biologie Physico-Chimique, 13 rue Pierre et Marie Curie, 75005 Paris, France

Abstract

The presence in the mitochondrial outer membrane of a cationic channel blocked by a 13 residue addressing peptide had been shown by the technique of "tip-dip". This channel was studied in planar bilayers and compared to the anionic porin, the voltage-dependent anion channel, observed in the same conditions. The two activities were clearly different. The peptide sensitive channel, present in both wild type and porin-deficient mutant yeast, is not a rescue channel, but an alternate independent permeability pathway. The channel interacted more strongly with a longer extension of the addressing peptide containing 22 residues, suggesting a physiological interaction of this type of molecule with the cationic channel.

Characteristics of the cationic channel from mitochondrial outer membrane in tip-dip records

Using bilayers formed at the tip of microelectrodes by the "tip-dip" technique (Coronado and Latorre, 1983; Suarez-Isla *et al.*, 1983; Wilmsen *et al.*, 1983), we have found, in mitochondrial outer membrane, a channel whose properties appeared different from those of the classical VDAC (Colombini, 1986). In mammalian mitochondria, such as those from bovine adrenal cortex, the channel is characterized by a slightly cationic selectivity ($P_{Na^+}/P_{Cl^-} \approx 4$), rapid kinetics with fast fluctuations between 3 main conductance levels separated by jumps of 220 pS (in 150 mM NaCl) and a clear asymmetrical voltage-dependence. It is permeant to large cations such as TEA as well as to large anions such as glutamate or glucuronic acid.

Interestingly, this channel is blocked by cationic peptides acting as mitochondrial addressing sequences. From this property, we proposed the name of PSC (Peptide Sensitive Channel) for the channel. We have mainly investigated the blocking effect using a 13 residue peptide containing the first 12 amino acids of the N-terminal extremity of the precursor of cyt-c oxidase subunit IV. This peptide (MLSLRQSIRFFKY), peptide M or pL4(1-12)Y

NATO ASI Series, Vol. H 83
Molecular Biology of Mitochondrial Transport Systems
Edited by M. Forte and M. Colombini

contains the minimal mitochondrial addressing sequence required to translocate the precursors synthesized in the cytosol (Hurt *et al.*, 1985).

The blockade, reversible by washing the peptide out, does not depend on the orientation of the channel in the bilayer. It is easier to observe using channels in which the voltage-dependent spontaneous fluctuations have been abolished by trypsin (Thieffry *et al.*, 1988; Henry *et al.*, 1989). The block duration has been analyzed under these conditions at 10 μM peptide concentration for different values of the applied voltage (Fig. 1).

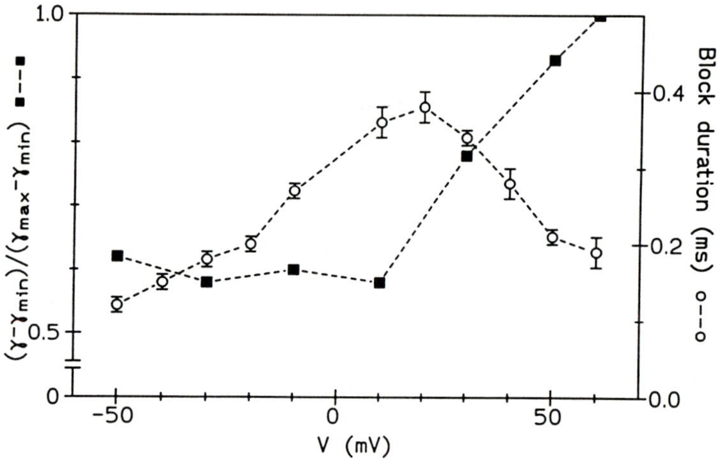

Figure 1. Voltage dependence of the conductance decrease (■) and block duration (O) induced by peptide M in cationic channels from bovine adrenal cortex. The channels were treated by trypsin to eliminate the spontaneous activity, washed and exposed to peptide M at the concentration of 10 μM (O) or 100 μM (■). The conductance decrease is evaluated by the ratio $(\gamma - \gamma_{min})/(\gamma_{max} - \gamma_{min})$ where γ, γ_{max} and γ_{min} are respectively the mean conductance in the presence of the peptide and the conductances of the open and closed states measured before proteolysis. The records were obtained by the "tip-dip" technique. The voltage refers to that of the electrode. The peptide, which carries 4 positive charges, was added to the bath. The driving force for the peptide thus increases from the right to the left of the voltage axis.

A bell shaped curve was obtained, the block duration first increasing then decreasing when the driving force applied to the peptide increased. This result is in good agreement with the voltage-dependence of the block intensity which stops increasing at a voltage corresponding to the peak of block duration (Fig. 1). It is in contradiction with a Woodhull type of block and strongly suggests that the peptide is going through the channel. Discussing the specificity of the peptide effect and the relevance of this observation to mitochondrial biogenesis is out of the scope of the present communication.

The same type of activity was observed in all mammalian mitochondrial preparations tested so far. Mitochondria fractionation experiments, in which the frequency of observation of the PSC was compared to the activity of enzymatic markers, indicated a channel localization in the outer membrane (Chich *et al.*, 1991).

Cationic channels are also found in mitochondria from Saccharomyces cerevisiae (Fèvre *et al.*, 1990) and Neurospora crassa (R. Lill, W. Neupert, F. Fèvre, J.P. Henry and M. Thieffry, unpublished observations). They are also characterized by three main conductance levels, but the jumps have an average value of 330 pS in 150 mM NaCl. Rapid closures of 140-250 pS may be present at potentials > 40 mV, but they are not always observed. When present, they occur rather by long bursts and are not sensitive to trypsin. The channels have slow kinetics closures and fast openings. When steady state voltages of either polarity are applied, long lasting closures to levels respectively 330 and 660 pS below the maximum conductance occur. This requires several tens of seconds around -30mV and a few seconds only around -80 mV. The fully open state can be restored by a brief pulse to 0 mV or to a voltage of reverse polarity. In spite of these differences, these channels are clearly related to the mammalian PSC. On one hand, they have the same cationic selectivity. On the other hand, they are blocked by basic peptides and the block properties are similar to those described above.

The PSC is different from the VDAC, but may be difficult to distinguish

Since our work points to a second channel of large conductance in the outer membrane, the relationship between the PSC and the VDAC have to be precised. This clarification is all the more required because different techniques have been used to characterize these two activities, "tip-dip" for the PSC and mostly planar bilayer for the VDAC.

The cationic channels from Saccharomyces or Neurospora have slow kinetics closures and fast openings which strongly remind those of VDAC (Colombini, 1989). In fact, it may be impossible to identify PSC or VDAC from the simple examination of current traces recorded at different potentials. Voltage ramps of low frequency are often applied to show the characteristic voltage-dependence of VDAC. Figure 2 shows a "tip-dip" record of the current flowing through cationic yeast mitochondrial channels sensitive to peptide M in response to a 5 mHz triangular voltage wave between -65 and +65 mV. By both the voltage-dependence and the magnitude of the current jumps, this record appears quite similar to those reported for VDAC reconstituted in giant liposomes and studied by the patch-clamp technique (Wunder and Colombini, 1991). However, the channels shown in Figure 2 were found in a porin-deficient mutant yeast (Guo and Lauquin, 1986). In this mutant, obtained by G. Lauquin in Bordeaux (Michedja *et al.*, 1989), the porin gene has been disrupted and it has been controlled that neither the corresponding mRNA nor the protein were expressed.

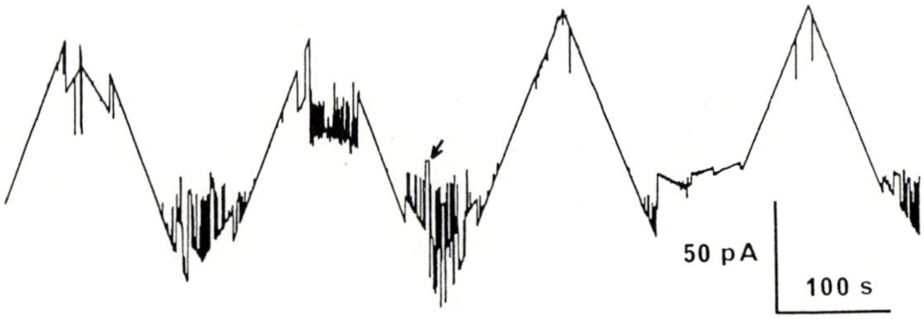

Figure 2. Current recorded in a "tip-dip" bilayer enriched in cationic channels from a porin-deficient mutant yeast. A 5 mHz triangular wave was used to vary the voltage between 65 and -65 mV. The arrow points to a control pulse indicating the zero level. The bath and the electrode contained 150 mM NaCl, 1 mM $MgCl_2$, 20 mM Hepes pH 7.3.

Characteristics of the PSC observed by the technique of planar bilayer

To evaluate the possibility of artefacts introduced by the "tip-dip" technique, we used as controls planar bilayers formed by the Mueller-Rudin method (Mueller *et al.*, 1963). Either membranes from bovine adrenal cortex mitochondria or proteoliposomes containing the same material were fused to PE : PS (7:3) planar bilayers in the presence of an osmotic gradient (150 mM NaCl, 20 mM Hepes cis, 20mM Hepes trans). In both cases, two types of insertion were mainly observed at 0 mV. In the first type, current was flowing from the cis to the trans compartment indicating a cationic selectivity (reversal potential -45 mV cis). After the two compartments were made symmetrical by adding 3 M NaCl to the trans side (Fig. 3, A), rapid fluctuations between three main conductance levels separated by jumps of 220 pS were observed at -30 mV. The voltage dependence of these levels (Fig. 3, B) was quite similar to that of the PSC observed in "tip-dip" records (Thieffry *et al.*, 1988). The maximum conductance was ≈ 750 pS and the minimum conductance, observed during brief closures 100 pS below the lowest of the three main conductance levels, was about 200 pS.

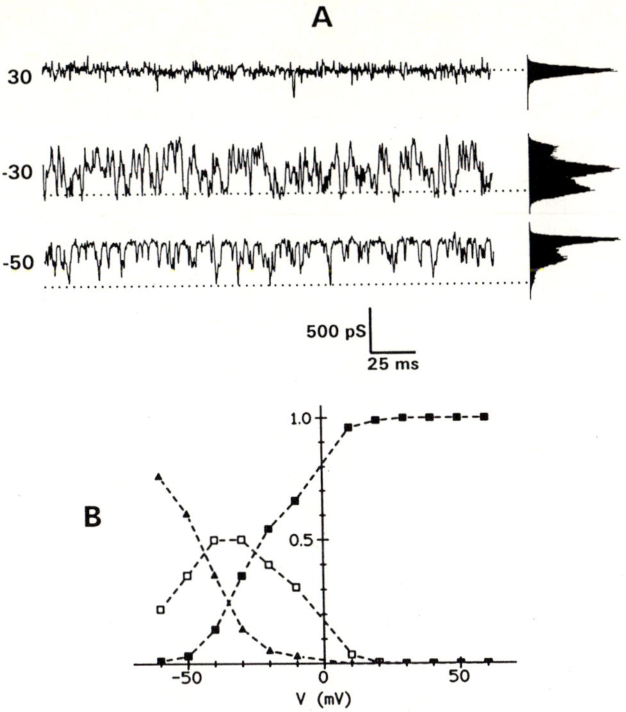

Figure 3. Characteristics of the adrenal cortex cationic channel recorded in a planar bilayer. The channel was inserted in a PE/PS (7:3) bilayer by fusion under osmotic gradient with proteoliposomes enriched in mitochondrial membranes. A : Samples of current recorded at different voltages indicated at the left of the traces. The distribution of the current amplitude, computed over periods of several tens of s, is shown at the end of each trace. The dotted line indicates the conductance of the open state. Symmetrical solutions : 150 mM NaCl, 20 mM Hepes, pH 7.5. Data filtered at 1.5 kHz. B: Voltage-dependence of the three main conductance levels corresponding to the peaks of the current histogram. The symbols ■, □ and ▲ correspond respectively to the open, intermediate and lowest conductance levels.

In the second type, the current flowing at 0 mV in the presence of the ionic gradient indicated a weak anionic selectivity (reversal potential +8 mV cis). Under symmetrical conditions (Fig. 4), this type of channel had a conductance of 560 pS. Slow kinetics closures of various amplitude occured following application of both positive and negative potentials and fast openings were triggered by return to 0 mV or to voltages of reverse polarity. Rapid fluctuations typical of the PSC were never present. These properties are typical of the VDAC. On some instances, changes in the direction of the current flow were observed in the presence of the ionic gradient used for fusion. These changes in selectivity were most likely due to transient closures of the channel, in good agreement with the cationic selectivity reported for the closed state of VDAC (Ludwig *et al.*, 1989; De Pinto *et al.*, 1989).

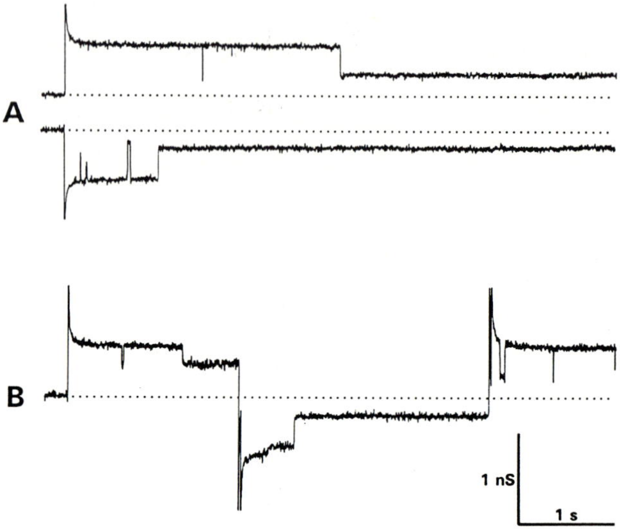

Figure 4. Conductance levels and kinetics of closing and opening transitions of the VDAC . A : Current traces recorded after a voltage jump from 0 mV to +50 mV (top trace) or -50 mV (bottom trace). B : Openings of the channel triggered by reversing the polarity of the applied voltage. The voltage was first switched from 0 mV to +40 mV, which induced a transition to one of the closed state. Reversing the voltage to -40 mV immediatly restored the open state and the channel closed to its lowest conductance level. Switching from -40 mV to +40 mV immediatly restored the open state. Channel from adrenal cortex mitochondria incorporated in proteoliposomes fused to a PE/PS (7:3) planar bilayer. Symmetrical solutions: 150 mM NaCl, 20 mM Hepes, pH 7.5. Data filtered at 250 Hz. The dotted lines indicate the 0 current level.

By contrast, when channels from porin-deficient mutant yeast were fused to planar bilayers, only one type of channel was found. From the direction of the current flowing at 0 mV, all the channels were clearly cationic. Their characteristics in the presence of symmetrical solutions were the same as those of yeast PSC observed by the "tip-dip" technique (Fèvre *et al.*, 1990). Maximum and minimum conductances were respectively about 900 pS and 100 pS in 150 mM NaCl.

In both adrenal cortex and yeast mitochondria, peptide M induced a reversible blockade of the PSC similar to that described above. By contrast, it was without any effect on the mammalian VDAC when added either to the cis or to the trans compartments.

Relative content of the membrane in PSC and VDAC

Studies of adrenal cortex mitochondria using planar bilayers clearly indicate the existence of two types of channels in the outer membrane, as suggested by various laboratories (Tedeschi *et al.*, 1987; Dihanich *et al.*, 1989; Moran *et al.*, 1990). A quantitative estimation of the PSC and VDAC relative abundance would be of great interest, but is in fact rather difficult to obtain because the methodology appears to affect differentially the two activities.

First, we never observed VDAC in "tip-dip" bilayers. Two steps of this method are particularly critical : the channels have to survive the freeze/thaw cycles used to prepare the proteoliposomes and they have to stay in a monolayer at the air-water interface. It is thus likely that the anionic porin is affected by one or both of these steps. Since we detected VDAC when proteoliposomes similar to those used in "tip-dip" experiments were fused to planar bilayers, some of these channels remain functionnal during the freeze/thaw cycles and are further inactivated at the air-water interface. However, a reliable estimate of content in VDAC and PSC cannot be obtained since sensitivity of the VDAC to freezing has been demonstrated by M. Colombini (Colombini, 1983), who noted that channels from samples stored in an inappropriate way had a cationic selectivity. This change in selectivity was interpreted as due to the insertion of partially denatured VDAC channels. In view of the results now available, it appears more likely that freezing inactivated VDAC and that the cationic permeability was due to the PSC.

Second, PSC is sensitive to detergents. The technique most often used for inserting VDAC in planar bilayers is the addition of a Triton X-100 extract containing the protein directly to the aqueous compartment. However, it appears unsuitable for PSC. This contention is based on our efforts to solubilize and reconstitute PSC from adrenal cortex (Fèvre *et al.*, in press). The solubilized material was reconstituted by dilution of the detergent and the particulate fraction was assayed by tip-dipping. Among various detergents and conditions tested, it was found that only solubilization by 1.5 % octyl ß-glucoside in the presence of 10 % glycerol preserved the activity. If glycerol was omitted, no PSC were observed and variable permeabilities were irreproducibly present for short times. Higher detergent concentrations resulted in important changes in the kinetics, the voltage-dependence and the sensitivity to peptide M of the channels which were then detected in the bilayers. Using a Triton X-100 extract of a porin deficient mutant yeast outer membrane, Benz and collaborators showed the insertion in the membrane of a cationic channel of large conductance (Dihanich *et al.*, 1989). From the comparison of the macroscopic conductance of the outer membrane in wild type and mutant and the values of the VDAC and cationic channel unitary conductance, they proposed that the former might be 10 times more abundant than the latter. However, also interacts irreversibly with yeast PSC (which is resistant to trypsin) and that the modified channel then shows fast fluctuations whose voltage-dependence resembles that of mammalian PSC.

The voltage-dependence of VDAC is modified by a soluble protein originating likely from the intermembrane space (Liu and Colombini, 1992). It has been proposed that the protein might act as polyanions such as dextran sulfate or the copolymer of methacrylate, maleate and styrene known as König's polyanion (Colombini *et al.*, 1987; Mangan and Colombini, 1987; Liu and Colombini, 1992). It is thus tempting to put together the effects of these compounds on VDAC and those of basic peptides on PSC. However, at the present time, no data exist which indicate that polyanions and basic peptides could modify the voltage-dependence by related mechanisms.

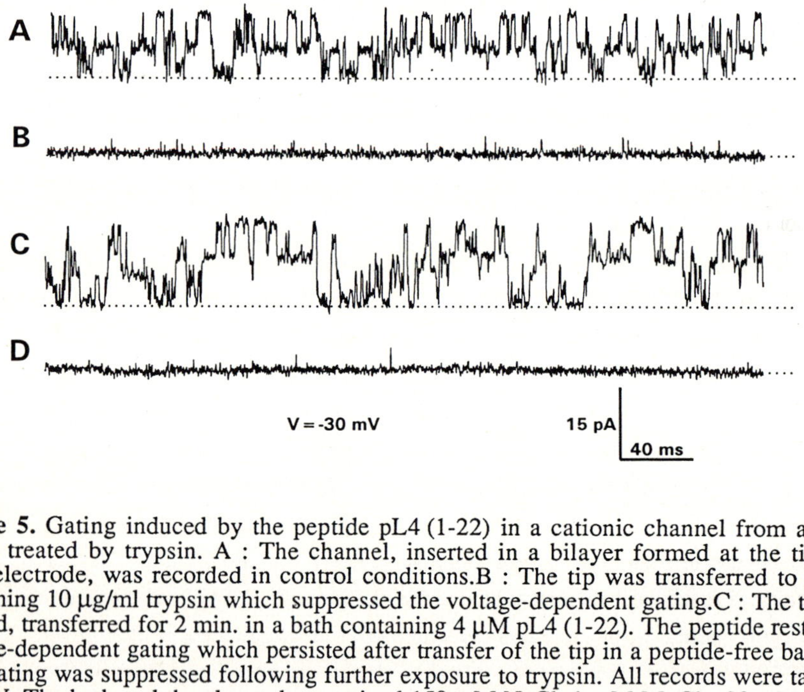

Figure 5. Gating induced by the peptide pL4 (1-22) in a cationic channel from adrenal cortex treated by trypsin. A : The channel, inserted in a bilayer formed at the tip of a microelectrode, was recorded in control conditions.B : The tip was transferred to a bath containing 10 μg/ml trypsin which suppressed the voltage-dependent gating.C : The tip was washed, transferred for 2 min. in a bath containing 4 μM pL4 (1-22). The peptide restored a voltage-dependent gating which persisted after transfer of the tip in a peptide-free bath. D : This gating was suppressed following further exposure to trypsin. All records were taken at -30 mV. The bath and the electrode contained 150 mM NaCl, 1 mM $MgCl_2$, 20 mM Hepes pH 7.3. The dotted line indicates the open state. Data filtered at 1.5 kHz.

since the channel they observed had a short lifetime (Dihanich *et al.*, 1989) and was devoid of voltage-dependence (Benz *et al.*, 1989), an effect of the detergent may be suspected and evaluation of the VDAC to PSC ratio deserves further investigation.

How different is the PSC from the VDAC ?

The difference between yeast VDAC and PSC (obtained from the porin deficient mutant) appears weak. Some data even show striking similarities (Fig. 2) and suggest that both activities could be carried by related molecules. Such an hypothesis is less likely for PSC and VDAC from adrenal cortex which, as illustrated in Figures 3 and 4, have clearly different properties (kinetics, voltage-dependence) permitting their easy identification. However, they also share common features such as their ability to translocate large molecules and the fact that they never completly close (a property which could not be demonstrated with the "tip-dip" technique).

One main difference between mammalian and yeast PSC is the presence of sustained and voltage-dependent rapid fluctuations characterizing the former. Recent experiments performed on addressing peptides longer than peptide M suggest an hypothesis on the possible mechanism of these fluctuations. Such an experiment, carried out on adrenal cortex mitochondria, is shown in Figure 5. The peptide used contained the 22 first residues of the precursor of cytochrome c oxidase subunit IV from Saccharomyces cerevisiae. This peptide, pL4 (1-22): MLSLRQSIRFFKPATRTLCSSR, is an extension of the peptide previously used: peptide M : MLSLRQSIRFFKY

The channel was recorded in control conditions (Fig.5, A), then exposed to trypsin which eliminated most of the spontaneous fluctuations (Fig.5, B). After washing out the protease, it was exposed to 4 μM pL4 (1-22) for about 2 min. The peptide, which induced numerous closures of heterogenous duration, was washed out for several minutes. Whereas the effect of peptide M is easily reversible by washing, that of pL4 (1-22) appeared to be irreversible. Sustained rapid fluctuations, reminiscent of the spontaneous activity observed before proteolysis, were then recorded (Fig.5, C). Moreover, the peptide restored a voltage-dependence strikingly similar to that observed in control conditions. When the bilayer was further exposed to trypsin, most of the spontaneous fluctuations were again eliminated (Fig. 5, D). Thus, peptide pL4 (1-22), but not peptide M, strongly interacts with proteolyzed adrenal cortex PSC and the induced gating mimicks that of intact channel.

Is it possible that the gating mechanism of the intact channel originates in a non covalent interaction between a "cationic porin" and some unidentified basic peptide ? According to such an interpretation, trypsin would cleave the interacting peptide and not the channel itself. Though this interpretation is highly speculative, it is worthnoting that the peptide pL4 (1-22)

Conclusion

We have shown that the electrical activity of the outer membrane that we had previously described using the "tip-dip" technique is not a methodological artefact. The same activity can be observed not only in planar bilayers formed by the Mueller-Rudin method, but also by patch-clamp of giant liposomes (M. Pelleschi, unpublished observations).

Several lines of evidence indicate that the PSC is different from the VDAC : electrical properties, presence in porin deficient mutants, sensitivity to peptides. The existence of different channels in the outer membrane has been previously proposed by various authors (Tedeschi *et al.*, 1987; Benz *et al.*, 1989; Moran *et al.*, 1990) and this hypothesis is strongly supported by our work. At the present time, the diversity of the recording and reconstitution techniques and the low frequency resolution of some reported data probably hamper the definition of a widely accepted repertoire for outer membrane channels. In addition, it is possible that interactions of the channels with various effectors such as "the modulator protein" of VDAC or cationic peptides associated with PSC might increase the apparent diversity of the observed activities. The existence of more than one porin in the mitochondrial outer membrane is an exciting hypothesis, though investigations by homology cloning using a porin probe have been unsuccessful (G. Lauquin, personal communication).

Acknowledgements

We are indebted to Dr J. Neyton (Ecole Normale Supérieure, Paris) for help in starting the planar bilayer experiments and to Dr G. Lauquin (Institut de Biochimie Cellulaire, Bordeaux) for discussion and preparation of yeast mutant mitochondria. We thank Dr M. Cumsky (University of California, Irvine) for the gift of pL4-(1-22) peptide.

This work was supported by the Centre National de la Recherche Scientifique (UPR 23 and URA 1112), by a grant from the Association Française contre les Myopathies and by the Direction des Recherches, Etudes et Techniques (Contrat n° 90/072).

References

Benz, R., Schmid, A. and Dihanich, M. (1989) Pores from mitochondrial outer membranes of yeast and porin-deficient mutant: a comparison. *J. Bioenerg. Biomembr.* **21**, 439-450.

Chich, J. F., Goldschmidt, D., Henry, J. P. and Thieffry, M. (1991) A peptide-sensitive channel of large conductance is localized on mitochondrial outer membrane. *Eur. J. Biochem.* **196**, 29-35.

Colombini, M. (1983) Purification of VDAC (voltage-dependent anion-selective channel) from rat liver mitotochondrial. *J. Membrane Biol.* **74**, 115-121.

Colombini, M. Voltage gating in VDAC. Toward a molecular mechanism. In: *Ion Channel Reconstitution*, edited by Miller, C. New York: Plenum, 1986, p. 533-552.

Colombini, M., Yeung, C. L., Tung, J. and König, T. (1987) The mitochondrial outer membrane, VDAC, is regulated by a synthetic polyanion. *Biochim. Biophys. Acta* **905**, 279-286.

Colombini, M. (1989) Voltage gating in the mitochondrial channel, VDAC. *J. Membrane Biol.* **111**, 103-111.

Coronado, R. and Latorre, R. (1983) Phospholipids bilayers made from monolayers on patch-clamp pipettes. *Biophys. J.* **43**, 231-236.

De Pinto, V., Benz, R., Caggese, C. and Palmieri, F. (1989) Characterisation of the mitochondrial porin from *Drosophila melanogaster. Biochim. Biophys. Acta* **987**, 1-7.

Dihanich, M., Schmid, A., Oppliger, W. and Benz, R. (1989) Identification of a new pore in the mitochondrial outer membrane of a porin-deficient yeast membrane. *Eur. J. Biochem.* **181**, 703-708.

Fèvre, F., Chich, J. F., Lauquin, G. J. M., Henry, J. P. and Thieffry, M. (1990) Comparison of mitochondrial cationic channels in wild-type and porin-deficient mutant yeast. *FEBS Lett.* **262**, 201-204.

Guo, X. J. and Lauquin, G. J. M. (1986) *Abstracts Fourth European Bioenergetics Conference, Prague* 292.

Henry, J. P., Chich, J. F., Goldschmidt, D. and Thieffry, M. (1989) Blockade of a mitochondrial cationic channels by an addressing peptide: an electrophysiological study. *J. Membrane Biol.* **112**, 139-147.

Hurt, E. C., Pesold-Hurt, B., Suda, K., Oppliger, W. and Schatz, G. (1985) The first twelve amino acids (less than half of the presequence) of an imported mitochondrial protein can direct mouse cytosolic dihydrofolate reductase into yeast mitochondrial matrix. *EMBO J.* **4**, 2061-2068.

Liu, M. Y. and Colombini, M. (1992) A soluble mitochondrial protein increase the voltage dependence of mitochondrial channel, VDAC. *J. Bioenerg. Biomembr.* **24**, 41-46.

Ludwig, O., Benz, R. and Schultz, J. E. (1989) Porin of Paramecium mitochondria isolation, characterisation and ion selectivity of the closed state. *Biochim. Biophys. Acta* **978**, 319-327.

Mangan, P. S. and Colombini, M. (1987) Ultrasteep voltage dependence in a membrane channel. *Proc. Natl. Acad. Sci. USA* **84**, 4896-4900.

Michedja, J., Guo, X.J. and Lauquin, G.J.M. . In: *Anion carriers of mitochondrial membranes*, edited by Azzi, A. Berlin: Springer, 1989, p. 225-235.

Moran, O., Sandri, G., Panfili, E., Stuhmer, W. and Sorgato, M. C. (1990) Electrophysiological caracterisation of contact sites in brain mitochondria. *J. Biol. Chem.* **265**, 908-913.

Mueller, P., Rudin, D. O., Tien, H. T. and Wescott, W. C. (1963) Methods for the formation of single bimolecular lipid membranes in aqueous solutions. *J. Phys. Chem.* **67**, 534-535.

Suarez-Isla, B. A., Wan, K., Lindstrom, J. and Montal, M. (1983) Single-channel recordings from acetylcholine receptors reconstituted in bilayers formed at the tip of patch pipets. *Biochemistry* .**22**, 2319-2323.

Tedeschi, H., Mannella, C. A. and Bowman, C. L. (1987) Patch-clamping of the outer mitochondrial membrane. *J Membrane Biol.* **97**, 21-29.

Thieffry, M., Chich, J. F., Goldschmidt, D. and Henry, J. P. (1988) Incorporation in lipid bilayers of a large conductance cationic channel from mitochondrial membrane. *EMBO J.* **7**, 1449-1454.

Wilmsen, U., Methfessel, W., Hanke, W. and Boheim, G. Channel current fluctuation studies with solvent-free lipid bilayers using Neher-Sakmann pipettes. In: *Physical Chemistry of Transmembrane Ion Motions*, edited by Spach, G. Amsterdam: Elsevier, 1983, p. 479-485.

Wunder, U. R. and Colombini, M. (1991) Patch clamping VDAC in liposomes containing whole mitochondrial membranes. *J. Membrane Biol.* **123**, 83-91.

ATP-SENSITIVE K^+ CHANNEL IN RAT LIVER MITOCHONDRIA: FUNCTIONAL CHARACTERISTICS

Adam Szewczyk[1], Sławomir Pikuła[1], Lech Wojtczak[2] and Maciej J.Nałęcz[1]

[1]Department of Muscle Biochemistry and [2]Department of Cellular Biochemistry
Nencki Institute of Experimental Biology
Polish Academy of Sciences
Pasteur str., 3
02-093 Warsaw
Poland

Introduction

Recently, ATP-sensitive potassium (K_{ATP}) channel has been described in rat liver mitochondria (Inoue *et al.*, 1991). This channel is blocked not only by ATP but also by the antidiabetic sulfonylurea, glibenclamide, giving evidence that mitochondrial K_{ATP} channel may belong to the well known family of ATP-dependent potassium channels found in plasma membranes of cardiac, smooth and skeletal muscle cells (Quast & Cook, 1989), pancreatic B-cells (Petersen & Dunne, 1989) and in the nervous system (Mourre *et al.*, 1990). In B-cells, K_{ATP} channels are specifically blocked by sulfonylureas which are the drugs capable to restore insulin secretion in patients affected by non-insulin-dependent diabetes mellitus (Petersen & Dunne, 1989). These channels are also blocked by some of non-sulfonylurea drugs like 8-methoxypsoralen (Szewczyk *et al.*, 1992a) and 8-(N,N-dimethylamino) octyl-3,4,5-trimethoxybenzoate (Szewczyk *et al.*, 1992b). K_{ATP} channels are activated by hormones such as somatostatin and galanin (De Weille *et al.*, 1988, 1989), free fatty acids (Müller *et al.*, 1992), as well as by K^+ channel openers (Cook & Quast, 1990; Duty & Weston, 1990).

It is not known how the mitochondrial K_{ATP} channel may be involved in the metabolism and function of liver mitochondria. Mitochondria in living cells swell and contract by changing their water content. Some of these changes are related to the action of hormones which cause an increase of cAMP (glucagon) or of calcium concentration

NATO ASI Series, Vol. H 83
Molecular Biology of Mitochondrial Transport Systems
Edited by M. Forte and M. Colombini

(vasopressin and α-adrenergic agonists) and to the respiratory chain activity. During respiration under phosphorylating conditions low water content is maintained in mitochondrial matrix; when adenine nucleotides or respiratory chain substrates are depleted mitochondria swell (Halestrap, 1989). An increase in intramitochondrial volume is sufficient to cause significant activation of mitochondrial metabolism (e.g. citrulline synthesis, pyruvate carboxylation, fatty acid oxidation). The increase in matrix volume is brought by influx of K^+. Hence, it remains to be determined whether the K_{ATP} channel is involved in the swelling-contraction cycle as an ATP sensor.

Addition of Me^{2+} ionophore A23187 to rat liver mitochondria in the presence of EDTA induces a uniport pathway for potassium ions, what results in mitochondrial swelling (Bernardi *et al.*, 1989; Nicolli *et al.*, 1991). It is believed that A23187 depletes magnesium ions from mitochondria, thus unmasking permeability for potassium ions. The study presented here focuses on the effects of A23187 and of different potassium channels openers and blockers on the energy-dependent swelling of rat liver mitochondria.

Materials and Methods

Rat liver mitochondria were prapared by the standard procedure involving centrifugation in sucrose/mannitol media and were used for experiments on the same day.

The kinetics of light scattering changes, reflecting mitochondrial volume changes linked to solute transport, were measured at 540 nm with Shimadzu UV-visible recording spectrophotometer (UV-160 A) at 25°C in the following media:

- *energy-dependent swelling induced by A23187 in Cl^- medium:* 100 mM KCl, 10 mM TRIS-MOPS, pH 7.4 or 8.6 (as indicated in the legends to the figures), 5 mM succinate-TRIS, 5 mM P_i-TRIS, 2 mM EDTA-TRIS, 2 μM rotenone and 5 μM cytochrome c; where indicated, A23187 from solution in ethanol was added to a final concentration of 5 nmoles per mg protein; protein concentration was 0.5 mg/ml;
- *energy-dependent swelling in CH_3COO^- medium:* 100 mM potassium acetate, 2 mM TRIS-acetate, pH 7.2, 2 μM rotenone and 10 mM succinate; protein concentration was 0.5 mg/ml;
- *passive swelling in SCN^- medium* - 150 mM KSCN, 20 mM TRIS-HEPES, pH 7.2, 2 μM rotenone and 2 mM EDTA; protein concentration was 0.5 mg/ml.

The drugs used in the present study were generous gifts from: Hoechst-Roussel

Pharmaceuticals (glibenclamide, glibonuride and tolbutamide), Leo Pharmaceuticals (pinacidil and P1060), Upjohn Company (minoxidil sulfate), Kirin Brewery Co. (KRN2391), and Rhone-Poulenc Rorer (RP 52891, i.e. aprykalim). A23187 and diazoxide were purchased from Sigma. All other chemicals were of the highest grade commercially available.

Results and Discussion

It has been postulated that rat liver mitochondrial K_{ATP} channel may play an important role in mitochondrial swelling (Inoue *et al.*, 1991). Me^{2+} ionophore A23187 is known to induce a rapid mitochondrial swelling via a mechanism involving also electrogenic uptake of potassium ions into mitochondria (K^+ uniport and K^+/H^+ exchanger), most likely due to the depletion of magnesium ions from mitochondria. Fig. 1 shows a typical recording of the absorbance changes of respiring mitochondria incubated in an isotonic medium, in the presence of potassium ions. Under these conditions, A23187 induced a rapid, low amplitude absorbance decrease. The K_{ATP} channel blocker, glibenclamide, abolished this effect in dose-dependent manner.

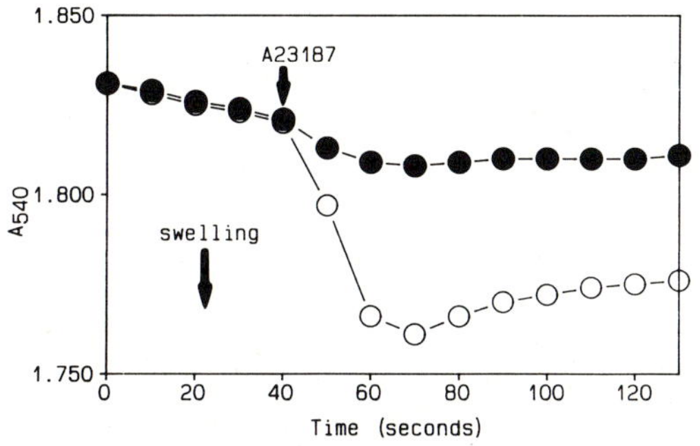

Fig. 1. Inhibition of A23187-induced swelling of rat liver mitochondria by glibenclamide. Energy-dependent swelling of mitochondria induced by A23187 was measured in 2 ml of the Cl^- medium described in Materials and Methods, pH 7.4, in the absence (open circles) or presence (closed circles) of 50 μM glibenclamide (added from 10 mM stock solution in dimethylsulfoxide, DMSO). At the time indicated, 5 nmoles of A23187 per mg protein was added. It is worthy mentioning that neither 1% DMSO nor 1% ethanol had an effect on swelling under these conditions.

The sulfonylurea receptor in the plasma membrane was shown to have a different affinity to various derivatives of antidiabetic drugs (Schmid-Antomarchi *et al.*, 1987). Moreover, an excellent correlation was found between the efficiency of blockage of K_{ATP} channel by sulfonylureas, degree of binding of sulfonylureas to their specific receptor and stimulation of insulin secretion in B-cells by these drugs (Schmid-Antomarchi *et al.*, 1987). In order to characterize mitochondrial K_{ATP} channel we used three sulfonylureas:

glibenclamide, glibonuride and tolbutamide to block mitochondrial swelling. Fig. 2 shows the concentration dependence of the inhibition of A23187-induced mitochondrial swelling by these three compounds. Their ability to inhibit this process turned out to be in the following order: glibenclamide > glibonuride > tolbutamide. Comparison of IC_{50} values derived from our experiments with K_d of listed sulfonylureas for half-maximal inhibition of specific [^{3}H]

Table 1. Effects of various sulfonylureas on K_{ATP} channel in B-cells, liver mitochondria, and as antidiabetic drugs. *Values are taken from Schmid-Antomarchi *et al.*, 1987.

Sulfonylurea	RIN5mF B-cells K_d* [nM]	IC_{50}* [nM]	Mitochondria IC_{50} (μM)	Therapeutic doses in man (mg/day)*
glibenclamide	0.3	0.06	20	2.5-15
glibonuride	90	120	300	12.5-75
tolbutamide	9000	22000	1100	1000-1500

glibenclamide binding to RINm5F (B-cells) microsomes, IC_{50} for specific efflux of $^{86}Rb^+$ from RINm5F B-cells and antidiabetic doses for man shows the same order of effectiveness of these sulfonylureas in the cases studied (Table 1). This suggests that the sulfonylurea receptor of the mitochondrial inner membrane may belong to the same family of receptors as found in the plasma membrane of B-cells.

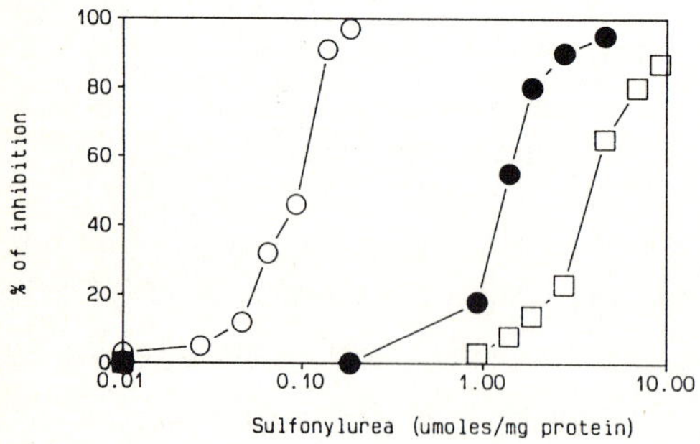

Fig. 2. Concentration dependence of inhibition of A23187-induced rat liver mitochondria swelling by antidiabetic sulfonylureas: glibenclamide (open circles), glibonuride (closed circles), and tolbutamide (open squares). Experimental conditions were the same as described in the legend to Fig. 1. The additions of antidiabetic sulfonylureas were made from concentrated stock solutions in DMSO, so the final concentration of the solvent did not exceed 1% (vol/vol). The results are expressed in % of inhibition of the maximal drop in absorbance in the presence of A23187 with no sulfonylureas (maximal swelling).

The A23187-induced swelling of mitochondria was blocked by glibenclamide with IC_{50} of 20 μM. On the other hand, it was shown that glibenclamide at nanomolar concentration range completely inhibited the activity of the reconstituted ATP-sensitive K^+ channel of rat liver mitochondria with IC_{50} of 62 nM (Woldegiorgis *et al.*, 1992). This apparent discrepancy could be explained as follows: as was shown (Woldegiorgis *et al.*, 1992) magnesium ions reduce inhibitory potency of glibenclamide in the reconstituted system, thus the relatively high concentration of glibenclamide (Table 1) needed to abolish the effect of A23187 on mitochondrial swelling could be ascribed to the presence of Mg^{2+} in preparation of mitochondria.

In pancreatic B-cells, using electrophysiological techniques, it has been observed that raising of pH of bath solution in the presence of a constant concentration of tolbutamide diminished the degree of block of ATP-dependent K^+ channel activity by sulfonylurea (Zunkler *et al.*, 1989). The conclusion is that the effective forms of sulfonylurea are undissociated forms which get access to their specific binding site(s) on the receptor molecule probably from the lipid phase of the B-cell plasma membrane (Zunkler *et al.*, 1989). To verify a possibility of a similar mechanism of action of sulfonylureas on mitochondrial K_{ATP} channel, we measured the effect of these drugs on the A23187-induced swelling at different pH values. The results are combined in Fig. 3. It was found that the pH of bath solution has no effect on A23187-induced swelling of mitochondria, while the inhibitory effect of glibenclamide observed at pH 7.4 was completely abolished at pH 8.6. This points to a similarity of interaction of antidiabetic drugs with K_{ATP} channel both in rat liver mitochondria and B-cell plasma membrane.

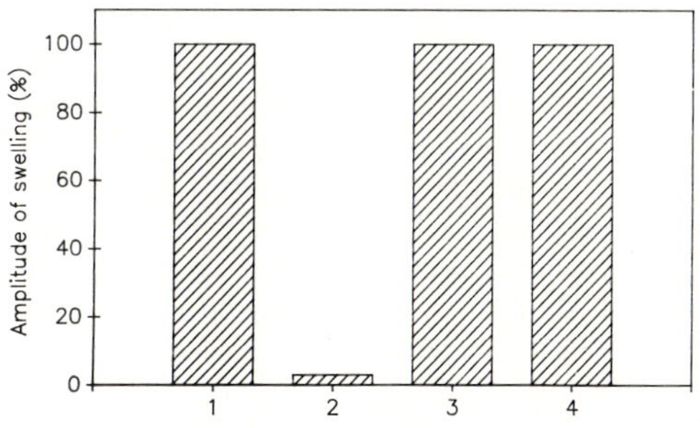

Fig. 3. Effect of glibenclamide on A23187-induced swelling of mitochondria at different pH values. Energy-dependent swelling of mitochondria was measured in the medium Cl^- described in Materials and Methods, at pH 7.4 (bars 1 and 2) or 8.6 (bars 3 and 4), in the presence of 50 μM glibenclamide (bars 2 and 4). Results are expressed in % of the maximal drop in absorbance (maximal swelling) caused by A23187 (5 nmoles/mg of protein) in the absence of drug (bars 1 and 3).

Despite low permeability of the mitochondrial inner membrane to charged and uncharged molecules, it is frequently reported that a variety of agents or conditions may increase this permeability (permeability transition) to solutes, leading to swelling of mitochondria, and to leakiness to K^+, Mg^{2+}, Ca^{2+}, etc. (for review see Gunter & Pfeiffer, 1990). Cyclosporin A, a cyclic peptide, is known to block permeability transitions in mitochondria. In our hands, however, cyclosporin A (5 μg/ml) was without effect on mitochondrial swelling induced by A23187 (not shown). This indicates that glibenclamide-sensitive, A23187-induced swelling of mitochondria does not involve the permeability transition mechanism.

Modulation of potassium channels activity by drugs, especially by those known as potassium channel openers, is a rapidly growing area of pharmacology. Recently, studies on the ability of different openers to influence K^+ channels activity focuses attention of many laboratories (Quast & Cook, 1989; Edwards & Weston, 1990; Weston & Edwards, 1992). Taking into account that the K_{ATP} channel openers are known to be tissue-specific, we applied a variety of compounds with different chemical structure to screen for their effect on mitochondrial swelling. The so-called passive swelling in SCN^- medium and the energy-dependent swelling in CH_3COO^- medium with mitochondria energized by the addition of succinate were checked (for further details see Materials and Methods). No effect of all the K_{ATP} channel openers on passive swelling of rat liver mitochondria was found (data not shown). Among six openers tested, 50 μM KRN2391 accelerated energy-dependent swelling, and this effect was fully reversed by the addition of 10 μM glibenclamide (Fig. 4). The same was observed with 50 μM diazoxide (data not shown), but not with the pinacidil, P1060, RP 52891 and minoxidil sulfate.

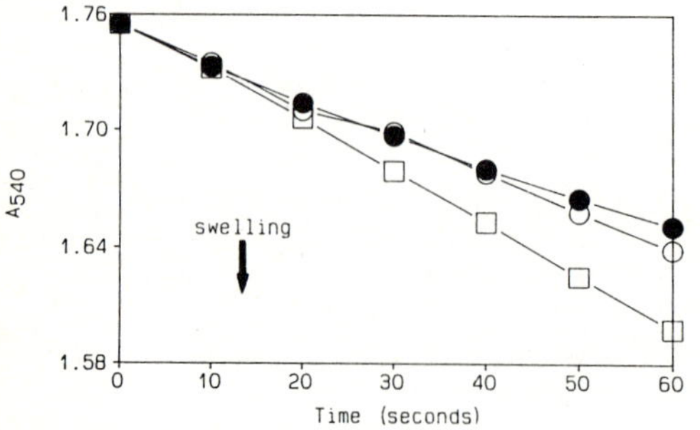

Fig. 4. Effect of potassium channel opener KRN2391 on energy-dependent mitochondrial swelling. Energy-dependent swelling of mitochondria was performed in CH_3COO^- medium (see Materials and Methods) with the following additions: none (open circles), 50 μM KRN2391 (open squares), and 10 μM glibenclamide + 50 μM KRN2391 (closed circles)

We believe that the present results reveal close similarities of the mitochondrial and the plasma membrane K_{ATP} channels. In addition, we describe here two pharmaceuticals that appear to be effective K_{ATP} channel openers in rat liver mitochondria. These conclusions, however, remain to be confirmed by observations coming also from other then measurements of mitochondrial swelling techniques.

Conclusion remarks

1. A23187-induced swelling of rat liver mitochondria is blocked by antidiabetic sulfonylureas: glibenclamide (IC_{50} 20 μM), glibonuride (IC_{50} 0.3 mM) and tolbutamide (IC_{50} 1.1 mM), but not by cyclosporin A (5 μg/ml).
2. The degree of inhibition of A23187-induced mitochondrial swelling by antidiabetic sulfonylureas depends on pH of the incubation medium: at pH 8.6 sulfonylureas were found to be not effective.
3. Among ATP-sensitive K^+ channel openers tested, KRN2391 (50 μM) and diazoxide (50 μM) stimulated energy-dependent mitochondrial swelling, and this effect was completely reversed by 10 μM glibenclamide. Other openers: RP 52891, P1060, pinacidil and minoxidil sulfate were found to be without effect.
4. Our results may suggest that K_{ATP} channel of the mitochondrial inner membrane is involved in energy-dependent swelling of liver mitochondria.

Acknowledgements

This study was supported by the Nencki Institute of Experimental Biology, Polish Academy of Sciences.

The following companies are gratefuly acknowledged for providing us with the drugs listed in Materials and Methods: Hoechst-Roussel Pharmaceuticals, Leo Pharmaceuticals, Upjohn Company, Kirin Brewery Co., and Rhone-Poulenc Rorer.

References:

Bernardi P, Angrilli A, Ambrosin V, Azzone GF (1989) Activation of latent K^+ uniport in mitochondria treated with the ionophore A23187. J Biol Chem 264: 18902-18906

Cook NS, Quast U (1990) Potassium channel pharmacology. In Potassium Channels: Structure, Classification, Function and Therapeutic Potential, Cook NS (ed). Ellis

Horwood Series in Pharmaceutical Technology Chichester.
De Weille JR, Schmid-Antomarchi H, Fosset M, Lazdunski M (1988) ATP-sensitive K^+ channels that are blocked by hypoglycemic sulfonylureas in insulin secreting cells are activated by galanin, an hyperglycemic hormone. Proc Natl Acad Sci USA 85: 1312-1316
De Weille JR, Schmid-Antomarchi H, Fosset M, Lazdunski M (1989) Regulation of ATP-sensitive K^+ channels in insulinoma cells: activation by somatostatin and kinase C and the role of cAMP. Proc Natl Acad Sci USA 86: 2971-2975
Duty S, Weston AH (1990) Potassium channel openers. Drugs 40: 785-788
Edwards G, Weston AH (1990) Structure-activity relationships of K^+ channel openers. Trends Pharmacol Sci 11: 417-422
Gunter TE, Pfeiffer DR (1990) Mechanisms by which mitochondria transport calcium. Am J Physiol 258: C755-C761
Halestrap AP (1989) The regulation of the matrix volume of mammalian mitochondria in vivo and vitro and its role in the control of mitochondrial metabolism. Biochim Biophys Acta 973: 355-382
Inoue I, Nagase H, Kishi K, Higuti T (1991) ATP-sensitive K^+ channel in the mitochondrial inner membrane. Nature 352: 244-247
Mourre C, Widmann C, Lazdunski M (1990) Sulfonylurea binding sites associated with ATP-regulated K^+ channels in the central nervous system: autoradiographic analysis of their distribution and ontogenesis, and of their localization in mutant mice cerebellum. Brain Res 519: 29-43
Müller M, Szewczyk A, De Weille J, Lazdunski M (1992) ATP-sensitive K^+ channels in insulinoma cells are activated by nonesterified fatty acids. Biochemistry 31: 4656-4661
Nicolli A, Redetti A, Bernardi P (1991) The K^+ conductance of the inner mitochondrial membrane. J Biol Chem 266: 9465-9470
Petersen OH, Dunne MJ (1989) Regulation of K^+ channels plays a crucial role in the control of insulin secretion. Pflüg Arch 414: S115-S121
Quast U, Cook NS (1989) Moving together: K^+ channel openers and ATP-sensitive K^+ channels. Trends Pharmacol Sci 10: 431-435
Schmid-Antomarchi H, De Weille J, Fosset M, Lazdunski M (1987) The receptor for antidiabetic sulfonylureas controls the activity of the ATP-modulated K^+ channel in insulin-secreting cells. J Biol Chem 262: 15840-15844
Szewczyk A, De Weille JR, Lazdunski M (1992a) 8-Methoxypsoralen blocks ATP-sensitive potassium channels and stimulates insulin release. Eur J Pharmacol 216: 323-326
Szewczyk A, De Weille JR, Lazdunski M (1992b) TMB-8 (8-(N,N-dimethylamino) octyl-3,4,5-trimethoxybenzoate) inhibits the ATP-sensitive K^+ channel. Eur J Pharmacol 226: 175-177
Weston AH, Edwards G (1992) Recent progress in potassium channel opener pharmacology. Biochem Pharmacol 43: 47-54
Woldegiorgis G, Paucek P, Mironova G, Mahdi F, Beavis A, Garlid K (1992) Purification and reconstitution of the ATP sensitive K^+ channel of rat liver mitochondria. EBEC Short Reports 7: 89 (Materials of 7th European Bioenergetics Conference, July 26-31, Helsinki, Finland, abstr. #IV-14)
Zunkler BJ, Trube G, Panten U (1989) How do sulfonylureas approach their receptor in the B-cell plasma membrane? Naunyn-Schmiedeberg's Arch Pharmacol 340: 328-331

PROBING FOR THE VOLTAGE SENSOR IN A MITOCHONDRIAL CHANNEL, VDAC, USING SITE-DIRECTED MUTAGENESIS

#L. Thomas, +E. Blachly-Dyson, *M. Colombini and +M. Forte
#Department of Biochemistry
University of Alberta
474 Medical Sciences Bldg.
Edmonton, Alberta
T6H 2H7

Abstract

The VDAC ion channel is voltage-dependent and forms a large pore across the mitochondrial outer membrane with a relatively small amount of protein. Thus, unravelling the molecular basis of voltage dependence for this channel should be an easier task than for the more complex voltage-gated channels of the plasma membrane. The ability to probe the channel for domains of the protein involved in voltage-gating, using site-directed mutations which change charge with respect to the wild type protein, has resulted in the identification of regions of the polypeptide which translocate through the transmembrane field during channel closure. The residues at which mutations alter the voltage dependence of the channel are not confined to discrete regions of the protein. Gating in VDAC seems to involve a substantial transmembrane movement of mass. There is at least one other membrane channel, colicin E1, where gating is associated with a substantial amount of mass translocated through the transmembrane field (Abrams *et al.*, 1991).

Introduction

A number of critical cellular functions, including membrane electrical excitability, secretion, egg fertilization and cell

*Laboratories of Cell Biology, Department of Zoology, University of Maryland, College Park, Maryland 20742; +Vollum Institute for Advanced Biomedical Research, Oregon Health Sciences University, Portland, Oregon 97201 USA

NATO ASI Series, Vol. H 83
Molecular Biology of Mitochondrial Transport Systems
Edited by M. Forte and M. Colombini

motility are controlled by voltage-dependent channels, which are intrinsic proteins forming aqueous pathways through membranes. These proteins sense and respond to very small changes in transmembrane potentials and through conformational changes modulate the transmembrane flux of ions and molecules. Two important structures of a voltage-gated channel are the selectivity filter and the voltage sensor. The portion of the protein that allows channels to discriminate between ions is referred to as the selectivity filter and that which responds to the applied field is referred to as the voltage sensor. This article focuses on attempts to discern the structural basis of voltage-dependence in a mitochondrial channel called VDAC.

Molecular Structure of VDAC

The mitochondrial voltage-dependent, anion-selective channel, VDAC, is the major permeability pathway through the mitochondrial outer membrane (Colombini, 1979). VDAC has been found in species of all eukaryotic kingdoms and its proposed function is regulation of the flux of metabolic intermediates into and out of the mitochondrion (Colombini, 1979; Smack and Colombini, 1985; Colombini, 1989). VDAC is in a high conductance (open) state at zero and low membrane potentials but adopts a low conducting (closed) state at higher negative and positive potentials (30 mV)(Schein, Colombini and Finkelstein, 1976). Open VDAC channels have a weak anion selectivity (2:1, $Cl^-:K^+$) (Colombini, 1989), while closed channels show a weak cation selectivity (Colombini, 1980; Benz, Kottke and Brdiczka, 1990; Zhang and Colombini, 1990). Electron microscopy and size exclusion studies indicate that the diameter the channel goes from 3 nm in the open state to 1.8 nm in the closed state (Colombini *et al.*, 1987; Mannella and Guo, 1990). As well as a large change in diameter, channel closure is accompanied by a large volume change, approximately 30 nm^3 (Zimmerberg and Parsegian, 1986). Scanning transmission electron microscopy (STEM) mass analysis indicates that VDAC is a monomeric channel (Thomas *et al.*, 1991). The fact that a large diameter channel

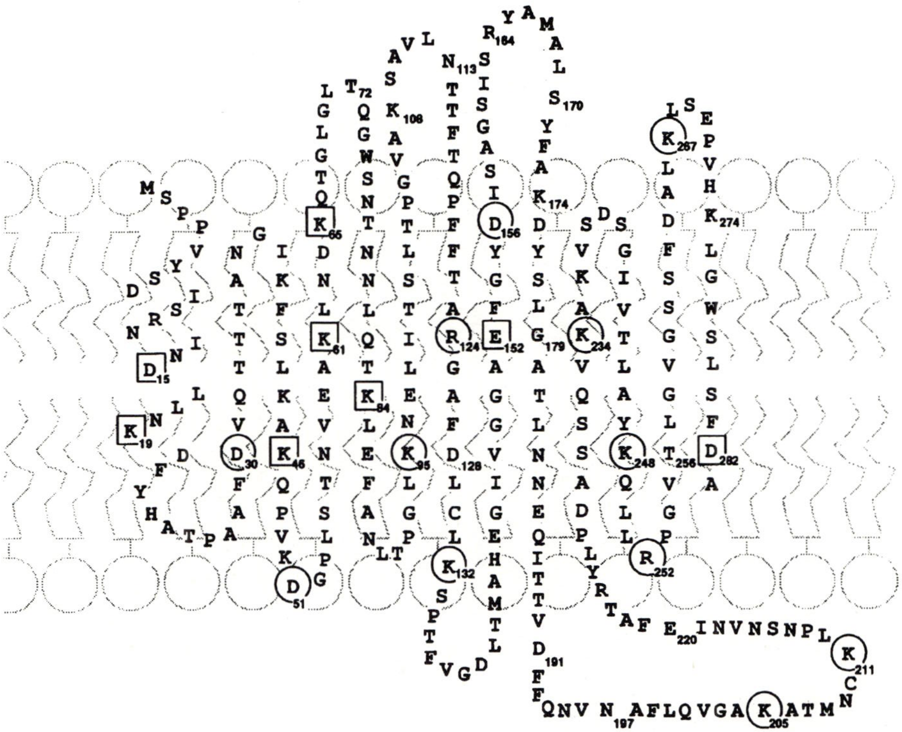

Figure 1. The proposed folding pattern of VDAC in a phospholipid membrane. The pattern, consisting of 12 transmembrane β strands and one α helix, is from Blachly-Dyson *et al.*, 1990. Here the circled residues are sites where a change in charge had no significant effect on voltage dependence. The boxed residues are sites where charge changes influenced the voltage dependence. Reprinted with permission from Thomas *et al.*, 1993.

is formed by a 282 or 283 amino acid polypeptide (Mihara and Sato, 1985; Kleene *et al.*, 1987; Thinnes *et al.*, 1989), restricts the wall of the channel to a single layer of protein.

A secondary structure for the yeast VDAC polypeptide has been proposed by Blachly-Dyson and coworkers (1990). This polypeptide is quite hydrophobic (45.5% charged and polar residues) which is consistent with it being a low molecular weight protein that forms a large aqueous pore. Polar or charged residues are positioned at every third or fourth amino acid in the first 20 N-terminal residues of VDAC. These residues, which are present in the mature protein, may form an amphipathic alpha helix typical of mitochondrial targeting sequences (Mihara and Sato, 1985). The remainder of the sequence contains many stretches of alternating hydrophobic and hydrophilic residues, which could form a "sided" beta sheet. VDAC contains no stretches of hydrophobic residues long enough to form a membrane-spanning hydrophobic alpha-helix. Twelve potential transmembrane beta strands were found using a computer algorithm designed to locate possible amphipathic transmembrane beta strands. Based on this information, a model was proposed in which the wall of the channel was formed by an N-terminal helix flanked by 12 strands of antiparallel beta sheet. This folding model was further refined using open channel selectivity data from single point mutations of the VDAC protein. Charged residues, which resulted in a net change in charge with respect to wild type, were introduced at positions predicted to line the channel in the alternating hydrophobic/hydrophilic pattern. If a mutation affected the open channel selectivity then the residue was assumed to comprise part of the aqueous pore. This analysis allowed the definition of protein domains forming the walls of the open channel (Figure 1).

Mapping VDAC's Voltage Sensor

Various mechanisms have been proposed for how channels sense electrical fields and how this ability is coupled with subsequent conformational changes. The simplest way for a protein to detect

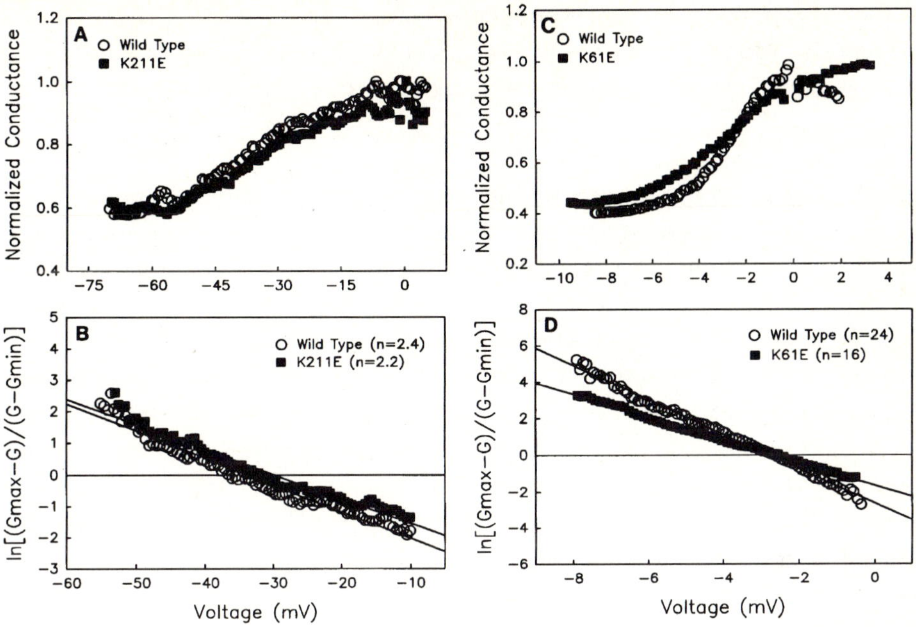

Figure 2. Comparison of wild-type VDAC with mutant channels containing single amino acid substitutions. Panels "A" and "C" illustrate the voltage dependence of the conductance of channels with substitutions of glutamate for lysine at position 211 (K211E) and 61 (K61E) resp. Panels "B" and "D" illustrate the same data after ln transformation and fitting to a two-state model. Both representations show that charge change at position 61 altered the steepness of the voltage dependence while charge change at position 211 had no significant effect. The properties of K61E and the wild type (panels C and D) were analyzed in the presence of 57 μg/ml 500 kDa dextran sulfate, sufficient to increase the steepness of the voltage dependence "n" by a factor of 10. To compare these to the results in "A" and "B" they should be divided by 10. By increasing the voltage dependence 10 fold, dextran sulfate must also reduce the voltage needed to close half the channels "V_0" by 10 fold. Using the ln transform (see text) the voltage dependence parameters (n and V_0) were estimated and tabulated (see Table).

an electric field would be to have charged residues that can move with respect to the field or dipoles that can change their dipole moments in the direction of the field. These charges or dipoles are referred to as the voltage sensors. The linkage of the conformational change responsible for the change in ion conductance to the movement of the voltage sensor results in voltage dependence. Voltage dependency could also arise from a change in field intensity at a fixed charge located on the walls of the pore, which occurs as a result of conformational change (Colombini, 1984). This would require that the sensor be located within the pore of the channel and the permeability filter not be uniformly distributed along the length of the channel.

Several lines of evidence indicate that the voltage sensor of VDAC involves the movement of positively charged residues through the transmembrane field. First titration of the channel to high pH neutralizes the sensor with an apparent pK of 10.6 (Bowen, Tam and Colombini, 1985). Second, low levels of succinic anhydride (which converts amino groups to carboxyl groups) result in the elimination of voltage dependence (Doring and Colombini, 1985). Third, lowering the pH to levels which would result in neutralization of the carboxyl groups increases the voltage dependence of the channel (Ermishkin and Mirzabekov, 1990). The apparent pK of the titratable groups and the succinic anhydride modification suggest that positive charge, possibly lysine residues, are involved in voltage sensing.

Since charged residues are involved in voltage sensing, mutants which differed in charge from the wild type were engineered by site-directed mutagenesis to probe for the VDAC's voltage sensor(s). Conductance data from two such mutations are presented in Figures 2a and 2c. To be able to quantify the effect a mutation had on VDAC's voltage dependence, conductance-voltage relationships were fit to a two state mathematical model:

$$\ln\frac{(G_{\max}-G)}{(G-G_{\min})}=\frac{(nFV-nFV_0)}{RT}$$

where G, G_{max}, and G_{min} are the conductance at any voltage V, the maximum conductance, and minimum conductance, respectively. F, R and T have their usual meanings. The free energy difference between the states (closed-open) is divided into a voltage-independent (nFV_0) and a voltage dependent (nFV) component, where n is the number of charges on the sensor, F is Faraday's constant, V is the potential difference across the membrane and V_0 is the voltage at which half the channels are open. The slope of the ln transformed data is proportional to n (Fig. 2b and 2d). A mutation may affect n or V_0, but only a change in n will be indicative of a change in the voltage dependence. A change in V_0 does not indicate that the voltage dependence of a channel has been changed. V_0 is influenced by many voltage-independent parameters including: surface charge, salt concentration, the stabilization of one state over another and changes in the height of the energy barrier (an apparent change due to kinetic delays).

A charge change in the sensing region was determined to have occurred when a change in n, that agreed in sign and magnitude with the charge engineered into the protein, occurred without perturbation of channel structure (assessed by calculating nFV_0).

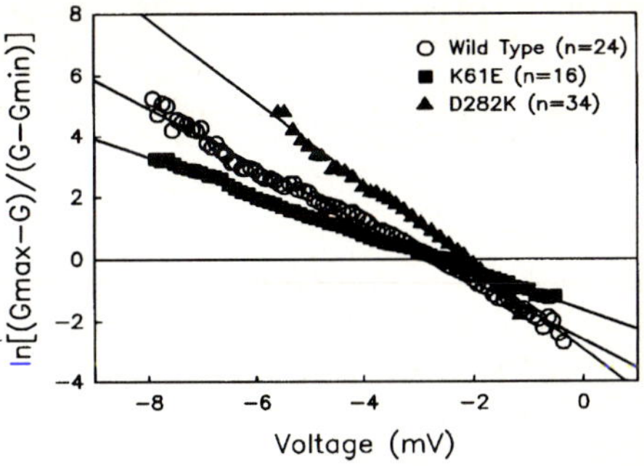

Figure 3. The changes in the steepness of the voltage dependence correlate with the sign of the changes in charge resulting from the engineered amino-acid substitutions. The data, collected in the presence of dextran sulfate, were plotted as the ln transform in order to clearly show the changes in the steepness of the voltage dependence as slope changes in the fitted straight line. The mutations were K61E, a glutamate substituted for a lysine at position 61 (a positive charge replaced by a negative charge), and D282K, a lysine substituted for an aspartate at position 282 (a negative charge replaced by a positive charge).

Increasing positive charge on the sensor results in an increased voltage dependence, while decreasing positive charge results in a decrease in the voltage sensitivity (Figure 3). The Table summarizes the voltage dependent parameters for the VDAC mutants. The results are divided into two groups, mutations which did not significantly affect voltage dependence (Group A), and mutations which affected voltage dependence (Group B). The residues at which mutations affect the voltage dependence (15, 19, 46, 61, 65, 84, 152, and 282) are not localized in a small portion of protein. The data support a large portion of protein translocating through the field during channel closure.

Table: Voltage-Dependence Parameters of Site-Directed Mutants

The values of "n" and V_0 were estimated as shown in Fig. 2. The results (mean ± S.D.) were grouped into "A": no significant effect on "n" (t test, $\alpha<0.05$) and "B": significant change in "n" (**:.01>α>.001; ***:.001>α). "Charge change" is the change in the charge on the sensor, estimated from the difference between the measured "n" values in the mutants and the wild type value. These are listed as the values in the absence of dextran sulfate. The values for experiments in the presence of dextran sulfate were divided by 10. Numbers in parenthesis are the engineered charge changes.

Species	No Dextran Sulfate		Plus Dextran Sulfate		Charge Change
	n	V_0 (mV)	n	V_0 (mV)	
Wild type	2.5 ± 0.4	30 ± 4	24 ± 2	2.4 ± 0.4	
Group A					
D30K	2.7 ± 0.2	32 ± 3			0 (+2)
D51K			26 ± 2	2.2 ± 0.2	0 (+2)
K95E			23 ± 2	2.1 ±0.2	0 (-2)
R124E			25 ± 5	2.6 ± 0.6	0 (-2)
K132E	2.3 ± 0.3	23 ± 3	23 ± 3	2.2 ± 0.4	0 (-2)
D156K	2.1	-27	25 ± 3	2.6 ± 1.2	0 (+2)
K205E	2.3 ± 0.4	29 ± 2			0 (-2)
K211E	2.6 ± 0.6	31 ± 7			0 (-2)
K234Q			23 ± 2	2.5 ± 0.3	0 (-1)
K248E	2.6 ± 0.2	19 ± 2			0 (-2)
R252E	2.6 ± 0.2	33 ± 3			0 (-2)
K267E	2.2 ± 0.4	29 ± 5			0 (-2)
Group B					
D15K			48 ± 9***	1.5 ± 0.5	+2 (+2)
K19E			10 ± 1***	-5 ± 1	-1 (-2)
K46E			17 ± 2***	3.0 ± 0.9	-1 (-2)
K61E			15 ± 3***	3.6 ± 0.8	-1 (-2)
K65E	1.6 ±0.4***	32 ± 8			-1 (-2)
K84E			17 ± 3**	3.1 ± 0.6	-1 (-2)
E152K			34 ± 7***	2.1 ± 0.6	+1 (+2)
D282K			34 ± 2***	1.9 ± 0.2	+1 (+2)

Correlation of Results from Voltage Dependent Analysis of VDAC Mutants with Selectivity Data

Two factors, steric considerations and interactions between the ion and the channel, are usually thought to contribute to selectivity in narrow channels where permeating ions must contact the wall of the protein. However, in channels with larger pore sizes, such as VDAC, it is unlikely that permeation of small ions requires an intimate interaction with the walls of the channel. Selectivity is likely be a function of the electric field generated by the distributed charges of amino-acid residues lining the wall of the pore. These should interact electrostatically with the permeating ions.

If some of the charged residues lining the wall of the pore were to move out of the pore, this motion would be associated with a voltage dependent energy change because charges would have moved through the transmembrane electric field. If this were the case, the voltage sensor of VDAC could be mapped directly by identifying residues which affect voltage dependence or indirectly by identifying residues which affect the open channel selectivity but not the closed channel selectivity. These residues are assumed to be removed from the wall of the pore during channel closure because they do not affect the electrostatic environment of the pore in the closed state. Thus, domains moving through the field during closure can be identified (Thomas *et al.*, 1993).

The same mutant channels were analyzed for the effects of specific sequence change on the channel selectivity in both the open and closed states (Peng *et al.*, 1992). Using an entirely different approach and different assumptions it is remarkable how well the results from these two studies agree. Residues identified as comprising the voltage sensor were: D15K, K19E, K46E, K61E, K65E, K84E, E152K, T236K, K248, and D282K. Residues that appeared to remain in the wall of the channel in the closed state were: D30K, K95E, R124E, D128R, G179D, and K234Q.

The results from the two approaches used to identify residues associated with the voltage sensor seem to disagree at two positions. A charge change at 252 did not affect the

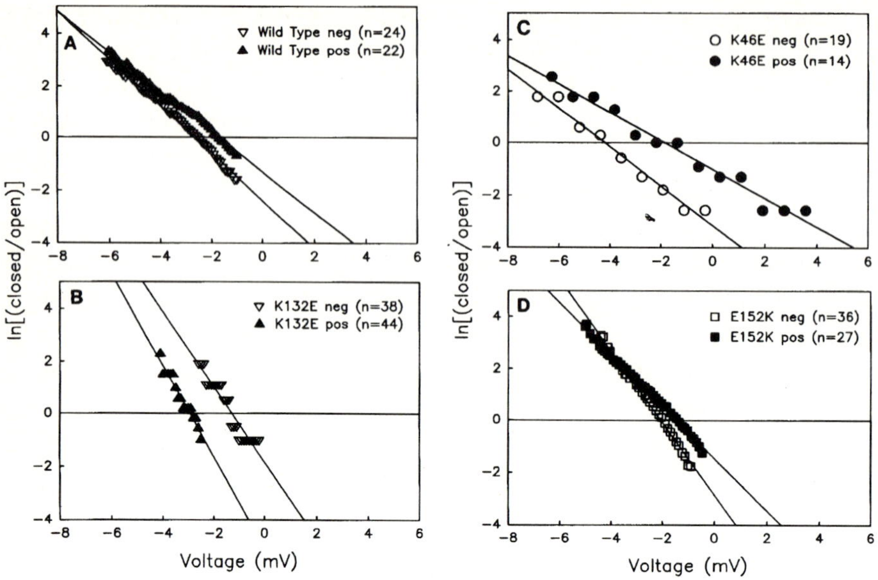

Figure 4. Comparison of the measured voltage-dependence parameters (n and V_0) for positive and negative transmembrane potentials. The data were collected with single-channel (panels B and C) and multi-channel (panel A and D) membranes. The data collected at positive potentials was plotted as if the applied potentials were negative in order to facilitate the comparison between the slopes of the two data sets. All experiments were performed in the presence of dextran sulfate. The standard amount of dextran sulfate (see Fig. 2) was used for panels (A, C, and D). In panel B the data was collected using an unknown (likely higher) amount of dextran sulfate. This does not affect the conclusion that this single channel had a similar voltage dependence at positive and negative potentials. The site-directed mutations were: K132E, glutamate for lysine at position 132; K46E, glutamate for lysine at 46; and E152K, lysine for glutamate at 152.

selectivity in the open channel but resulted in a change of selectivity in the closed state. It may be that with the reduction in pore diameter and volume that accompanies closure, this residue's location changes by a small amount. Its proposed location at the edge of the pore (Fig. 1) is consistent with a small change in location resulting in this site influencing ion selectivity. Similar results were observed for the mutant K248E. From the voltage dependent parameters this residue does not appear to be moving in response to the field, while the selectivity studies indicate that this residue forms a portion of the voltage sensor. During channel closure this residue may move a short distance and be removed from the channel, no longer influencing channel selectivity. Translocation of charge a short distance with respect to the electric field may not be resolvable using the techniques employed in this study.

Further Considerations

Symmetry of Voltage Dependence

A characteristic of VDAC is that the voltage dependent closure at both positive and negative potentials is fairly symmetric (Figure 4a). While single channels were known to close at both positive and negative potentials (Colombini, 1979; Colombini, 1989) the symmetry in the quantitative value of the steepness of the voltage dependence could be due to an equivalent amount of effective charge translocation for gating at both negative and positive potentials or it could be an artifact of random insertion of channels into planar phospholipid membranes. Single channel analysis of K132E, a mutant whose multichannel voltage dependent parameters were indistinguishable from those of the multichannel wild type (Figure 4b), indicated that the symmetry of the steepness of voltage dependence is an inherent property of the channel and not an artifact of random insertion. The steepness of the voltage dependence in response to both negative and positive potentials (under the conditions used) is almost identical (n values are within the standard deviation observed with multichannel wild type membranes). Recalling that each channel is made by one 30 kDa polypeptide and looking at the proposed secondary structure, it is clear that VDAC must be asymmetric and thus it is likely that the symmetry is a conserved property retained by natural selection.

Single or Multiple Sensor(s)?

A single VDAC channel responds to both positive and negative fields. The voltage dependent closure of the channel in response to opposite potentials may be due to one gating mechanism or two distinct gating mechanisms. While there is some kinetic evidence for two distinct gating processes (Colombini, 1976), these processes could share the same sensor and/or gate. Single channel analysis of a mutant whose multichannel steepness of voltage dependence was reduced with respect to wild type, K46E, indicates there is at least a common element to the voltage sensor(s) responding to positive and negative potentials. A decrease in charge at one position in the protein affected the voltage sensitivity of the protein in response to both negative and positive potentials (Figure 4C). From the folding pattern (Figure 1), one would expect some difference in the region of protein translocating because of the way the protein chain is folded in the membrane. Results from succinic anhydride modification of VDAC (Doring and Colombini, 1985) also supports this view of a common element to the voltage sensor. In these experiments anhydride was added under very asymmetric conditions (applied voltage, side of addition, state of the channel) resulting in quite different degrees of inhibition of voltage dependence. Yet, in general, the resulting channels behaved in the same way to positive and negative potentials.

Sidedness of the Channel

VDAC is a monomeric channel, therefore, residues which are positioned at one end of the channel and involved in voltage sensing, might be expected only to affect gating on one side. These residues would translocate through the entire field in response to one potential and would not move through the field in response to the opposite potential. A mutation at residue 152 results in wild type behavior in response to negative potentials and an increased voltage dependence in response to positive potentials (Figure 4D). This residue is positioned at one end of the channel in the proposed secondary structure (Figure 1).

Are the Selectivity Filter and the Voltage Sensor the Same Structural Entity?

Two lines of evidence support that some of the same charges responsible for selectivity may also be involved in the voltage sensor. First, changes in charge at residues throughout the primary sequence of yeast VDAC (15, 19, 46, 61, 65, 84, 152 and 282) affected both the open channel selectivity and the voltage dependence of the channel. Second, succinic anhydride modification affected voltage dependence and ion selectivity (Doring and Colombini, 1985; Adelsberger-Mangan and Colombini, 1987). This indicates that the charge on the amino groups modified by the anhydride are involved in forming portions of the selectivity filter and the sensor.

Model of the Mechanism of Voltage Dependent Gating

The fact that many of the mutations altered open-channel selectivity eliminates the hypothesis that voltage dependency in VDAC could arise from a change in electrical potential at a fixed charge located on the walls of the pore, which occurs as a result of conformational change. This would require a discrete position for the selectivity filter in the pore of the channel. The selectivity filter in VDAC appears to be the result of uniformly distributed fixed charges within the cylindrical aqueous pore. Succinic anhydride modification (Adelsberger-Mangan and Colombini, 1987) affected the voltage dependence and the selectivity filter suggesting that these structures were not different entities. More persuasive is the fact that specific residues that influence VDAC selectivity (based on the site-directed mutations) also affect the voltage dependence. Thus the voltage sensor and selectivity filter are not totally different entities but share common protein domains.

Figure 1 summarizes the results of the voltage dependence studies. Residues for which a mutation altered the voltage dependence are boxed; residues in which the mutation left the voltage dependence unchanged are circled.

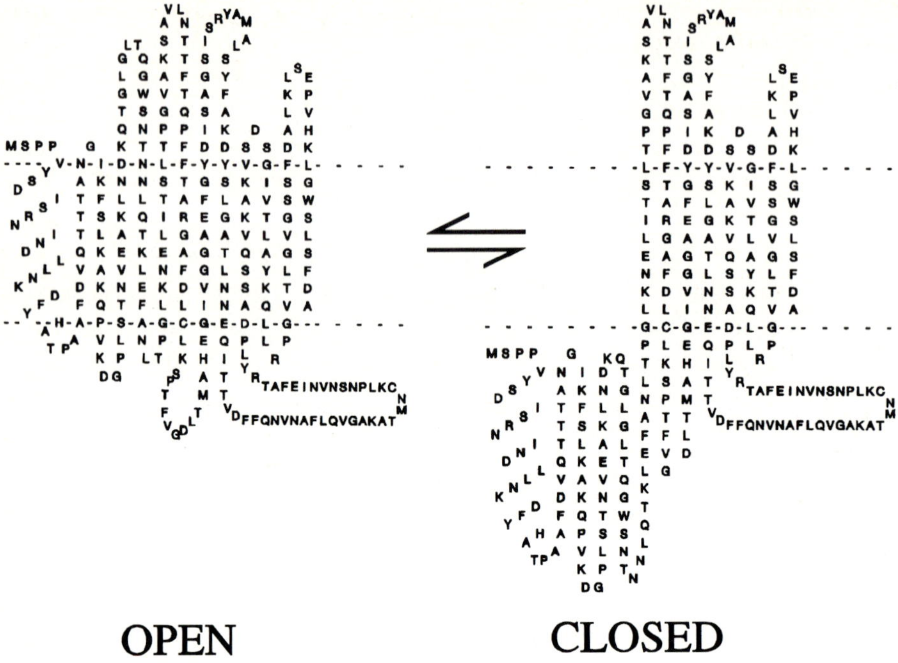

Figure 5. Working model for the conformational change resulting in VDAC channel closure. The open and closed channels are illustrated simply as the transmembrane folding pattern of the amino-acid chain using the single-letter code. Presumably the strands would be tilted and wrapped into a cylinder. A membrane potential, negative in the lower part of the illustration, would attract the positively-charged voltage sensor (the N-terminal α-helix and adjacent 4 β strands) out of the membrane resulting in a smaller cylindrical pore with a net negative charge.

A large portion of the protein appears to be translocated through the field. Our working model for the molecular rearrangements involved in gating includes the translocation of an alpha helix and the four neighboring beta strands through the field (Figure 5). The movement of the alpha helix-beta sheet region would move enough charges through the field, approximately 2.5, to account for the voltage dependence of the channel. Note that movement of the alpha helix into the channel during closure unaccompanied by beta strand movement (as some have proposed) would not account for enough charge movement through the field to account for the voltage dependence of VDAC. The sensor could then orient with the hydrophobic residues towards the lipid bilayer and the hydrophilic residues facing the aqueous environment. Alternatively, the non-polar regions of the mobile protein domain could face each other (away from the water) forming a self-contained aqueous domain. Either way exposure of the non-polar surface to the aqueous phase would be avoided thus reducing the energy difference between the open and closed states. This large movement of mass would account for the large change in 1) pore diameter and volume accompanying channel closure and 2) the large change in electrostatic environment within the pore. From the proposed secondary structure of VDAC, Figure 1, clearly the same set of strands cannot translocate in both directions. What we propose is that most of the strands comprising the voltage sensor are common but, depending on the field polarity, the strands immediately adjacent to the sensor that form part of the pore in the closed channel may be different (see Fig. 5).

Inconsistent Results

The results from analysis of the voltage dependence of point mutations at residues 30 and 51 were unexpected. These mutations are residues which are flanked by regions which appear to translocate through the membrane field. Conversely, charge change at residue 152 affects voltage dependence while residues on either side, 132 and 156, appear to be immobilized with respect to the transmembrane field. Translocation of the protein domains comprising the voltage sensor can be understood (see Fig. 5) in response to potentials of one sign. However, it is difficult to conceive of a conformational change, consistent with the data, for potentials of the opposite sign. It is possible that data gathered from multi-channel membranes confuses diametrically opposite effects of a mutation of the two gating processes.

Topological mapping of bovine heart VDAC, using peptide-specific antibodies indicates that the α helix may be located outside the membrane proper (Mannella *et al.*, 1992). This seems inconsistent with the observation that charged residues on this helix influence the selectivity of the open state and have reduced or no effect on the selectivity of the closed state (Peng *et al.*, 1992). The identification of the α helix as part of the voltage sensor argues for a transmembrane motion of this helix during channel gating. It is possible that the presence of antibodies trap the α helix outside the membrane when the channels spontaneously close even in the absence of a membrane potential (the probability of closure is low but not zero). Thus, this apparent conflict may be the result of the methodological approach taken to probe the system.

Conclusions

Site-directed mutations, which change the net charge of the mutant proteins with respect to the wild type, have been used in combination with electrophysiological techniques to define regions of VDAC involved in voltage sensing. Unlike the neuronal Na^+ and the K^+ channel, where the voltage sensor appears to be comprised of charged amino acid residues localized in discrete protein domains, translocating through only a fraction of the transmembrane field (W. Stuhmer *et al.*, 1989; Papazian *et al.*, 1991; Liman *et al.*, 1991), VDAC's voltage-sensing charge seems to be distributed over a large region of the protein. Residues affecting the voltage dependence are found in the amino and carboxyl terminus. Movement of large amounts of protein would account for the large reduction in channel diameter and volume that occur during channel closure. The removal of positively charged residues from the conducting pathway during closure is consistent with the marked change in selectivity of the channel upon closure. Translocation of the proposed alpha-helix and the four adjacent beta-strands would result in charge movement, accounting for the steepness of voltage dependence. Although this model is not able to exactly define the molecular mechanism of voltage dependence, it has given us a framework to think about the voltage-induced conformational transitions within the channel.

Outlook for the Future

Further point mutations in regions immediately flanking residues 51, 30 and 152 should be generated and analyzed with respect to their voltage dependent parameters to see if the ambiguities in this region can be resolved. The major limitation of the approaches used so far to define structure-function relationships has been the absence of a topological marker for VDAC. Without this information the orientation of the protein in the membrane is unknown. If the information acquired from site-directed mutagenesis can be combined with biochemical techniques (protease sensitivity, cysteine engineering, sulfhydryl chemistry, circular dichroism or accessibility to peptide-specific antibodies) which map the orientation of the channel in the outer membrane and the position of proposed membrane spanning domains with respect to each other, the working model can be refined and the ultimately the molecular structure-function relationship of this simple ion channel may be elucidated.

Acknowledgement: This work was supported by grants from NIH (GM 35759) and ONR (N00014-90-J-1024).

References

Abrams CK, KS Jakes, A Finkelstein, and SL Slatin (1991) Identification of a translocated gating charge in a voltage-dependent channel: colicin E1 channels in planar phospholipid bilayer membranes. *J Gen Physiol* **98:**77-93

Adelsberger-Mangan DM, and M Colombini (1987) Elimination and restoration of the voltage dependence in the mitochondrial channel, VDAC, by graded modification with succinic anhydride. *J Membr Biol* **98:**157-168

Benz R, M Kottke, and D Brdiczka (1990) The cationically selective state of the mitochondrial outer membrane pore: a study with intact mitochondria and reconstituted mitochondrial porin. *Biochim Biophys Acta* **1022:**311-318

Blachly-Dyson E, SZ Peng, M Colombini and M Forte (1990) Alteration of the selectivity of the VDAC ion channel by site-directed mutagenesis: implications for the structure of a membrane ion channel. *Science (Wash DC)* **247:**1233-1236

Bowen KA, K Tam, and M Colombini (1985) Evidence for titratable gating charges controlling the voltage dependence of the outer mitochondrial membrane channel, VDAC. *J Membr Biol* **86:**51-59

Colombini M (1979) A candidate for the permeability pathway of the outer mitochondrial membrane. *Nature (Lond)* **279:**643-645

Colombini M (1980) Structure and mode of action of a voltage dependent anion-selective channel (VDAC) located in the outer mitochondrial membrane. *Ann NY Acad Sci* **341:**552-563

Colombini M (1984) A novel mechanism for voltage control of channel conductance. *J Theor Biol* **110:**559-567

Colombini M (1989) Voltage gating in the mitochondrial channel, VDAC. *J Membr Biol* **111:**103-111

Colombini M, CL Yeung, J Tung, and T König (1987) The mitochondrial outer membrane channel, VDAC is regulated by a synthetic polyanion. *Biochim Biophys Acta* **905:**279-286

Doring C, and M Colombini (1985) Voltage dependence and ion selectivity of the mitochondrial channel, VDAC, are modified by succinic anhydride. *J Membr Biol* **83:**81-86

Ermishkin LN, and TA Mirzabekov (1990) Redistribution of the electric field within the pore contributes to the voltage dependence of mitochondrial porin channel. *Biochim Biophys Acta* **1021:**161-168

Kayser H, H Kratzin, Thinnes FP, H Gotz, WE Schmidt, K Eckart and N Hilschmann (1989) Charakterisierung und primarstruktur eines 31-kDa-porins aus menschlichen B-lymphozyten (Porin 31HL). *Biol Chem Hoppe-Seyler* **370:**1265-1278

Kleene R, N Pfanner, R Pfaller, TA Link, W Sebald, W Neupert, and M Tropschug (1987) Mitochondrial porin of *Neurospora crassa*: cDNA cloning, *in vitro* expression and import into mitochondria. *EMBO J* **6:**2627-2633

Liman ER, P Hess, F Weaver, G Koren (1991) Voltage-sensing residues in the S4 region of mammalian K^+ channel. *Nature* **353:**752-

Mannella CA, and X-W Guo (1990) Interaction between the VDAC channel and a polyanionic effector: an electron microscopic study. *Biophys J* **57:**23-31

Mannella CA, M Forte, and M Colombini (1992) Toward the molecular structure of the mitochondrial channel, VDAC. *J Bioenerg Biomembr* **24:**7-19

McCormack K, MA Tanouye, LE Iverson, J-W Lin, M Ramaswami, T McCormack, JT Campanelli, MK Mathew, and B Rudy (1991) A role for hydrophobic residues in the voltage-dependent gating of Shaker K^+ channels. *Proc Natl Acad Sci (USA)* **88:**2931-2935

Mihara K, and R Sato (1985) Molecular cloning and sequencing of cDNA for yeast porin, an outer mitochondrial membrane protein: a search for targeting signal in primary structure. *EMBO J* **4:**769-774

Papazian DM, LC Timpe, YN Jan, and LY Jan (1991) Alteration of voltage-dependence of *Shaker* potassium channel by mutations in the S4 sequence. *Nature* **349:**305-310

Peng SZ, E Blachly-Dyson, M Forte, and M Colombini (1992) Large scale rearrangement of protein domains is associated with voltage gating of the VDAC channel. *Biophys J* **62:**123-135

Schein SJ, M Colombini, and A Finkelstein (1976) Reconstitution in planar lipid bilayers of a voltage-dependent anion selective channel obtained from *Paramecium* mitochondria. *J Membr Biol* **30:**99-120

Smack DP, and M Colombini (1985) Voltage-dependent channels found in the membrane fraction of corn mitochondria. *Plant Physiol* **79:**1094-1097

Stuhmer W, F Conti, H Suzuki, X Wang, M Noda, N Yahagi, H Kubo and S Numa (1989) Structural parts involved in activation and inactivation of the sodium channel. *Nature* **339:**597-603

Thomas L, E Kocsis, M Colombini, E Erbe, BL Trus, and AC Steven (1991) Surface topography and molecular stoichiometry of the mitochondrial channel, VDAC, in crystalline arrays. *J Struct Biol* **106:**161-171

Thomas L, E Blachly-Dyson, M Colombini, M Forte (1993) Mapping of residues forming the voltage sensor of the voltage-dependent anion-selective channel. *Proc Natl Acad Sci USA* **90:**5446-5449

Zhang DW, and M Colombini (1990) Group IIIA-metal hydroxides indirectly neutralize the voltage sensor of the voltage-dependent mitochondrial channel, VDAC, by interacting with a dynamic binding site. *Biochim Biophys Acta* **1025:**127-134

Zimmerberg J, and VA Parsegian (1986) Polymer inaccessible volume changes during opening and closing of a voltage-dependent ionic channel. *Nature (Lond)* **323:**36-39

INSIGHTS INTO THE STRUCTURE OF THE MITOCHONDRIAL CHANNEL, VDAC, PROVIDED BY ELECTRON MICROSCOPY

Carmen A. Mannella
Wadsworth Center for Laboratories and Research
New York State Department of Health
and Department of Biomedical Sciences
School of Public Health
State University of New York at Albany
Empire State Plaza, Box 509
Albany, NY 12201-0509

Introduction

It has been ten years since publication of the finding that outer membranes isolated from mitochondria of the fungus *Neurospora crassa* may bear periodic arrays of stain-accumulating subunits, which were later identified as the voltage-dependent anion-selective channel, VDAC (Mannella, 1982). In the ensuing years, considerable progress has been made both in controlling the crystallization process and in characterizing the structure of the arrays. The aim of the present report is to summarize information about the structure of the VDAC channel that has been obtained from high-resolution electron microscopy and image analysis of these two-dimensional (2D) crystals.

Relevance of studies of 2D crystals

Proteins may adopt numerous conformational states in solution or in bilayers. There is general consensus that the structures obtained by X-ray

NATO ASI Series, Vol. H 83
Molecular Biology of Mitochondrial Transport Systems
Edited by M. Forte and M. Colombini

diffraction from crystals of water-soluble proteins are "photographs" of the lowest-energy (most likely) states. The same is presumably true for membrane proteins, although high-resolution data are much more limited. Atomic structures of only three different membrane proteins have been solved to-date by x-ray and/or electron crystallography: photosynthetic reaction center (Deisenhofer *et al.*, 1984), bacteriorhodopsin (Henderson *et al.*, 1990), and bacterial porin (Jap *et al.*, 1991; Weiss *et al.*, 1991).

In the case of VDAC, the technique used to induce 2D crystallization (lipid depletion of native membranes by phospholipase A_2 treatment) has no more (and more likely less) potential for denaturing proteins than other commonly used protocols involving detergent extraction and lipid reconstitution. Since our introduction of phospholipase/dialysis with VDAC (Mannella, 1984), the technique subsequently has been used for 2D crystallization of other membrane protein complexes, including Na/K-ATPase (Mohraz *et al.*, 1985), bacterial porin (Massalski *et al.*, 1987), and bacterial flagellar L-P rings (Akibi *et al.*, 1991). In each case, the structure of the membrane protein in phospholipase-induced arrays is consistent with information obtained from other sources. Likewise, the basic structural parameters of the VDAC channel in phospholipase-induced crystals are compatible with those derived from other techniques. For example, center-to-center distances between adjacent VDAC channels in the crystalline arrays (4.3-5.0 nm, see below) are consistent with the mean outer diameter of the prominent in-plane subunits formed by the 31-kDa trypsin-resistant VDAC protein in plant mitochondrial outer membranes (4.5-5 nm), as determined by x-ray scattering (Mannella and Bonner, 1975). Also, the projected width of VDAC's lumen in ordered arrays falls between within the range inferred from different measures of this channel's permeability (see below).

Comparisons with bacterial porins

Bacterial porins, like VDAC, form large pores with a single copy of a 30-40 kDa polypeptide. Thus, it is not surprising that VDAC is also commonly referred to as mitochondrial or eukaryotic porin. While there is no significant homology between available VDAC sequences and those of the porin family (Forte *et al.*, 1987), this does not preclude possible structural relatedness, since numerous examples of "structural homologues" have been found involving proteins with no extensive sequence homology. Electron and X-ray crystallography (Jap *et al.*, 1991; Weiss *et al.*, 1991) have shown recently that the lumens of bacterial porins are cylindrical β-sheets or β-barrels. It has been suggested (Forte *et al.*, 1987; Blachly-Dyson *et al.*, 1990) that VDAC might have the same structural motif, based on the occurrence in its sequences of several runs of alternating polar-nonpolar amino acids, 10-14 residues long. These polypeptide segments might traverse the membrane as strands in a β-barrel, with polar groups lining the lumen interior and nonpolar residues facing the bilayer's hydrophobic interior.

Electron microscopic and image analysis techniques

The environment inside the electron microscope is hostile to biological molecules. Because electrons are scattered by gases, the specimen is kept in a vacuum (typically 10^{-6} torr), where it is exposed to high radiation doses during conventional imaging procedures. Many studies of VDAC crystals have involved specimens that were either air-dried in a heavy-metal salt (negatively stained; Mannella *et al.*, 1986) or freeze-dried and metalshadowed (Thomas *et al.*, 1991). In both cases, the metal coatings provide enhanced contrast as well as resistance to the damaging effects of radiation. In effect, it is the metal cast and not the organic material that is being imaged. While these techniques have provided considerable information about the VDAC crystals, there are legitimate concerns about denaturing effects of dehydration and specimen-metal interactions. Also, the attainable resolution is no better (and usually considerably worse) than 2 nm.

This resolution limit has been overcome in recent years by avoiding the use of metal coatings, and by application of cryo-techniques, in which specimens are fast-frozen in thin layers of vitreous ice and directly imaged in the electron microscope. The low (liquid nitrogen) temperatures also provide a 3-5-fold reduction in radiation damage, compensating in part for the absence of protective metal coats. Images of frozen-hydrated VDAC arrays can be obtained with resolution (after averaging, see below) approaching 1.0 nm (Mannella *et al.*, 1989). Since there is no stain, these projections are true 2D density maps of the protein, lipid, and water distribution in the arrays. However, contrast in the images is very low and technical difficulties (controlling ice thickness, mechanical stability of the cryo-stages) complicate data collection.

The 2-nm barrier also has been overcome with VDAC crystals by embedding in aurothioglucose (a molecule of glucose with one gold atom attached), which yields image resolution in the 1.2-1.3 nm range (Guo and Mannella, 1992). This is a compromise between frozen-hydrated and conventional negative-stain approaches, *i.e.*, the small gold content provides modest contrast enhancement (and essentially no protection against radiation damage), while the hydroxyl-rich sugar provides a water-like environment for biomolecules. However, images obtained with aurothioglucose are primarily projection maps of the gold distribution and not of inherent array density.

Computer-based approaches used to generate averages of projection images of 2D VDAC crystals are described in detail elsewhere (Mannella *et al.*, 1986; Mannella and Guo, 1990). The basic technique is correlation averaging, by which the overlapped crystalline layers in collapsed tubular membranes may be separated and corrected for translational lattice disorder. Three-dimensional reconstruction involves merging correlation-averaged projections from crystals tilted in many different directions with respect to the electron beam. Since radiation damage limits the number of projections that can be recorded from one field, data must be collected from several different crystals.

Geometry of crystalline arrays of VDAC.

Figures 1A,B are correlation averages of untilted projections of the most common crystalline form of *N. crassa* VDAC (called the "oblique" array), embedded in vitreous ice and aurothioglucose, respectively. The pore lumens are white (low density) in Figure 1A because they are filled with water, and black (high density) in Figure 1B because they contain gold. The crystalline array is composed of hexameric groups of pores repeated on a parallelogram lattice with characteristic repeating distances (vectors) of a=13.3 and b=11.5 nm, separated by an angle of 109°. The repeat unit has twofold rotational symmetry, meaning that it will superimpose on itself if rotated by 180°. Other polymorphs of the lattice are also observed, the most common being a "contracted" form in which the b vector is decreased to 10 nm and the lattice angle to 99° (Mannella and Guo, 1990).

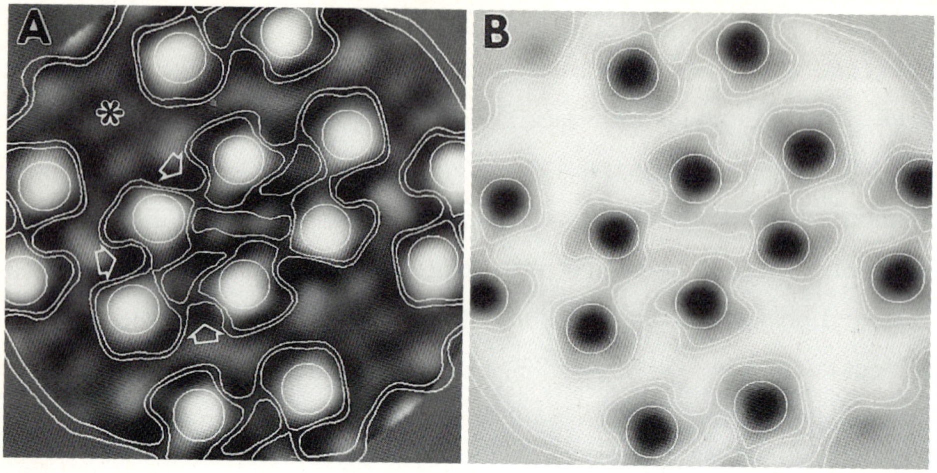

Figure 1 Correlation averages of (A) frozen-hydrated and (B) aurothioglucose-embedded arrays of the VDAC channel. Density contours from (B) are superimposed on (A) to illustrate that gold-glucose is excluded from the densest loci in the latter map (arrows), indicating surface protrusions (see text). The "arms" in one corner region are marked by "*". Bar = 5 nm.

Projected lumen shape.

Figure 2 is a comparison of the parallelogram VDAC array with an orthorhombic 2D array formed by *E. coli* PhoE porin (Jap *et al.* 1991). The actual projected trace of the Cα backbone is shown for porin, whose structure has been solved at atomic resolution. The β-barrels are roughly ellipsoidal in projection, with major and minor axes of 3.8 and 2.8 nm. In the case of VDAC, the resolution attained in projection images is too low to unambiguously trace the carbon backbone. However, modeling studies (Mannella *et al.*, 1989) indicate that projected density maps like Figure 1A may be fit by an ideal (uniform, circular) β-barrel with a diameter of 3.8 +/- 0.1 nm. This diameter is consistent with the functional size of the VDAC pore, which is variously estimated to fall in the range 2-4 nm. Assuming

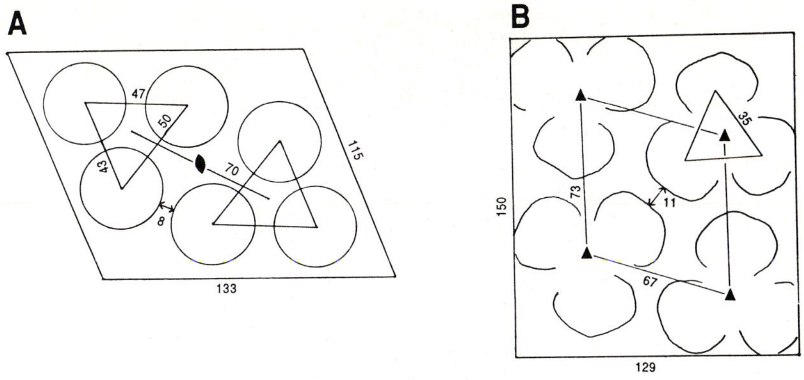

Figure 2 Comparison of the packing of VDAC and PhoE pores in 2D arrays. Distances are indicated in angstroms. (Reproduced from Mannella, 1992b with permission of the American Crystallographic Association.)

that the amino acid residues lining VDAC's putative β-barrel extend about 0.5 nm from the carbon backbone, the cylindrical lumen would have a mean inner diameter of about 2.8 nm. However, there are indications (see below) that the inner diameter of the VDAC pore in some types of crystals is smaller than that expected for a 3.8-nm diameter β-barrel.

While we assumed for simplicity that VDAC's pore is circular in the model computations noted above (and in Figure 2), projection images of the 2D crystals indicate that VDAC's pore is neither circular nor as elliptical as porin's, but rather polygonal in outline. Straight edges and sharp corners are especially evident in contour representations of the projected density map of frozen-hydrated arrays (Figure 3), suggesting that the lumen may have four straight sides. A non-circular pore shape would have important implications for understanding functional properties of VDAC. For example, electrostatic interactions between charged amino-acid residues on a radially nonsymmetric lumen wall and ions inside the

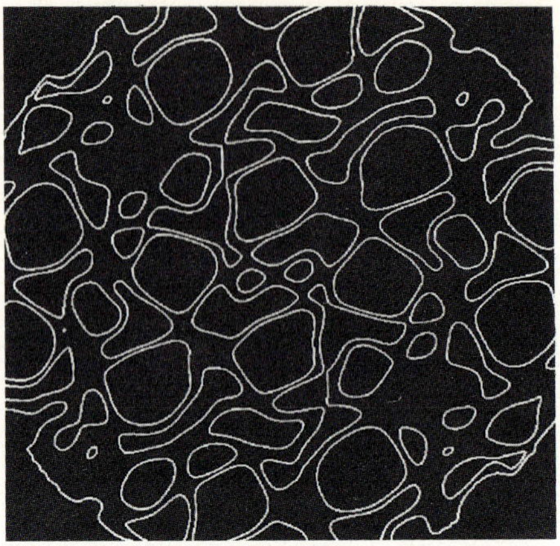

Figure 3 Contour map of the projected density of the frozen-hydrated VDAC array of Figure 1A.

lumen will be affected by residue location, *e.g.*, positioning of a residue in an acute corner may attenuate its effect on ion selectivity.

Height of the lumen wall

Recently completed 3-D reconstruction of the VDAC array embedded in aurothioglucose (Guo *et al.*, 1992) indicates that the main gold-glucose-filled compartment has a height of 3-3.5 nm, which is just sufficient to traverse the hydrophobic interior of a phospholipid bilayer. The height of porin's wall varies from about 4 nm along the outside of the trimer (the region in contact with the bilayer) to only about 2 nm in the interior of the complex (the region of contact between the barrels). Because of this uneven height, the lumens within a porin trimer appear to merge when filled with negative stain. VDAC's lumens show no such tendency to merge in 3-D reconstructions, but there are other indications that the wall defining

VDAC's lumen is not radially uniform in height. In particular, the projected density surrounding the water-filled lumen in the frozen-hydrated array (Figure 1A) has maxima and minima, indicating uneven distribution of mass around the pores. The strongest maxima in Figure 1A coincide with minima in Figure 2B, consistent with the occurrence of aurothioglucose-excluding surface protrusions at these points (arrows in Figure 1A). The image of the aurothioglucose-embedded array also shows numerous subsidiary density maxima on the periphery of the pores. 3-D reconstruction of the aurothioglucose-embedded array indicates that these maxima are superpositions of features on both sides of the membrane crystal, some of which extend well into the interior. Assuming that the density in these projection images arises primarily from the gold in the aurothioglucose, these features are probably narrow aurothioglucose-filled grooves or indentations in the walls of the channel. These would occur, for example, if some transmembrane strands in a β-barrel were shorter than the others.

Lateral protein arms: the N-terminal helix.

Recent immunolabelling experiments have shown that Fab fragments from antibodies directed against the first 20 amino acids of the fungal VDAC polypeptide bind to VDAC crystals in the region between the channel hexamers (*i.e.*, at the corners of the unit cell) (Mannella *et al.*, 1992b). In the image of the frozen-hydrated array (Figure 1A), linear densities (protein "arms") can be seen criss-crossing this same region. Thus, it is likely that the linear densities are the N-terminal region of the VDAC polypeptide, which is predicted to fold as an amphiphilic α-helix (Kleene *et al.*, 1987). In fact, recent computer modeling has shown that the pattern of linear densities in the normal array can be mimicked very closely by six 20-amino-acid-long α-helices, one extending from each of the channels bounding the corner region in the array (Mannella, 1992a). This model is consistent with the known tendency of amphiphilic α-helices to extend laterally near the surface of phospholipid bilayers, at the interface between the acyl-chain and polar-headgroup regions (Tamm *et al.*, 1991).

The corner regions of the VDAC crystalline array have been found to be "sticky": both cytochrome c and the targetting sequence of subunit IV of cytochrome oxidase bind there (Mannella *et al.*, 1992a). This is consistent with identity of the arms in this region as amphiphilic α-helices, which tend to form stable complexes with other such helices (*e.g.*, Cox *et al.*, 1985). Mitochondrial targetting sequences probably fold as amphiphilic α-helices (Roise *et al.*, 1986), while cytochrome c contains amphiphilic α-helices at both its N- and C-termini that are implicated in mitochondrial binding (Jordi *et al.*, 1989). Whether there is any physiological significance to the binding of import polypeptides to the VDAC protein is unclear at this time. If the N-termini of VDAC proteins were exposed on the cytosol-facing side of the mitochondrial outer membrane, it is almost inevitable that precursors would attach to them, given the large surface density of VDAC (10^3 to 10^4 per square μm).

Crystal polymorphism: influence of channel effectors

Two kinds of systematic structural variations occur in 2D crystals of fungal VDAC. The first variation, noted above, is in lattice geometry, *i.e.*, contraction of the arrays. The second variation, detected by multivariate statistical analysis of images of aurothioglucose-embedded arrays, is in the projected diameter of the gold-glucose accumulation at each pore, which can vary from 2.2 to 1.4 nm (Guo and Mannella, 1992).

Both kinds of structural variations are observed in "control" VDAC arrays, but their occurrence is strongly influenced by exposure of the membrane crystals to chemical effectors of the channel (Guo and Mannella, 1992). For example, the polyanion that reduces VDAC's gating voltage favors crystals with smaller gold-glucose accumulations, while aluminum and high pH, which inhibit VDAC's voltage-dependent closure, favor larger accumulations in the pores. This suggests that the structural state of the channels in the crystals (wide *vs.* narrow pores) may correlate with the expected functional state induced by the effectors (open *vs.* closed). That both wide and narrow pore diameters are found normally in the crystals may reflect the low energy difference (less than one hydrogen bond) between the open and closed states (Bowen *et al.*, 1985).

The modulator polyanion also triggers contraction of the VDAC lattice, which involves shrinkage of the corner regions in the crystal's unit cell that is normally occupied by the lateral protein arms. This suggests that lattice contraction involves a conformational change in the VDAC protein, *i.e.*, movement of this part of the channel protein (the N-terminal α-helix) off the bilayer.

Implications for the mechanism of gating.

The effector-associated changes in the structure of VDAC crystals have important implications for the kinds of conformational changes that the channel may undergo (Mannella, 1990). That the modulator polyanion triggers lattice contraction suggests that it causes detachment of the N-terminal helix from the bilayer. This effect could be due to the amphiphilic nature of the polyanion; another macromolecule that has been found to induce contraction is the targetting peptide of cytochrome oxidase subunit IV (Mannella *et al.*, 1992a), an amphiphilic helix that binds to the arm region. It is possible that the lowest energy state of the channel is the conformation with N-terminal arms interacting with the bilayer. Anything that causes the arm to be displaced from the bilayer might destabilize this open state of the channel, leading to reduced gating voltage and easier closure.

That both wide and narrow gold-glucose-filled pores are observed in arrays with the same "oblique" (*i.e.*, not contracted) lattice parameters suggests that the inner diameter of the channel can change significantly without altering the packing of the channels. This suggests that the underlying conformational change does not involve a significant rearrangement of the outside of the lumen. One way that the pore's inner diameter might be decreased without affecting its outer diameter is by entry of external parts of the protein into the lumen. The N-terminal arm, once detached from the bilayer, might be one such moveable domain but there could well be others. In the case of porin, the β-barrel lumen is occupied by a 44-amino-acid loop of the protein (Weiss *et al.*, 1991) which may be responsible for reducing porin's permeability below that expected for an "empty" cylinder of the same dimensions.

Acknowledgements

This work was supported by a research grant from the National Science Foundation (DMB-8916315) and by instrumentation grants from the NSF (PCM-8313045, DIR-8914757) and the National Institutes of Health (1S15-GM4344301).

References

Akiba T, Yoshimura H, Namba K (1991) Monolayer crystallization of flagellar L-P rings by sequential addition and depletion of lipid. Science 252:1544-1546

Blachly-Dyson E, Peng S Z, Colombini M, Forte M (1990) Selectivity changes in site-directed mutants of the VDAC ion channel: structural implications. Science 247:1233-1236

Bowen K A, Tam K, Colombini M (1985) Evidence for titratable gating charges controlling the voltage dependence of the outer mitochondrial membrane channel, VDAC. J Membr Biol 86:51-59

Cox J A, Conte M, Fitton J E, DeGrado W F (1985) The interaction of calmodulin with amphiphilic peptides. J Biol Chem 260:2527-2534

Deisenhofer J, Epp O, Miki K, Huber R, Michel H (1984) X-ray structure analysis of a membrane protein complex. Electron density map at 3A resolution and a model of the chromophores of the photosynthetic reaction center from Rhodopseudomonas viridis. J Mol Biol 180:385-398

Forte M, Guy H R, Mannella C A (1987) Molecular genetics of the VDAC ion channel: structural model and sequence analysis. J Bioenerg Biomembr 19:341-350

Guo X W, Mannella C A (1992) Classification of projection images of crystalline arrays of the mitochondrial channel, VDAC, embedded in aurothioglucose. Biophys J 63, 418-427

Guo X W, Smith P R, Radermacher M, Mannella C A (1992) Structure of crystalline VDAC, the voltage-gated channel in the mitochondrial outer membrane. In: Proc. 50th Annual Meeting of the Electron Microscopy Society of America. Bailey G W, Bentley J, Small J A (eds) San Francisco Press, Inc. San Francisco, p. 440-441

Henderson R, Baldwin J M, Ceska T A, Zemlin F, Beckmann E, Downing K H (1990) Model for the structure of bacteriorhodopsin based on high-resolution electron cryo-microscopy. J Mol Biol 213:899-929

Jap B K, Walian P J, Gehring K (1991) Structural architecture of an outer membrane channel as determinined by electron crystallography. Nature 350:167-170

Jordi W, De Kruijff B, Marsh D (1989) Specificity of the interaction of amino- and carboxy-terminal fragments of the mitochondrial precursor protein apocytochrome c with negatively charged phospholipids. A spin-label electron spin resonance study. Biochemistry 28:8998-9005

Kleene R, Pfanner N, Pfaller R, Link T, Sebald W, Neupert W, Tropschug M (1987) Mitochondrial porin of *Neurospora crassa*: cDNA cloning, in vitro expression and import into mitochondria. EMBO J 6:2627-2633

Mannella C A (1982) Structure of the outer mitochondrial membrane: ordered arrays of porelike subunits in outer-membrane fractions from *Neurospora crassa* mitochondria. J Cell Biol 94:680-687

Mannella C A (1984) Phospholipase-induced crystallization of channels in mitochondrial outer membranes. Science 224:165-166

Mannella C A (1992a) The ins and outs of mitochondrial membrane channels. Trends Biochem Sci, 17:315-320

Mannella C A (1992b) Model for a gated mitochondrial channel based on electron microscopy of 2D crystal polymorphs. Trans Am Crystal Assn, in press

Mannella C A, Bonner W D,Jr. (1975) X-ray diffraction from oriented outer mitochondrial membranes: Detection of in-plane subunit structure. Biochim Biophys Acta 413:226-233

Mannella C A, Guo X W (1990) Interaction between the VDAC channel and a polyanionic effector. An electron microscopic study. Biophys J 57:23-31

Mannella C A, Guo X W, Cognon B (1989) Diameter of the mitochondrial outer membrane channel: evidence from electron microscopy of frozen-hydrated membrane crystals. FEBS Lett 253:231-234

Mannella C A, Guo X W, Dias J (1992a) Binding of a synthetic targeting peptide to a mitochondrial channel protein. J Bioenerg Biomembr 24:55-61

Mannella C A, Ribeiro A, Frank J (1986) Structure of the channels in the outer mitochondrial membrane. Biophys J 49:307-318

Mannella C A, Stanley S, D'Arcangelis D, Dias J A (1992b) Location of the N-terminal region of the channel protein, VDAC, in the outer membrane of *Neurospora* mitochondria. Biophys J 61:A14

Massalski A, Sass H J, Zemlin F, Beckmann E, van Heel M, Buldt G, Dorset D L, Zeitler E, Rosenbusch J P (1987) High-resolution, low-dose electron cryomicroscopy of negatively stained matrix porin, a transmembrane protein from *Escherichia coli* outer membranes. In: Proc. 45th Annual Meeting of the Electron Microscopy Society of America. Bailey G W(ed) San Francisco Press, Inc. San Francisco, p. 788-789

Mohraz M, Yee M, Smith P R (1985) Novel crystalline sheets of Na,K-ATPase induced by phospholipase A_2. J Ultrastruct Res 93:17-26

Roise D, Horvath S J, Tomich J M, Richards J H, Schatz G (1986) A chemically synthesized pre-sequence of an imported mitochondrial protein can form an amphiphilic helix and perturb natural and artificial phospholipid bilayers. EMBO J 5:1327-1334

Tamm L K (1991) Membrane insertion and lateral mobility of synthetic amphiphilic signal peptides in lipid model membranes. Biochim Biophys Acta 1071:123-148

Thomas L, Kocsis E, Colombini M, Erbe E, Trus B L, Steven A C (1991) Surface topography and molecular stoichiometry of the mitochondrial channel, VDAC, in crystalline arrays. J Struct Biol 106:161-171

Weiss M S, Kreusch A, Schiltz E, Nestel U, Welte W, Weckesser J, Schulz G E (1991) The structure of porin from *Rhodobacter capsulata* at 1.8 A resolution. FEBS Lett 280:379-382

EXPERIMENTAL SUPPORTS TO A SIXTEEN-STRANDS MODEL OF TRANSMEMBRANE ARRANGEMENT OF MITOCHONDRIAL PORIN AND PRELIMINARY RESULTS CONCERNING A MULTIGENE FAMILY IN *Drosophila melanogaster* RELATED TO HUMAN MITOCHONDRIAL PORIN

Vito De Pinto[1], Ruggiero Caizzi[2], Jalal Ahmad Aljamal[1], Corrado Caggese[2] and Ferdinando Palmieri[1]

[1]Department of Pharmaco-Biology, Laboratory of Biochemistry and Molecular Biology, University of Bari, Bari, Italy
and
[2]Institute of Genetics, University of Bari, Bari, Italy

Introduction

Mitochondrial porin or VDAC (voltage-dependent anion channel) is the intracellular aqueous pore which allows the very high permeability of the outer mitochondrial membrane (Colombini,1979). It was shown to be present in every tissue investigated and in many organisms with approximately the same functional features, the same molecular size and a similar quantitative expression (De Pinto et al,1987). The primary structure of porins from various sources has been determined. A great deal of interest is nowadays focussed in the structural arrangement of the polypeptide chain(s) of porin and its/their role in the properties of the channel. In the first part of this paper we give further experimental supports to a sixteen-stranded beta-barrel model of the transmembrane topology of the mammalian mitochondrial

NATO ASI Series, Vol. H 83
Molecular Biology of Mitochondrial Transport Systems
Edited by M. Forte and M. Colombini

porin we have previously published (De Pinto et al,1991b). The results reported here concern with the localization of the hydrophobic region responsible for the DCCD-binding and the characterization of porin cysteines with respect to their position in the channel unit. At the end our model and other models for mitochondrial porin appeared in the literature are discussed and compared to the bacterial porin structures determined by crystal analysis.

In the second part of this paper we were concerned with the working hypothesis that more porin-related genes could justify the growing evidences for the molteplicity of roles attributed to the mammalian porins (Mannella, 1992, Thinnes, 1992, Gelb et al,1992, McEnery, 1992, Blachly-Dyson et al,1992). This hypothesis is currently investigated in the genome of *Drosophila melanogaster*, an organism very convenient for its relative simplicity and for the genetic facilities available.

1.1 Porin cysteines are localized in the channel wall, in an environment accessible both to hydrophilic and hydrophobic reagents

Porin from bovine heart mitochondria contains one methionine (position 154), and two cysteines (position 126 and 231) (Kayser et al,1989 and unpublished results). Two peptides, each containing one cysteine, can easily be obtained by CNBr cleavage. This allowed us to clearly spot the chemical reactivity of each cysteine. The polarity of the environment in which each cysteine is located was exploited by means of maleimides with residues of opposite polarity. The reaction products of maleimides with sulfhydryl groups in protein are very stable, which means that it is possible to localize sulfhydryls after labelling with radioactive or fluorescent maleimides. We used eosin-5-maleimide, which is fluorescent and impermeable to membranes and N-(1-pyrenyl)-maleimide, which is also fluorescent but it is highly hydrophobic. The labelling was performed in intact mitochondria. After labelling porin was isolated from the mitochondria and the protein fraction was run on SDS-gel electrophoresis. The fluorescent bands were cut out of the gel and the gel slices were exposed to CNBr cleavage. Surprisingly, both CNBr peptides were fluorescent when porin was labelled either with eosin-5-maleimide or with N-(1-pyrenyl)-maleimide. This means that both cysteines are in an environment accessible either to a large hydrophilic or to a strongly hydrophobic reagent (De Pinto et al,1991a).

The porin-channel contains two possible localizations for water-accessible amino acid residues: the external surface, surrounding the mouth of the pore, and the walls of the pore itself, facing the water inside the channel. Only the latter possibility is allowed to amino acid residues able to react both with hydrophobic and with hydrophilic reagents. To confirm this conclusion we used commercially available chromatographic matrices (Affi-Gel 501 and Thiopropyl-Sepharose 6 B) as probes for -SH groups. Both are specific for free -SH groups but have spacer arms with different lengths (0.51 and 1.75 nm). The LDAO-porin was bound only by the material with the longer arm (De Pinto et al,1991a). Consequently it may be suggested that at least one of the two cysteines is localized between 0.51 and 1.75 nm away from the surface of the protein-detergent micelle. The other -SH group should have a similar localization or it is located even deeper in the protein-detergent micelle. As both sulphydryl groups are labelled by eosin-5-maleimide they may face the water phase anywhere inside of the channel. In our model the localization of the two cysteines is exactly as expected by these last experimental results: they are in the seventh and thirteenth transmembrane beta strands, thus in the region constituting the walls of the beta-barrel (Fig. 1C).

1.2 Localization of the dicyclohexylcarbodiimide-reactive residue in porin.

Hexokinase is one of the rate limiting enzymes of glycolisis. In the cell it was shown that hexokinase can be free in the cytosol or bound to the outer mitochondrial membrane. The binding of hexokinase to mitochondria is assumed to be important in the regulation of glucose metabolism. DCCD is a potent inhibitor of the transmembrane proton conductivity catalyzed by the F_OF_1 ATPase complex. Moreover, treatment of intact mitochondria with DCCD resulted in inhibition of their ability to bind hexokinase (Nakashima et al,1986). Since mitochondrial porin was demonstrated to be identical with the mitochondrial hexokinase-receptor, the DCCD inhibition of hexokinase binding to mitochondria has been attributed to modification of porin (Nakashima et al,1986). The identification of the amino acid residue(s) modified by DCCD can thus give new insights into the elucidation of the interaction between porin and hexokinase.

The localization of the DCCD-binding residue in the primary structure of porin was performed by amino acid sequencing of peptides obtained from porin purified from [^{14}C]DCCD labelled mitochondria. Many peptides were obtained after cleaving porin with CNBr, trypsin and *Staphilococcus aureus* V8 protease. Their N-terminal sequence and the

^{14}C-radioactivity content were determined. At the end, the region containing the [^{14}C]DCCD labelled residue(s) was confined to the sequence 40-88. In this part of the sequence there are six glutamic acid and one aspartic acid residues (Kayser et al,1989). We thus sequenced the whole region. Aliquots of the PTH-derivatives released at each cycle of the sequence were analysed for ^{14}C-radioactivity. The [^{14}C]DCCD radioactivity was released at position 72. No significant radioactivity was found to be associated to the other cycles of the Edman degradation. In order to identify the amino acid present at position 72, two peptides encompassing this residue, obtained by cleaving the unlabelled porin, were sequenced and in this way the presence of glutamic acid at position 72 was shown. We could thus conclude that there is only one DCCD-modified residue in the sequence of mitochondrial porin and this is identified as glutamic acid 72 (De Pinto et al,1993).

In our topological model the hydrophobic DCCD-reactive residue Glu 72 is localized inside the membrane, in the fourth of the sixteen amphipatic beta-strands which cross the membrane (Fig. 1C). This is thus a further experimental result which agrees with the proposed model.

1.3 Transmembrane arrangement of bacterial and eukaryotic porins

Fig. 1A and 1B show the topologies of *Rhodobacter capsulatus* and *E. coli* OmpF porin monomers obtained by X-ray diffraction analysis (Weiss et al,1991, Cowan et al,1992). The two bacterial porins are structurally very similar. They are composed of three subunits, but the channel-forming subunit of both porins is a 16-stranded anti-parallel beta-barrel. These 16 beta strands have a length ranging from 6 to 17 residues. In the OmpF porin the amino and carboxy termini are linked by a salt-bridge within the 16th strand. The tilt of the strands relative to the barrel axis varies between 30° and 60°. The beta-strands are connected by loops. One end of the barrel (the smooth end) is defined by short loops (two to five residue) whereas the "rough end" contains longer irregular loops. One of the long loops folds into the barrel and constricts the inner diameter of the pore at about half the height of the barrel. There is an electrostatic asymmetry between the two sides of the pore which should have important functional consequences.

Fig. 1C shows the model of transmembrane arrangement of mammalian mitochondrial porin we have proposed (De Pinto et al,1991b). It was based i) on computer prediction analysis of the secondary structure of the sequence of the protein and ii) on experimental results, namely the accessibility of water-exposed protein domains to the action of specific proteases and of antibodies against

---

Fig. 1. A. The sequence of porin from *R. capsulatus* displayed in the 16-stranded beta-barrel established by X-ray diffraction analysis. The transmembrane arrangement of the 16-stranded beta-barrel is viewed from the barrell outside. The dashed lines indicate the hydrogen-bonding scheme between the beta-strands. Appropriate marks indicate the secondary structure assignments (quadrangle= b; rectangle= a; triangle= hydrogen bond reverse turn). Reproduced with permission from (Weiss et al,1991).
B. Topology of *E. coli* OmpF porin sequence obtained by X-ray analysis. The view is from outside the unrolled 16 beta-strands barrel. Secondary structure elements (bold if their side-chains are external) as above. Reproduced with permission from (Cowan et al,1992).
C Model of the transmembrane arrangement of the polypeptide chain of porin from mammalian mitochondria. The starting amphipathic alpha-helix lies on the external surface of the membrane. Sixteen amphipathic beta-strands cross the membrane. The arrows indicate the putative cleavage sites of protease V 8, trypsin and chymotrypsin and the localization of Glu 72 and cysteines (in boxes).

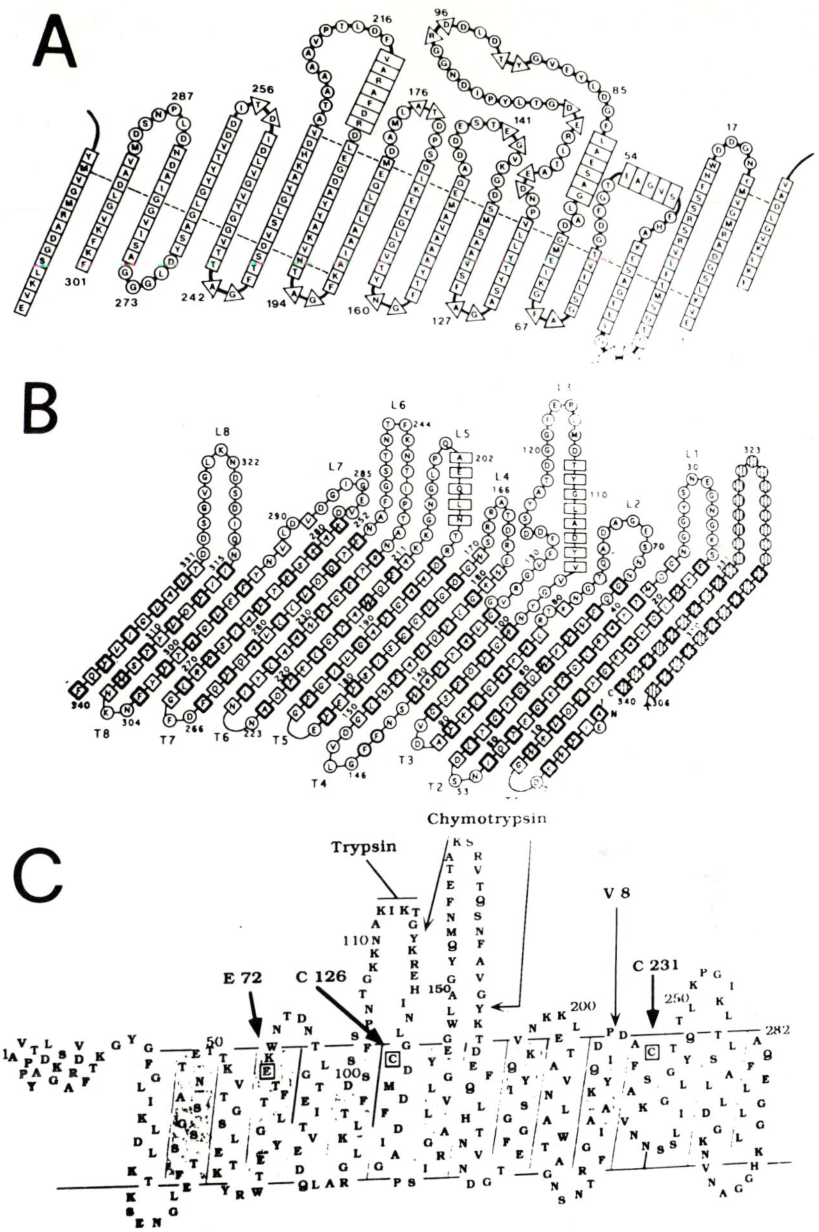
A
B
C
L8
L7
L6
L5
L4
L2
L1
T8
T7
T6
T5
T4
T3
T2
Chymotrypsin
Trypsin
V 8
E 72
C 126
C 231
282

synthetic peptides with the same sequence of the protein. In this paper we have reported further experimental results which support the model drawn by us, and exactly iii) the localization of porin cysteines in the channel-walls, an environment accessible to both hydrophilic and hydrophobic reagents (De Pinto et al,1991a) and iv) the localization of a highly hydrophobic DCCD-binding site inside the fourth transmembrane beta strand (De Pinto et al,1993). Our topological model has several similarities with the bacterial porin structures described above. The pore would be formed by 16 beta-strands connected by loops: on one side all the loops are short, while on the other side the loops are longer and irregular. An asymmetryc distribution of charged residues is present also in this model, even if with an opposite sign. No tilt of the strands is indicated in the figure, but the number of 16 strands and EM and STEM observations agree with a monomeric beta-barrel structure (Mannella et al,1989, Thomas et al,1991).

The main difference between the bacterial and mammalian porin is the amphipathic alpha-helix located at the beginning of the mitochondrial sequence. This alpha-helix looks very conserved and likely is a distinctive structural feature of the eukaryotic porin (Kleene et al,1987). In other papers 12, 15 or 19 beta-strands were predicted from the analysis of yeast and *N. crassa* mitochondrial porin. In particular, a model of transmembrane arrangement was proposed after site-directed mutagenesis experiments (Blachly-Dyson et al,1990). In this model 12 beta-strands and the amino-terminal alpha-helix cross the membrane. This 12-strands model omits large part of the sequence from the membrane since mutations in this part did not alter the channel behaviour of the corresponding mutant-proteins. Comparison with our model shows that important residues are also within the membrane in mammalian porin, but most of them are not conserved although the channel properties seem to be identical.

One of the major arguments against the similarity between mitochondrial and bacterial porins is that they have nothing in common in their sequence. It was recently shown that, according to sequences alone, bacterial and eukaryotic porins can be classified so far into ten groups: if porins from different groups are compared, identities of 13-14% are obtained (Schiltz et al,1987). Since porins appear a typical example of a function preserved during evolution in spite of dramatic changes in the primary structure, we believe reasonable to hypothesize that the sixteen-stranded beta-barrel is the secondary structure motif that can be in common to most, if not all, porins.

2.1 Fishing porin-related *D.melanogaster* DNAs.

We have looked for porin-related genes in the genome of D. melanogaster, the fruit fly largely studied by genetists, with an EcoRI/EcoRI fragment of the B-lymphocite human porin cDNA (HVDAC1; a kind gift of Mike Forte, Portland, Oregon), starting from the 5' end, 633 bp long and containing about two third of the coding frame. The rest of the cDNA, about 1.2 kbp in size, containing the 3' end of the coding frame and a very long uncoding tail, was seldom used for comparison.

With the 0.6 kbp probe we performed a Southern analysis and screened a genomic library of *D. melanogaster*. Figure 2 shows a Southern blot of bovine and fly genomic DNAs. The two panels of hydrization refer to the different probes used: 0.6 kbp in panel B and 1.2 kbp in panel C.

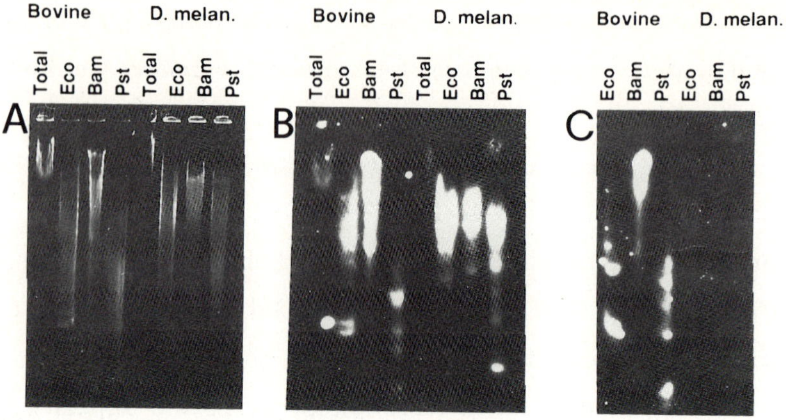

Figure 2. Southern blots of genomic DNA from *D. melanogaster* and bovine heart probed by HVDAC1 cDNA 0.6 kbp (panel B) and HVDAC1 cDNA 1.2 kbp (panel C). For further details see text.

Hybridization was performed at 55°C and washings at 55°C and 0.5 X SSC. We observed several and very strong signals not only in the bovine panels, as it was expected, but also in the fly panel hybridized with the 0.6 kbp probe. The presence of several hybridizing bands was a clear indication that several genes were identified by the probe. A Northern analysis on purified mRNAs at different developmental stages was then performed, but the human probe gave signals very faint (not shown). We thus decided to screen a genomic library. The hybridization conditions were the same described for the Southern analysis: we were again astonished by the large number of positive spots detected. Twenty positive clones were randomly choosen and their genomic inserts analysed. DNA fragments identified by hybridization with the human probe were then subcloned.

Most of the cloned genomic fragments were used as probes for Northern blots analysis using mRNA purified from the different developmental stages of the insect. In addition, the same fragments were used to isolate the corresponding cDNA from opportunely selected cDNA libraries and to look for their chromosomal localization by *in situ* hybridization. Since each genomic fragment showed more transcripts in different developmental stages (see below) the work soon appeared very large and complex and it is far to be completed. We'll describe the results obtained until now.

2.2 Northern analysis

Polyadenylated RNA from various stages of embryonic, larval, pupal and adult development was subjected to Northern blot analysis using six different genomic fragments as a probe (fragments called: A5, B1, B3, C1, D1, D3). The hybridization and washing conditions were at high stringency (65°C, 0.5x SSC). The results showed three main possible patterns. In the first case a single, possibly constitutive band was detected in every developmental stage (i.e. fragment D1). The second group showed a single, major transcription product at a definite developmental stage. This was for example the case of B1 and C1 whose signals, almost identical, were very strong and present at the same late pupal stage, the key stadium in the tissue differentiation of the fruit fly. The third group, the most frequent, showed a mixed pattern with some stage-specific and some constitutive messages (A5, B3 and D3). We'd like to point out that every genomic fragment fished by HVDAC1 cDNA 0.6 kbp has one or more transcripts, thus ruling out the possibility of being untranslated sequences.

2.3 Chromosomal localization and characterization of genomic fragments.

Reciprocal cross hybridization and restriction maps were performed to group the DNAs. The location of nine inserts on the *D. melanogaster* polytenic chromosomes was determined by *in situ* hybridization.

The genomic fragments under investigation mapped in the chromosomes 2, 3 and X. In the chromosome X region 14BC the genomic DNAs B3 and A5 mapped. In this region several loci encoding for genes of importance in the neurological development (*para, disco*) are localized. B3, a genomic 1 kbp DNA, was completely sequenced. Surprisingly its 5' starting 200 bp were present in a pen repeat sequence already in the EMBL database (clone p19, Haynes et al,1987). The small fragment A3, mapping at the chromosome 3R region 100DE, was also sequenced and was identical with the GCR-1 (Flavell et al,1987). p19 and GCR-1 belong to the group of *Drosophila* DNA termed pen repeats (Haynes et al,1987) and GC-rich sequences (Flavell et al,1987) which have in common an unusually high content of GGN triplets, where N represents any nucleotide and was found to occurr as a C with the highest frequency. These sequences are short, middle repetitive and are found dispersed in the fly genome. In the many clones selected the only regions common consisted of clusters of the triplets GGN, relatively short and interspersed with other nucleotides (Haynes et al,1987, Flavell et al,1987). The pen repeats and GC-rich sequences are transcribed, since their hybridization to Northern blots show complex transcript patterns. Some of the sequences analyzed probably encode proteins, since large ORFs include pen repeat (Haynes et al,1987). These predicted proteins show an extremely high glycine content. The pen repeat are not transposable elements, because they are present in the same genomic location in different strains. Therefore pen repeats and GC-rich sequences could be similar to repetitious sequences like the *opa* repeat, contained in several genes, which specifies the development of spatial pattern and the nervous system of *D. melanogaster*, or the homeobox. Haynes et al. hypotesized that pen repeats are a sequence motif: when transcribed they would encode for glycine-rich domains whose role could be to provide flexible loops or regions of tight polypeptide chains packing (Haynes et al,1987).

The fragments A2, C3 and D3 mapped in the almost uninvestigated region 47DE in the chromosome 2R. Preliminary sequencing obtained from D3 gave data unknown in the databases. D1 mapped at chromosome X region 18A and D4 at chromosome 3R region 92EF, both regions almost completely unknown.

C1 mapped in the chromosome X at 1B, a very well-known region where the locus ASC is located. ASC codes for the *Achaete-scute Complex*, a large complex of 100 kbp in which at least 9 trascription units have been identified. It is one of the best studied example of developmental regulation of a neurological structure, i.e. the cuticular sensory organs (Campunzano & Modolell, 1992). Interestingly B1, which mapped in the chromosome 3R at 97DE, a region where is located the neurogenic locus *Beaded*, showed a pattern on Northern blots almost identical to that typical of C1, namely a very strong signal at the late pupae stage (see above).It was suggestive to find that most of the regions where our genomic fragments mapped had relationships with neurological important loci.

```
B 1  ATGGTTACACCTCTGGCGGCTCCGGCGGCGGTTTCGGCGGAGGTGGCAGCATTGGCGGATTCGGAGGCGGCGGCGGTGTG    80
0.6  ---GCCGC-TCGCT--CGGCTCCGCTCCCTGGCTCGGC-----TCCCTGCCTCCGCG---TCGCAGCCCCCGCCG-T---    63
        *   *  * **  ********    * *  *****     *  * ** *  ***   *** ** *  ** ** *

B 1  TGGTCGAGCCACCTTGTGCAGAAGCACATCTACGTGCACGTGCCACCACCAGAGCAGGAAGAAGTGCGCAGCGCCCTAACC 161
0.6  ------AGCCGCCTC-CG----AGCCCGCC---GC-CACAT-CC----TCTGA----GAAGA--TG-GCTGTGCC---ACC 115
           **** ***   *    *** *  *   *  *** * **     * **    *****  ** ** * ***   ***

B 1 TGCCCATTGGCCAGTCCCAGAAGCACTACCAAGATCA-TCTTCATCAAGGCCCCATCGCCACCATCGTACCAGGCTCCGGT  241
0.6 CACGTAT--GCCGATCTTGGCAAATCTGCCCAGGGATGTCTTCACCAAGGGCT-ATGG-----ATTTG-----GCT----T  179
      *  **  ***  **   * *   ** ** **     ****** ***** *  ** *     **        ***    *
```

Figure 3. Sequence comparison between 0.6, part of the human VDAC cDNA, and the 1.05 kbp cDNA fished by means of B1 as a probe. Underlined are triplets starting with GG or GC.

2.4 cDNA libraries screenings.

Four genomic fragments have been used for screenings of cDNA libraries obtained from mRNA of *D.melanogaster* at the appropriate developmental stage. The most advanced analysis was performed on a group of cDNA fished in a late pupae library by means of B1 as a probe. 15 cDNA of various sizes have been subcloned. They were all part of the same messenger. The longest cDNA (about 1.05 kbp) was sequenced.: Unfortunately it does not contain the whole message, which should be about 500 bp longer. The nucleotide sequence and three alternative ORFs of about 200, 125 and 100 amino acids did not fish any porin-related sequence in the databases. A comparison between the 5' termini of this cDNA and the 0.6 kbp cDNA probe obtained from human VDAC, is shown in Fig. 3. A certain degree of homology is apparent between the sequences, enough to account for their cross-hybridization. Part of the homology is confined to the pre-coding region of the human VDAC (up to base 99). The sequences shown in Fig. 3 have several GGN or GCN triplets.

2.5 What role for GC-rich sequences?

When this work begun, we were intrigued by the large number of *D. melanogaster* sequences recognized by the human porin cDNA. It could just be due to hybridization conditions, not stringent enough to select the right clones. On the other hand the large evolutionary distance between the fly and human justified the use of medium stringency conditions.

The picture we have now, after one year of work is still puzzling. We were not yet able to find the *D. melanogaster* fragment homologous to the human porin, which was our primary goal. This will probably need more efforts and possibly screenings of cDNA libraries with more probes (HVDAC1 and HVDAC2, for example; see Blachly-Dyson et al,1992). Anyway the results presented here are in progress and we do not yet know if more information, more pertinent to porin will come out. The presence of GC-rich sequences in all the fragments analyzed until now looks the connection among HVDAC1 0.6 kbp and the genomic and cDNA sequences under investigation. Also the sequence of a putative mammalian mitochondrial carrier associated with Grave's disease is unusually rich in GCN triplets, expecially in the 5' terminus of the message (Fiermonte et al,1992). Role and importance of them are still to be recognized. Since pen repeats have been demonstrated to be part of polypeptides, a suggestion for the similarity of these sequences with the 5'-end untranslated HVDAC1 cDNA is that also in mammals these repeated sequence can be part of other transcription units.

References

Blachly-Dyson, E., Peng, S., Colombini, M. and Forte, M. (1990) Science, 247, 1233-1236

Blachly-Dyson, E., Zambronicz, E.B., YU, W.H., Adams, V., McCabe, E.R.B., Adelman, J., Colombini, M. and Forte, M. (1992) in NATO ARW on "Molecular Biology of Mitochondrial Transport Systems" Abstract Book, p. 39

Colombini, M. (1979) Nature 279, 643-645

Cowan, S.W., Schirmer, T., Rummel, G., Steiert, M., Ghosh, R., Pauptit, R.A., Jansonius, J.N. and Rosenbusch, J.P. (1992) Nature 358, 727-733

De Pinto, V., Ludwig, O., Krause, J. Benz, R. and Palmieri, F. (1987) Biochim. Biophys. Acta 894, 109-119

De Pinto, V., Aljamal, J.A., Benz, R., Genchi, G. and Palmieri, F. (1991a) Eur. J. Biochem. 202, 903-911

De Pinto, V., Prezioso, G., Thinnes, F., Link, T.A. and Palmieri, F. (1991b) Biochemistry, 30, 10191-10200

De Pinto, V., Aljamal, J.A. and Palmieri, F. (1993) J. Biol. Chem., in press

Fiermonte, G., Runswick, M.J., Walker, J.E. and Palmieri, F. (1992) DNA Sequence 3, 71-78

Flavell, A.J., Dyson, J. and Ish-Horowicz, D. (1987) Nuc. Ac. Res. 15, Campunzano, S. and Modolell, J. (1992) T. Genet. 8, 202-208

Gelb, B.D., Adams, V., Jones, S.N., Griffin, L.D., MacGregor, G.R. and McCabe E.R.B. (1992) Proc. Natl. Acad. Sci. USA 89, 202-206

Haynes, S.R., Rebbert, M.L., Mozer, B.A., Forquignon, F. and Dawid, I.B. (1987) Proc. Natl. Acad. Sci. USA 84, 1819-1823

Kayser, H., Kratzin, H.D., Thinnes, F.P., Götz, H., Schmidt, W.E., Eckart, K. and Hilschmann, N. (1989) Hoppe-Seyler 370, 1265-1278

Kleene, R., Pfanner, N., Pfaller, R., Link, T.A., Sebald, W. Neupert, W. and Tropschung, M. (1987) EMBO J. 6, 2627-2633

Mannella, C.A., Guo, X.-W. and Cognon, C. (1989) FEBS Lett. 253, 231-234

Mannella, C.A. (1992) Trend Biochem. Sci. 17, 315-320

McEnery, M.W. (1992) J. Bioenerg. Biomembranes 24, 63-69

Nakashima, R.A., Mangan, P.S., Colombini, M. and Pedersen, P.L. (1986) Biochemistry 25, 1015-1021

Schiltz, E., Kreusch, A., Nestel, U. and Schulz, G.E. (1991) Eur. J. Biochem. 199, 587-59420. Haynes, S.R., Rebbert, M.L., Mozer, B.A., Forquignon, F. and Dawid, I.B. (1987) Proc. Natl. Acad. Sci. USA 84, 1819-1823

Thinnes, F. (1992) J. Bioenerg. Biomembranes 24, 63-69

Thomas, L., Kocsis, E., Colombini, M., Erbe, E., Trus, B.L. and Steven, A.C. (1991) J. Struct. Biol. 106, 161-171

Weiss, M.S., Kreusch, A., Schiltz, E., Nestel, U., Welte, W. Weckesser, J. and Schulz, G.E. (1991) FEBS Lett. 280, 379-382

STRUCTURE AND FUNCTION OF THE VDAC ION CHANNEL

Marco Colombini
Department of Zoology
University of Maryland
College Park, MD 20742
U.S.A.

Introduction

Mitochondria function within the cytoplasm of eukaryotic cells separated from the cytoplasm by a membranous barrier, the mitochondrial outer membrane. Permeability pathways in this membrane, such as the VDAC channels, can be viewed both as a means by which metabolites may travel from one compartment to the other and as a means of controlling the flux of such metabolites. Both of these functions depend on the properties of the permeating pathways, such as the VDAC channels. By examining in detail the properties of VDAC one should obtain insights into its role in mitochondrial function.

The need for mitochondrial regulation may have its roots in the origin of mitochondria. Primordial bacteria that survived the phagocytic process could only be useful to the cell if useful products could be harvested and the growth and reproduction of the bacterium could be controlled. The ADP/ATP exchanger allows ATP to travel from the bacterial cytoplasm (mitochondrial matrix) to the intermembrane space. VDAC allows flux between the intermembrane space and the host's cytoplasm. The ability to close VDAC restricts the exchange of metabolites and should restrict the growth and replication of the bacterium (mitochondrion).

NATO ASI Series, Vol. H 83
Molecular Biology of Mitochondrial Transport Systems
Edited by M. Forte and M. Colombini

VDAC was discovered in 1975 (Schein et al., 1976) because its voltage dependence made it stand out from undefined conducting elements in the reconstituted membrane. Voltage-dependent channel closure is certainly one way in which the permeability of the outer membrane could be reduced. The characteristics of this voltage dependence in channels reconstituted into planar membranes (Schein et al., 1976; Colombini, 1989) and in liposomes by path clamping (Wunder and Colombini, 1991), have been well described. However, the characteristics of this voltage-gating *in vivo* may be rather different due to local effects.

Outer membrane permeability and its regulation

Early work on the permeability of the mitochondrial outer membrane showed a size-dependence on the permeability to non-electrolytes (Werkheiser and Bartley, 1957). Electron microscopy revealed the presence of channel-like structures in the mitochondrial outer membrane (Parsons et al., 1965). By measuring the ability of non-electrolytes to permeate VDAC-containing vesicles, a non-electrolyte molecular weight cut-off of about 4000 was estimated (Colombini, 1980, Zalman et al., 1980) for VDAC, in line with permeability estimates for the mitochondrial outer membrane. Non-electrolytes were purposely used for these estimates to avoid electrostatic effects. Yet, physiologically, charge effects may have a large influence on the permeability of metabolites. A net positive charge in VDAC that accounts for the weak anion selectivity may contribute to the outer membrane's impermeability to the positively-charged protein, cytochrome C (Wojtczak and Zaluska, 1969).

Upon channel closure, the effective radius of the channel is reduced such that the cyclic molecule, gamma cyclodextrin, just barely permeates (Colombini et al., 1987). From the size of this non-electrolyte, the radius of this closed state was estimated to be 0.9 nm. A similar estimate was obtained from electron microscopy of negatively-stained channels pre-treated in a way that should induce closure (Mannella and Guo, 1990, Mannella et al., 1992). Thus, even in a closed state, the physical size of the pore is large. If this is the case, how can closure of VDAC be used as a mechanism of regulation of mitochondrial function by regulating the permeability of the outer membrane?

The closed states of VDAC that are most commonly observed in electrophysiological recordings are those most readily accessible at the voltages used, in the environment used, and at short times. The conditions in the cell may be quite different and certainly the time scale is much longer. However, even in the reconstituted system, VDAC channels can be observed (Colombini, 1986; Holden and Colombini, 1988) to enter very low conducting states with presumably (but not necessarily) smaller pore radius. These tend to be observed at higher potentials and at longer times after the application of a membrane potential. The very-low conducting states are also observed more frequently in the presence of specific modulating factors (Holden and Colombini, 1988). Thus, these states may play a more important role in regulating flow across the outer membrane.

From another point of view, even the higher-conducting closed states may be effective barriers to the flow of metabolites if electrostatic and not merely steric constraints are considered. Channel closure involves not only a reduction in pore radius but a dramatic inversion of ion selectivity (Colombini, 1980; Benz et al., 1990; Zhang and Colombini, 1990). Since most metabolites crossing the outer membrane are anions, a change to cation preference upon channel closure would help to reduce the permeability of these anions.

Experiments with isolated mitochondria with a mostly intact outer membrane indicate that VDAC's gating mechanism can regulate the access of externally added metabolites to the mitochondrial compartments. König's polyanion, a negatively-charged polymer shown (Colombini et al., 1987) to induce VDAC closure and increase VDAC's voltage dependence, was effective at inhibiting ADP-dependent respiration only if the outer membrane was not damaged (Benz et al., 1988). While this polymer acts on other proteins in addition to VDAC, the results strongly indicated that VDAC closure was responsible for the restricted flow of ADP. Similar experiments performed using the VDAC modulator (a soluble mitochondrial protein capable of augmenting VDAC's voltage dependence (Holden and Colombini, 1988; Liu and colombini, 1991; Liu and Colombini, 1992b) and inducing VDAC closure) showed a similar result. This natural regulator restricted ADP translocation through the outer membrane as monitored by inhibition of both ADP-dependent respiration and adenylate kinase activity (Liu and Colombini, 1992a). This inhibition was eliminated by damaging the outer membrane either by hypotonic treatment or the action of digitonin. Thus, while the readily-accessible closed state has a physical radius that should be large enough to allow ADP to penetrate, the combination of the steric and electrostatic barriers may effectively block the flow of ADP. The possibility that lower-conducting states are responsible for the blockage of ADP flow has not been eliminated.

Physiological situations have been shown to exist in which the outer membrane is a rate-limiting step to the flow of adenine nucleotides. The establishment of the existence of dynamic compartmentation of adenine nucleotides in the intermembrane space indicates that the outer membrane can be an important permeability barrier in the functioning of the intact cell

(Gellerich and Kunz, 1987). The gating of VDAC may be involved in this compartmentation.

Molecular basis for the selectivity of the open state

The large size of VDAC's open pore (about 1.5 nm in radius) makes it likely that electrostatic effects might dominate the channel's selectivity. While it has been speculated that the selectivity might be determined by the unusual properties (such as high charge density) of a particular region of the pore, the data collected by site-directed mutations indicate otherwise (Blachly-Dyson et al., 1990).

Site-directed mutations were used to change the charge at a particular location in VDAC by substituting the amino-acid side chain at that location with one differing in charge. By making conservative changes (substituting polar or charged residues by similar residues) the chances of inducing undefined conformational changes were minimized. This expectation was confirmed in most of cases by observing little or no change in single-channel conductance in the altered proteins and selectivity changes of the appropriate sign and magnitude. In two cases (K234E and K236E), the substitution resulted in such changes in structure that channel reconstitution was not achieved.

The mutations that resulted in changes in VDAC selectivity were all located in regions that should form good transmembrane beta strands by analysis of the primary sequence (Blachly-Dyson et al., 1989). They are found distributed over the entire surface of the protein wall proposed to form the aqueous pore. Save for one unusually large selectivity change (mutation K19E), the selectivity change normalized for the engineered charge change was rather similar among the large majority of the mutations. As impressive was the observation that the selectivity change observed in a channel with multiple substitutions was almost equal to the summation of the changes observed in channels with the corresponding single mutation (Blachly-Dyson et al., 1990). This shows that the selectivity properties of the channel are the result of ionic interactions between the mobile ion and charges all over the inner surface of the protein wall forming the pore. This view of the molecular basis for the ion selectivity of VDAC can be understood in terms of an overall theory of ion flow through large channels (Peng et al., 1992b). This large channel theory (LCT) is based on a theory by Teorell of ion flow through ion-exchange membranes (Teorell, 1957). A cylindrical shell of solution next to the charged wall of the pore is considered to behave like solution in a matrix of fixed charges (Fig. 1). The cylindrical core of the channel is treated as if it is unaffected by the charge on the walls of the channel. These two pathways are in parallel and both conduct ions. The channel's ion-selective properties are therefore a combination of the

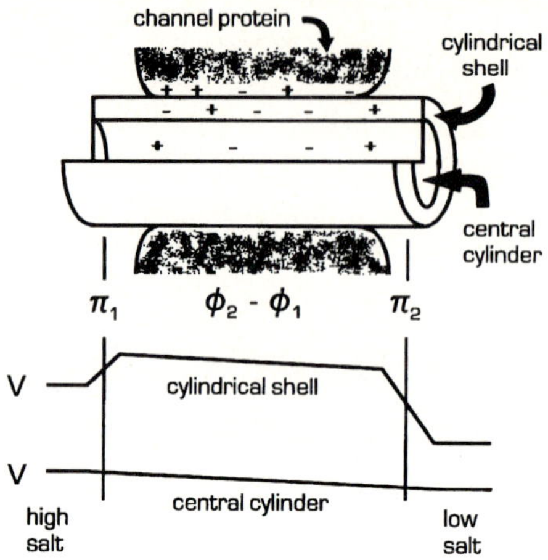

Figure 1. The large channel theory (LCT) proposes that the aqueous phase within the pore is composed of a cylindrical shell of solution containing the fixed charges extending from the surface of the protein and a neutral central cylinder of solution. Donnan potentials (π_1, π_2, for the cylindrical shell) at the interfaces with the bulk phase and diffusion potentials (ϕ_1, ϕ_2, for both aqueous compartments) within the pore contribute to the measured zero-current potential.

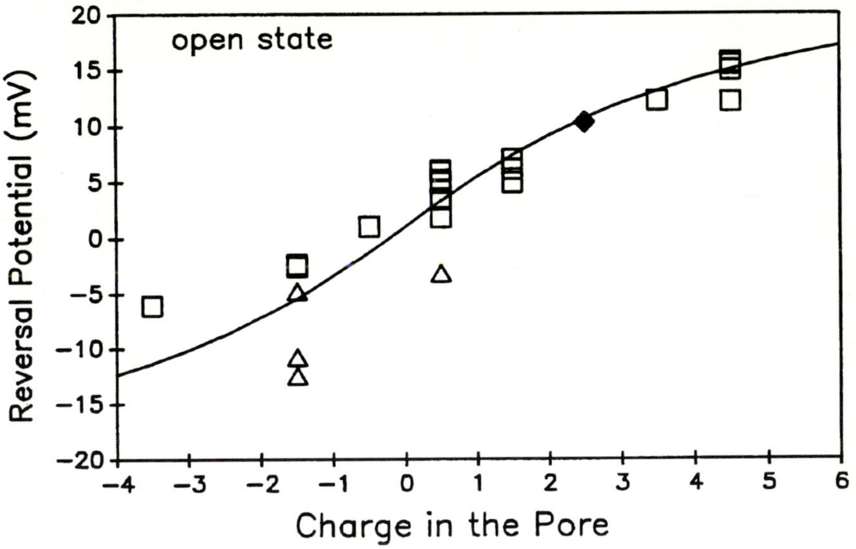

Figure 2. Selectivity changes in VDAC resulting from site-directed mutations compared to theoretical expectations. Selectivity measurements (reversal potentials in the presence of a KCl gradient across the membrane, 1M vs 0.1 M; the sign refers to the high-salt side) were made on wild-type VDAC from *S. cerevisiae* (filled diamond) and VDAC channels in which specific amino acids had been substituted (other symbols) so as to generate the indicated charge change in the molecule (the net charge in the pore of the wild-type channel was set at +2.5 to fit the data to the theoretical line). Only sites thought to be located on the inner wall of the pore are included. The solid line is that obtained from the modification of Teorell's theory. Reproduced with permission from Peng et al., 1992b.

properties of these two aqueous domains.

Considering its simplicity, LCT does remarkably well in accounting for the experimental observations. With just one adjusted value (the net charge on the wall of the pore), LCT can account not only for the effect on selectivity of the site-directed mutations (Fig. 2) but also effects of changing ionic strength, ion mobility, and salt gradient (Zambrowicz and Colombini,

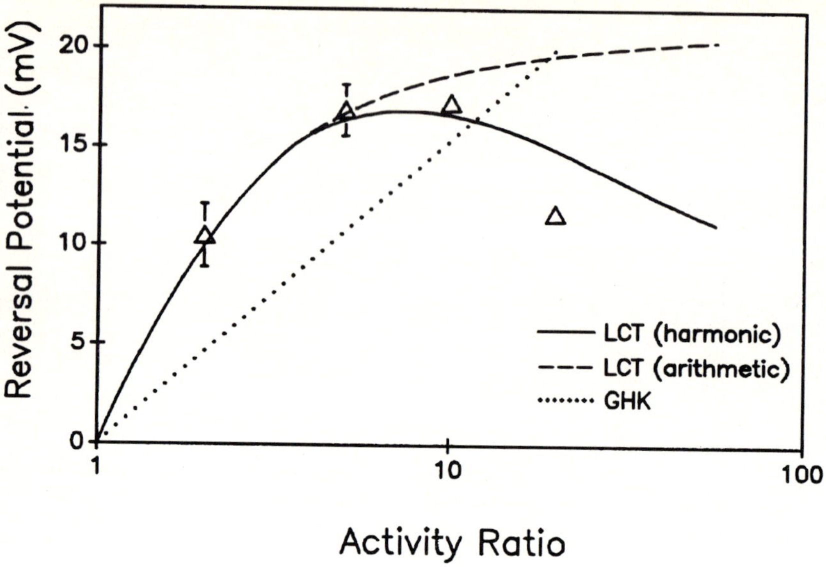

Figure 3. Variation in the reversal potential of VDAC channels from *N. crassa* with changes in the salt gradient across the membrane. The reversal potential is a measure of the selectivity of the channels. If the selectivity did not change, as assumed by the used of the Goldman-Hodgkin-Katz (GHK) equation the reversal potential would increase linearly with the ratio of the ion activities across the membrane (dotted line). The measured values (triangles ± S.D.) follow a curve consistent with LCT using either the harmonic mean or the arithmetic mean when combining the effects of the cylindrical shell and the central cylinder. The curve indicates that the selectivity changes with ionic concentration. Salt screening reduces the influence of the charges on the walls of the channel on the motion of ions within the channel. In LCT this translates into a thinned cylindrical shell and a larger central cylinder.

1993). It also predicts totally unexpected effects such as current flow in opposite directions in different parts of the channel at the same time. The results of this effect is a peak in the value of the zero-current potential as a function of the magnitude of the transmembrane salt gradient (Fig. 3).

Other classical approaches only predict continuous increases in the zero-current potential with increasing salt gradient.

Molecular basis for the selectivity change upon channel closure

The closure of VDAC channels results in not only a reduction in conductance but also a marked change in ion selectivity. This is apparent even with small ions such as KCl. Channels that showed a preference for Cl^- over K^+ when the channels were in the open state now show the opposite preference. The major change then is a selective reduction in the permeability to anions (Colombini, 1980; Peng et al., 1992b).

According to LCT, such a change requires a change in the net charge on the protein walls of the channel from a net positive to a net negative. For yeast VDAC, an estimated change in net charge of 5 units is required (according to LCT) to account for the change in zero-current potential. The effect of charge substitutions on the closed-channel selectivity are consistent with this (Fig. 4).

Such a change in the nature of the protein surface in close proximity with the ion flow through the channel requires a large conformational change. The results are consistent with movement of a portion of the wall of the channel out toward the membrane surface. The region that moves must have a net positive charge so as to leave a net negative charge in the pore. The movement out of the channel of such a positively-charged region is consistent with the voltage dependence of the gating process.

Site-directed mutations have shown that some of the sites that affected the selectivity of VDAC in the open state had a weaker or no effect in the closed state (Fig. 4 open symbols). These sites are primarily located on the transmembrane strands close to the N-terminus: the alpha helix and the first

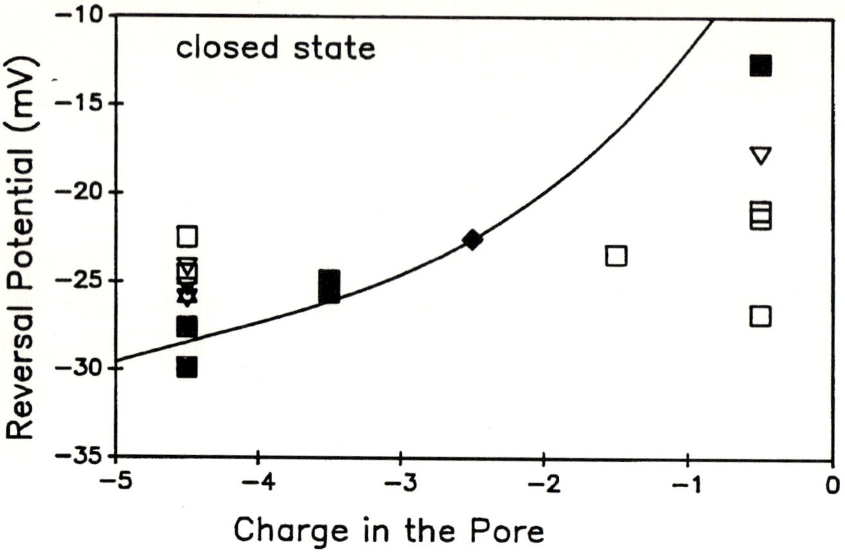

Figure 4. Selectivity changes, in VDAC in the easily-accessible low-conducting "closed" state resulting from site-directed mutations, compared to theoretical expectations. The mutants used to generate the data in figure 2 were also examined to measure the selectivity of the closed states obtained by applying a membrane potential. The *filled symbols* represent the values for the wild-type VDAC (diamond) and mutants at locations judged to be at sites within the pore. The others (open symbols) represent domains that are believed to have moved out of the pore. The theoretical line fits the data (only filled symbols) it the wild-type channel is assumed to have a net charge of -2.5 in the pore. These results were published in Zambrowicz and Colombini, 1993. Reproduced with permission from Peng et al., 1992b

4 beta strands (Peng et al., 1992b; Thomas et al., 1993). If such a region were to move out of the channel, the pore radius would be reduced, the pore volume would be reduced, and the net charge on the wall of the pore would be inverted. All these are consistent with experimental observations (Colombini et al., 1987; Mannella and Guo, 1990; Zimmerberg and Parsegian, 1986; Peng et al., 1992b; Thomas et al., 1993). This mobile region

is that portion of the channel that is responsible for the voltage dependence and can thus be referred to as the voltage sensor.

Unfortunately, the situation is not so simple. VDAC must form a fundamentally asymmetric structure since it is composed of *one* 30kDa polypeptide (Thomas et al., 1992; Peng et al., 1992a). Two gating processes, one at positive and the other at negative potentials, require that the region that moves at positive potentials differ, at least to some extent from the one that moves at negative potentials. Yet, the fundamental mechanism seems sound.

Alternate voltage-gating mechanisms have been proposed for VDAC. In one mechanism the alpha helix is proposed to lie on the surface of the membrane in the open state and thus the walls of the channel would be composed solely of beta strands. Closure would involve the alpha helix entering the channel and thus reducing its conductance. While this model has the elegance of a simple beta barrel without the problem of interfacing with an alpha helix, it is also inconsistent with experimental observations. First of all, site-directed mutations at two residues on the alpha helix show that these sites influence selectivity in the open state but have a weaker (K19E) or no effect (D15K) on the closed state selectivity. This demonstrates that the alpha helix interacts much more strongly with the ions flowing through the channel in the open state than in the closed state. Secondly, the alpha helix in yeast has no net charge and thus its movement into or out of the channel should not influence the selectivity of the channel. This is in contrast to the marked change in selectivity upon channel closure. Thirdly, the alpha helix for yeast VDAC has no net charge and almost no dipole moment since the distribution of charged residues on the helix tends to counteract the dipole moment of the backbone hydrogen bonds. Thus it is hard to see how this motion would be voltage dependent. Finally, site-

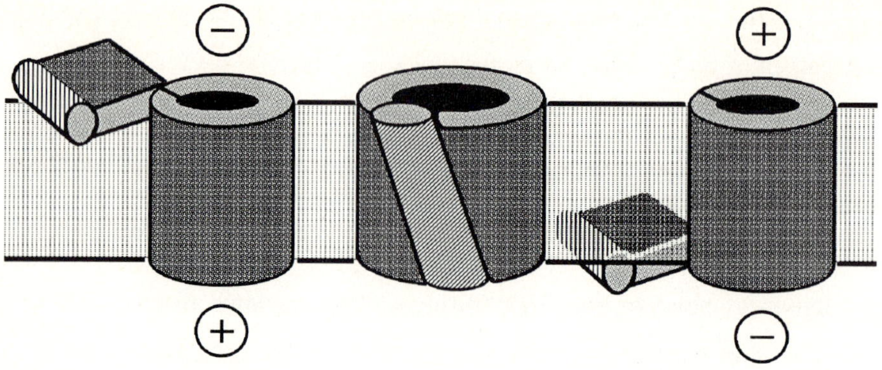

Figure 5. Working model for the voltage gating of VDAC. The open state is a 12-strand β barrel plus 1 α helix (tilted cylinder). Upon application of a membrane potential the α helix and perhaps 4 β strands are forced out of the channel toward one membrane surface depending on the sign of the applied potential.

directed mutations of residues on the alpha helix and elsewhere influence the voltage dependence of VDAC. Thus the alpha helix is part but not the entire voltage sensor. Other regions have a net positive charge and the entire mobile structure, not just the alpha helix, must account for the voltage gating.

Conclusions

The highly-conserved properties of VDAC (Colombini, 1989; Blachly-Dyson et al., 1993) are consistent not only with a passive pathway for metabolite flow through the mitochondrial outer membrane, but also with a site for the regulation of flow of anionic metabolites. The molecular mechanism (Fig. 5) for voltage dependence allows for a natural coupling between charge translocation, reduction in channel radius, and a drastic change in ion selectivity. Closure of VDAC should greatly reduce the flux of anionic metabolites through the outer membrane and thus restrict mitochondrial function.

Although no estimates are available for an electrical potential across the outer membrane, unless the charge on macromolecules in the cytoplasm and intermembrane space are identical, a Donnan potential must exist. The potential could easily be altered by such processes as protein phosphorylation, changes in mitochondrial state from the condensed to the orthodox configuration, local changes in pH or concentration of Ca^{++}, and so forth. Thus voltage dependence *in vivo* need not be discounted solely because VDAC channels are highly conductive.

The properties of VDAC in the cell may indeed be rather different for those in the isolated system. Yet they must be derived from the fundamental properties of the channel. We know that VDAC's properties can be amplified and tuned by such things as colloidal osmotic pressure, binding of the VDAC modulator, and interacting with polyanions. The ability of the channel to be tuned by different agents reinforces the notion that VDAC will be found to be an important factor in the regulation of cellular metabolism.

Acknowledgement: This work was supported by grants from NIH (GM 35759) and ONR (N00014-90-J-1024).

References

Benz R, Kottke M, Brdiczka D (1990) The cationically selective state of the mitochondrial outer membrane pore: a study with intact mitochondria and reconstituted mitochondrial porin. *Biochim Biophys Acta* 1022:311-318.

Benz R, Wojtczak L, Bosch W, Brdiczka D (1988) Inhibition of adenine nucleotide transport through the mitochondrial porin by a synthetic polyanion. *FEBS Lett* 231:75-80

Blachly-Dyson E, Peng SZ, Colombini M, Forte M (1989) Probing the structure of the mitochondrial channel, VDAC, by site-directed mutagenesis: a progress report. *J Bioenerg Biomembr* 21:471-483

Blachly-Dyson E, Peng SZ, Colombini M, Forte M (1990) Alteration of the selectivity of the VDAC ion channel by site-directed mutagenesis: implications for the structure of a membrane ion channel. *Science* 247:1233-1236

Blachly-Dyson E, Zambrowicz EB, Yu WH, Adams V, McCabe ERB, Adelman J, Colombini M, Forte M (1993) Cloning and functional expression in yeast of two human isoforms of the outer mitochondrial membrane channel, the voltage-dependent anion channel. *J Biol Chem* 268:1835-1841

Colombini M (1980b) Structure and mode of action of a voltage-dependent anion-selective channel (VDAC) located in the outer mitochondrial membrane. *Ann NY Acad Sci* 341:552-563

Colombini M (1986) Voltage gating in VDAC: toward a molecular mechanism. *In* "Ion Channel Reconstitution" (C. Miller, ed.), pp. 533-552, Plenum, New York.

Colombini M (1989) Voltage gating in the mitochondrial channel, VDAC. *J Membr Biol* 111:103-111

Colombini M, Yeung CL, Tung J, König T (1987) The mitochondrial outer membrane channel, VDAC, is regulated by a synthetic polyanion. *Biochim Biophys Acta* 905:279-286

Gellerich FN, Kunz W (1987) Cause and consequences od dynamic compartmentation of adenine nucleotides in the mitochondrial intermembrane space in respect to exchange of energy rich phosphates between cytosol and mitochondria. *Biomed Biochim Acta* 46:S545-S548

Holden MJ, Colombini M (1988) The mitochondrial outer membrane channel, VDAC, is modulated by a soluble protein. *FEBS Lett* 241:105-109

Liu M, Colombini M (1991) Voltage gating in the mitochondrial outer membrane channel, VDAC, is regulated by a very conserved protein. *Am J Physiol* 260:C371-C374

Liu M, Colombini M (1992a) Regulation of mitochondrial respiration by controlling the permeability of the outer membrane through the mitochondrial channel, VDAC. *Biochim Biophys Acta* 1098:255-260

Liu MY, Colombini M (1992b) A soluble protein increases the voltage dependence of the mitochondrial channel, VDAC. *J Bioenerg Biomembr* 24:41-46

Mannella CA, Gou XW (1990) Interaction between the VDAC channel and a polyanionic effector. *Biophys J* 57:23-31

Mannella CA, Forte M, Colombini M (1992) Toward the molecular structure of the mitochondrial channel, VDAC. *J Bioenerg Biomembr* 24:7-19

Parsons DF, Bonner WD Jr., Verboon JG (1965) Electron microscopy of isolated plant mitochondria and plastids using both the thin-section and negative-staining techniques. *Can J Bot* 43:647-655

Peng S, Blachly-Dyson E, Forte M, Colombini M (1992a) Determination of the number of polypeptide subunits in a functional VDAC channel from *Saccharomyces cerevisiae*. *J Bioenerg Biomembr* 24:27-31

Peng S, Blachly-Dyson E, Forte M, Colombini M (1992b) Large scale rearrangement of protein domains is associated with voltage gating of the VDAC channel. *Biophys J* 62:123-135

Schein SJ, Colombini M, Finkelstein A (1976) Reconstitution in planar lipid bilayers of a voltage-dependent anion-selective channel obtained from *Paramecium* mitochondria. *J Membr Biol* 30:99-120

Teorell T (1957) Transport processes and electrical phenomena in ionic membranes. *Prog Biophys Chem* 3:305-369

Thomas L, Kocsis E, Colombini M, Erbe E, Trus BL, Steven AC (1991) Surface topography and molecular stoichiometry of the mitochondrial channel, VDAC, in crystalline arrays. *J Struct Biol* 106:161-171

Thomas L, Blachly-Dyson E, Colombini M, Forte M (1993) Mapping of residues forming the voltage sensor of the voltage-dependent anion-selective channel. *Proc. Natl. Acad. USA* 90:5446-5449

Werkheiser WC, Bartley W (1957) The study of steady-state concentrations of internal solute of mitochondria by rapid centrifugal transfer to fixation medium. *Biochem J* 66:79-91

Wojtczak L, Zaluska H (1969) On the impermeability of the outer mitochondrial membrane to cytochrome c. I. Studies on whole mitochondria. *Biochim Biophys Acta* 193:64-72

Wunder UR, Colombini M (1991) Patch clamping VDAC in liposomes containing whole mitochondrial membranes. *J Membr Biol* 123:83-91

Zambrowicz EB, Colombini M (1993) Zero-current potentials in a large membrane channel: a simple theory accounts for complex behavior. *Biophys J* (in press)

Zhang DW, Colombini M (1990) Group IIIA-metal hydroxides indirectly neutralize the voltage sensor of the voltage-dependent mitochondrial channel, VDAC, by interacting with a dynamic binding site. *Biochim Biophys Acta* 1025:127-134

Zalman LS, Nikaido, H, Kagawa, Y (1980) Mitochondrial outer membrane contains a protein producing nonspecific diffusion channels. *J Biol*

Chem 255:1771-1774
Zimmerberg J, Parsegian VA (1986) Polymer inaccessible volume changes during opening and closing of a voltage-dependent ionic channel. *Nature* (Lond.) 323:36-39

THE FUNCTION OF THE MITOCHONDRIAL OUTER MEMBRANE PORE IN THE REGULATION OF PERIPHERAL KINASES AND ENERGY METABOLISM.

Dieter Brdiczka, Ulrike Wicker, and Frank Gellerich*
Faculty of Biology,
University of Konstanz,
D-7750 Konstanz

Introduction

The mitochondrial outer membrane pore protein has recently been detected in the cell membrane of human lymphoblasts by light and electron microscopy and isolation [Thinnes et al. 1989]. Taking into account the known impermeability of the cell membrane for charged compounds we have to presume efficient regulatory mechanisms of the pore permeability which may principally be appropriate also for the regulation of the pore in the outer mitochondrial membrane. Studies of the mitochondrial porin reconstituted in artificial membranes, so far revealed three possible mechanisms of regulation: i) a transmembrane voltage above 30 mV reduced the pore permeability [Colombini 1979] and changed anion to cation selectivity [Benz and Brdiczka 1992], ii) maromolecules such as dextran decreased the pore response to a transmembrane voltage of 5-10 mV [Zimmerberg and Parsegian 1986], and iii) proteins present in the intermembrane space modulated the voltage sensitivity [Holden and Colombini 1988]. All three mechanisms are physiologically relevant: the inner membrane potential may affect the outer membrane by capacitive coupling between the two envelope membranes in the contact sites [Benz et al. 1990], the mitochondria are embedded in a 30% protein solution [Wojcieszyn et al. 1981] and the intermembrane space contains proteins such as kinases which are capable to interact with the isolated porin as will be shown. In view of these findings a reduced permeability of the pore can be anticipated under physiological conditions.

*Institute of Biochemistry, Medical Academy Magdeburg, Germany

NATO ASI Series, Vol. H 83
Molecular Biology of Mitochondrial Transport Systems
Edited by M. Forte and M. Colombini

Effect of macromolecules on the permeability of the outer membrane pore in intact mitochondria.

When cytosolic proteins were substituted by 10% dextran [Gellerich et al. (1992), see accompanying lecture], we observed a reduced outer membrane permeability for adenine nucleotides which was explained by a direct effect of the macromolecules on the pore and by an indirect effect because of the observed induction of contact sites by dextran and presumably correlated changes of the transmembrane voltage [Wicker et al. 1992].

Effects of dextran on the structure of the outer mitochondrial compartment in liver mitochondria.

Isolated liver mitochondria were fixed by rapid freezing techniques [Knoll et al. 1982] in state 4 and state 3 and analyzed in freeze fractured samples. In agreement with the *in situ* observed structure [Malhotra and Van Harreveld 1965], the substitution of macromolecules by 10 % dextran led to a significant reduction of the space between the two envelope- and the crista membranes, whereas in cross fractured control mitochondria the outer compartment was enlarged [Wicker et al. 1992]. Thus it appeared that the presence of dextran was keeping the structure of the outer mitochondrial compartment close to that observed *in vivo*. Supposing that persistence of the physiological intermembrane space structure would provide a prerequisite for contact formation led us to investigate this aspect.

Effect of dextran on the frequency of boundary membrane contact sites.

Liver mitochondria were subjected to rapid freezing in different functional states (state 4 and state 3). As described previously [Knoll and Brdiczka 1983] it was assumed that contact sites are responsible for the frequent jumping of the fracture plane between the two adjacent envelope membranes in freeze fractured mitochondria. By looking closely to the structure where the fracture planes deflected, it appeared that particles (either proteins, or phospholipid micelles) were responsible for attachment between the membranes. In view of these findings we determined the frequency of fracture plane deflections as a means to estimate changes in contact site formation. By applying this method [Bücheler et al. 1991] we observed that phosphorylating mitochondria in state 3 exhibited a higher frequency of fracture plane deflections indicating an increase of contact sites (Table I). In the presence of 10 % dextran T 70, mitochondria in state 4 had a 2.4 fold higher frequency of contacts than control mitochondria. When in the presence of dextran mitochondrial respiration was fully activated (state 3), a 1.6 fold increase of the fracture plane deflections compared to the control was observed, suggesting additional formation of contact sites [Wicker et al. 1992] caused by the macromolecules. On the whole, electron microscopy showed that dextran reduced the space between boundary and crista membranes and induced contact site formation.

Regulation of contact sites and correlation with bound hexokinase activity.

By analyzing the frequency of contacts in freeze fractures of intact hepatocytes we observed that the contacts were subjected to hormonal and metabolite control (Table I). They were decreased by glucagon and increased by epinephrine, acting as a1 agonist [Brdiczka et al. 1986]. Free fatty acids led as well to a reduction of contacts *in vitro* and *in vivo* [Klug et al. 1984]. In all these cases hexokinase activity bound to the mitochondrial surface correlated positively with the frequency of contact sites, suggesting that hexokinase might be a marker enzyme for contact sites. This agreed with electron microscopic observations, applying immuno-gold labelling. In liver and brain mitochondria hexokinase was found to be concentrated in areas where both envelope membranes were closely attached [Weiler et al. 1985, Kottke et al. 1988]. The clustered distribution of hexokinase pointed to a higher enzyme affinity of the pore in the contact sites, because the pore protein, which is known to be the binding site for the enzyme [Fiek et al. 1982, Lindén et al. 1982], was found to be randomly distributed in the outer membrane [Adams et al. 1989, Ohlendieck et al. 1986].

Table I Regulation of contact sites and bound hexokinase activity in the liver.

Effector	Increase	Decrease	Hexokinase
State 4 [a)]		0.26	1
State 3 [a)]	1		4-5
ADP [b)]	1		4-5
Atractyloside [b)]	1		nd
10% Dextran 70 [c)]	1.6		6
DNP [a)]		0.04	0.2
20% Glycerol [a)]		0.03	nd
Fatty acids [d)]		0.35	0.24
Glucagon (b-receptor) [e)]		0.58	0.64
Epinephrine (a1-receptor) [e)]	1.7		nd
HT29 highly glycolytic [f)]		0.21	10
HT29 low glycolytic [f)]	1		8

The frequency of contact sites was determined in freeze fractured liver mitochondria according to the method described [[a)]Knoll and Brdiczka 1983]. The values are based on the frequency of contacts in phosphorylating (state 3) mitochondria. The data for hexokinase represent the activity which remained bound during isolation of mitochondria or after washing of isolated mitochondria respectively. The hexokinase values are given as activity relative to that bound to non-phosphorylating mitochondria in the presence of substrate (state 4). References: [b)][Bücheler et al. 1991] [c)][Wicker et al. 1992],[d)][Klug et al. 1984],[e)][Brdiczka et al. 1986],[f)][Denis-Pouxviel et al. 1987].

Topology of peripheral mitochondrial kinases.

Based on the electron microscopic localization and the correlation between bound hexokinase and contact site frequency, this enzyme was chosen as a contact fraction marker enzyme during the attempts to enrich the contacts from osmotically disrupted liver mitochondria by density gradient centrifugation [Ohlendieck et al. 1986]. Applying this method also to brain and kidney mitochondria, a density gradient fraction was obtained which was distinct from the main outer and inner membrane fractions and was characterized by a high hexokinase activity. In addition, it contained inner and outer membrane components. The putative contact fraction was subsequently removed from the density gradient and treated with specific antibodies against hexokinase followed by decoration with protein-A-gold. Although this procedure exclusively increased the density of hexokinase, upon recentrifugation on a second discontinuous density gradient, the inner and outer membrane components migrated to the same higher density as the hexokinase activity, supporting the proposed existence of a complex between hexokinase and the two boundary membranes [Adams et al. 1989].

The organization of other kinases was studied in brain, kidney and liver mitochondria by the same method described above for hexokinase. The activity profiles in the gradients showed a concentration of nucleoside diphosphate kinase and creatine kinase activities in the fraction of intermediate density, like hexokinase, whereas the adenylate kinase activity remained on top of the gradients and thus appeared to reside outside the contact site fraction [Adams et al. 1989].

Specific binding of hexokinase in the contact sites as induced by macromolecules

The observation of a correlation between bound hexokinase activity and the frequency of contact sites and the induction of contacts by dextran provided the basis for using the effect of dextran to study the specific interaction between hexokinase and the pore in the contact sites.

When we analyzed the binding of isolated hexokinase I to liver mitochondria, we observed a co-operative behaviour, the mean value of the calculated Hill-coefficients was 3 (Fig 1A). Addition of 10% dextran shifted the half maximal saturation constant to a significantly lower concentration of free hexokinase than compared to the control, suggesting an effect of the induction of contact sites by dextran. This suggestion was further supported by the observation that suppression of the contacts by addition of 20% glycerol or 50 µM dinitrophenol [Knoll and Brdiczka 1983] (Fig 1B) or destruction of the membranes by 0.5% Triton-X-100 (not shown) led to an approximately 10 fold decrease of activity and the cooperativity disappeared completely. The results indicated that the induction of contacts by dextran had two effects on hexokinase: it increased the binding affinity to the membrane surface and the enzyme activity.

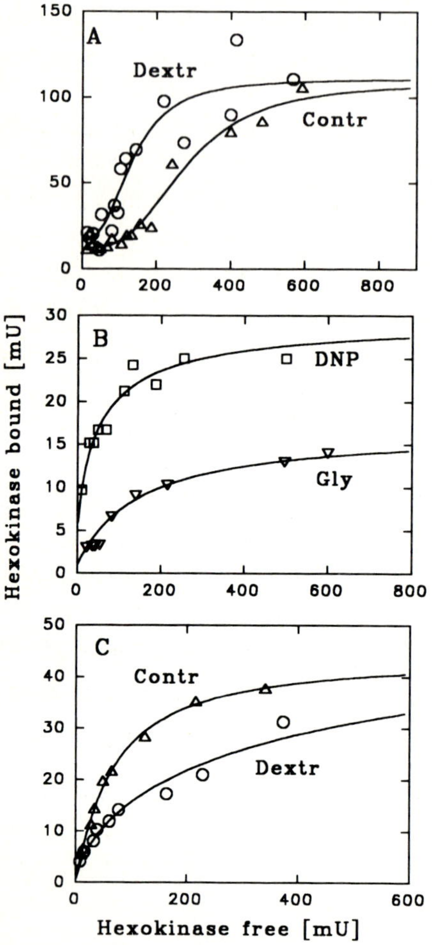

Figure 1
Liver mitochondria were incubated for 10 min at room temperature with increasing amounts of isolated hexokinase, isozyme I, in isotonic medium in the presence of 1 mM ADP, 5 mM glucose and 5 mM $MgCl_2$. After centrifugation, the sediments were resuspended in isolation medium and the activity was determined under hypotonic conditions (0.1 M triethanol-amine pH 7.4) in the supernatant and sediment.

Binding of hexokinase to the isolated outer membrane in the presence of dextran

In order to prove the suggestion that the induction of contacts was responsible for the specific effects on hexokinase binding and activation we studied the binding of the enzyme to the isolated outer membrane (Fig. 1C). The experiments were performed in the presence of 10% dextran 20 because it was difficult to sediment outer membrane vesicles in dextran 70. In contrast to the investigations with intact mitochondria, described above, the binding to the outer membrane did not show cooperativity. The binding efficiency (Umax/Ks) of 0.42 to the outer membrane was about the same compared to control mitochondria (0.33 Fig 1A). The presence of dextran had a negative effect on the binding of hexokinase to the outer membrane. The binding efficiency of 0.1 was significantly lower compared to that of 0.64, which was observed in intact mitochondria.

Taken together, these findings suggested two regulatory effects of contact formation on hexokinase namely activation and increase of the binding affinity. The direct effect of dextran on the pore structure according to Zimmerberg and Parsegian (1986), which was presumably existent in the outer membrane, had no or a negative consequence on the affinity of the enzyme as shown by the binding studies with isolated outer membrane (Fig 1C). Recent results of Xie and Wilson (1990) might provide a basis to explain the co-operative binding to the contacts and the activation. By employing cross-linking, the authors found that hexokinase I forms tetramers when bound to the membrane of liver mitochondria.

Regulatory meaning of hexokinase binding in the contact sites

The results of the present study showed that the binding of hexokinase to the contact sites may have two regulatory functions: i) the regulation of the total cellular activity of glucose phosphorylation via the activation by binding as described earlier [Weiler et al. 1985, Adams et al. 1988] and ii) the saturation of the adenine nucleotide translocator in the presence of high extramitochondrial ATP/ADP quotients because of microcompartmentation of the ADP between hexokinase and the adenylate translocator.

Functional coupling to the AdN-translocator.

The changes of the kinetic properties of originally bound hexokinase in brain mitochondria, pointed to a functional coupling with the AdN-translocator. In agreement with Bessman and co-workers [Gots et al 1974, Bessman et al. 1980] we noted a significantly lower KM for ADP (ATP) when the enzyme was utilising internal ATP as compared to external ATP (Table II). As observed by others [Kabir and Nelson 1991, BeltrandelRio and Wilson 1991] the Vmax with internal ATP was about halve of that with external ATP. The addition of dextran had no effect on these parameters, suggesting that contact formation did not improve the supply of hexokinase with internal ATP. On the other hand, when we studied the ADP supply of the oxidative phosphorylation by the activity of originally bound hexokinase, we noticed a significant decrease in KM and a duplication of the catalytic efficiency in the presence of dextran (Table II). This points to the importance of contacts in optimizing the AdN exchange. Externally added ADP (to produce ATP for hexokinase) led to saturation of the AdN translocator and unimpaired uptake, while the saturation of the translocator with ADP, produced by hexokinase, was difficult in the presence of increasing concentrations of external ATP which competes for ADP binding sites. In this case, induction of contacts by dextran significantly improved the saturation of the translocator because of facilitated communication with hexokinase.

Table II Kinetic properties of original bound hexokinase in brain mitochondria.

Sample	Vmax [mU/ml]	KM [µM]	Vmax/KM
External ATP			
Dextran	1671 ±802	173 ±30.5	9.6 ±3.9
Control	1942 ±920	166 ±33.4	12.3 ±6.5
ATP via OxPhos			
Dextran	610 ±186	39 ±21	16.1 ±1.3
Control	919 ±238	66 ±13.1	14.0 ±1.2
OxPhos via HK	[nMol ADP/min x ml]		
Dextran	1902 ±163	66 ±2.4	30.6 ±1.5
Control	1983 ±5.3	144 ±46.3	14.8 ±0.8

The activity of originally bound hexokinase was analyzed in isotonic sucrose Hepes medium in the presence 2 mM glucose and 10% dextran 70 as indicated. The velocity of the activity was determined from the formation of glucose-6-phosphate by optical test in experiment 1-2, while the stimulation of state 3 respiration was measured by the oxygen electrode in experiment 3. In experiment 2, ATP was formed via activity of oxidative phosphorylation in the presence of 5 mM succinate and 5 mM phosphate and 2 mM di-adenosine 5'-pentaphosphate. In experiment 3 the ADP to stimulate the oxidative phosphorylation was produced by the activity of hexokinase in the presence of 5 mM succinate and phosphate. Hexokinase activity was calculated from the ADP/O ratio and the oxidation rate per min and ml mitochondrial suspension. Mean of 3 experiments.

Function of peripheral kinases in view of the reduced permeability of the outer membrane pore.

A physiologically restricted permeability of the outer membrane for adenine nucleotides as anticipated above would lead to a **third compartment** [Gellerich et al. 1989, see also Gellerich et al., accompanying paper] for these substrates between the two mitochondrial envelope membranes. This would provide the option to saturate the adenine nucleotide translocator in the presence of a constantly high phosphorylation potential during high ATP turnover as was observed in muscle and brain while activity. Specific ADP and ATP levels in the intermembrane space would be created by the activity of several kinases which were found to be organized at the mitochondrial periphery by formation of functional complexes with components of the inner and outer membrane, the contact sites. These complexes, as described above, were characterized by electron microscopy and were isolated by subfractionation of mitochondrial membranes. The work resulted in the definition of three types of kinases: i) type **"energy consuming" kinases** such as hexokinase and glycerol kinase which bound with high affinity to the pore protein in the contact sites, ii) type **"energy transmitting" kinases** such as creatine kinase and nucleoside diphosphate kinase which were concentrated in the contact sites between the two envelope membranes, and iii) type **"energy buffering" kinases** such as adenylate kinase which was located in the outer mitochondrial compartment but not organized in the contact sites [Brdiczka D 1991].

Formation of complexes between kinases and porin.

There is ample evidence that hexokinase binds to the pore in the outer membrane and thus is able to form complexes with this protein [Felgner et al. 1979, Fiek et al. 1982, Lindén et al. 1982]. For creatine kinase we have recently provided indirect evidence that this enzyme might also interact with the pore. After treatment of brain mitochondria with digitonin, creatine kinase was localized by electron microscopy between the inner membrane and attached outer membrane vesicles [Riesinger et al. 1992]. The activity of the enzyme after digitonin treatment stayed not accessible for negatively charged inhibitors such as in intact mitochondria [Kottke et al 1991] and could be inhibited by Königs polyanion [König et al 1977] suggesting that the pore remained attached to creatine kinase, controlling the activity, although the outer membrane was disrupted.

The hexokinase-porin complex

Hexokinase I was isolated according to Wilson (1989) and incubated with porin prepared from liver mitochondria in lauryl dimethyl amino oxide (LDAO) as described by DePinto (1989). Hexokinase was activated 5-10 fold, when the complex formed under similar conditions as learned from binding of hexokinase to intact mitochondria: pH- (6.5) and Mg^{2+}- (10 mM) dependence, suppression by glucose-6-P (5 mM). Hexokinase was not activated (or did not interact) with porin when prepared in Triton X-100. This suggested that the porin N-terminus must be available for interaction with hexokinase, which agreed with the fact that antibodies against the porin N-terminus suppressed the activation. In intact mitochondria the porin N-terminus is accessible [DePinto et al. 1991].

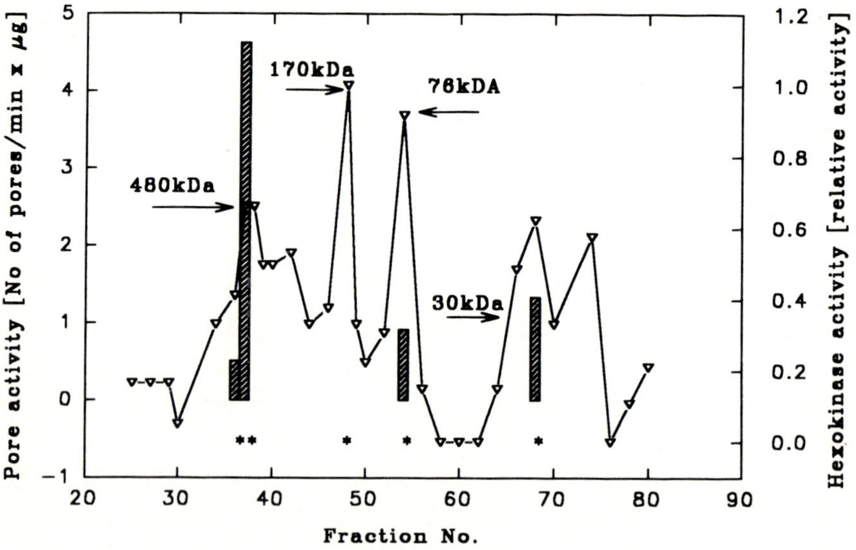

Figure 2: 200µg porin isolated with LDAO and 100µg of hexokinase-I were incubated for 10 min at 20°C in 0.25 M sucrose, 10 mM Hepes pH 6.5 and 10 mM $MgCl_2$. The complex was sedimented by centrifugation (90 min, k-factor 12) and chromatographed on Superose 12 in Hepes sucrose medium pH 7.4. The fractions marked by asterisk where analyzed for pore activity (bars) in artificial membranes.

Analysis by column chromatography (Fig 2) revealed activity of porin and hexokinase in a protein fraction of 480 kDa, suggesting the formation of a complex composed of a porin dimer and a hexokinase tetramer. Hexokinase dimer (170 kDa) and hexokinase fragments (76 kDa) as well as porin (30 kDa) were additionally eluted from the column. As indicated by SDS-gel electrophoresis and immuno blots with antibodies against hexokinase I and the N-terminal end of porin [these were a gift of Dr. Thinnes, Göttingen] a porin dimer and hexokinase were present in the 480 kDa fraction from the column. The pore in the complex with hexokinase was reconstituted in artificial membranes, and the voltage dependence of the conductance was measured in the presence of 10 % dextran (Fig 3A).

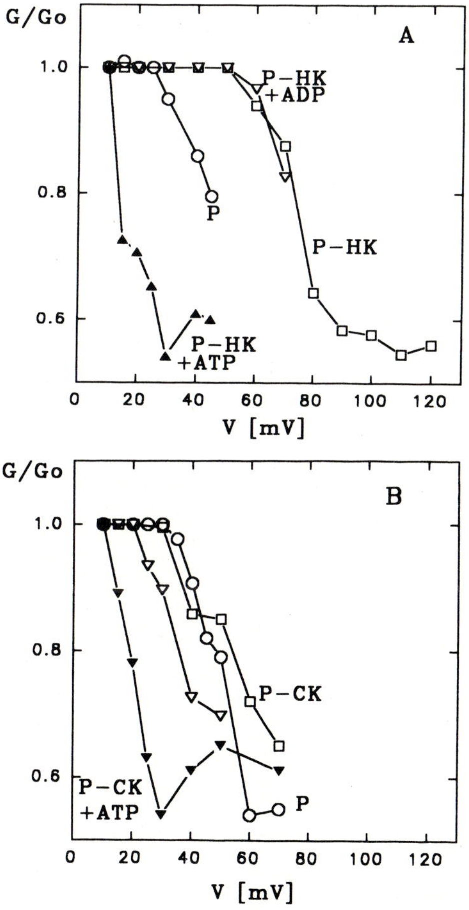

Figure 3
Relation between the voltage independent current and the maximal conductance of the pore protein (P) prepared from liver mitochondria with LDAO (upper panel) and with Triton X-100 from brain mitochondria. Approx. 2 µg of pore protein or 20µg of the complex with hexokinase (P-HK) and creatine kinase (P-CK) was added to both sides of a membrane formed of diphytanoyl phosphatidyl choline/ n-decane in 1 M KCl solution pH 6.5. Additions of ADP or ATP were 1 mM pH 7. The complexes were measured in the presence of 10% dextran. The maximal conductance Go was determined measuring the instantaneous current response to the transmembrane voltage which was varied between 10 and 120 mV. G represents the voltage independent current after approximately 30 sec.

The pore was found to be less voltage sensitive compared to the free LDAO pore. In the presence 1mM ATP the voltage sensitivity of the pore in the complex increased significantly. The results suggested that porin and hexokinase are able to form a complex *in vitro* with properties observed in binding studies to the pore in intact mitochondria. The complex formation reduced the voltage sensitivity of the pore. ATP appears to be a regulator of the pore conductance.

The creatine kinase-porin complex

The octameric form [Rojo et al. 1991] of the mitochondrial creatine kinase was extracted from rat heart muscle by the method of Schlegel et al. (1988). The octamer (50 µg) were incubated with 100 µg porin, which was isolated from brain mitochondria in Triton X-100 according to DePinto (1989). The incubation was performed for 10 min at 20° C in 20 mM phosphate buffer pH 8.0 and led to a 1,7 fold activation of the enzyme. Comparable results were obtained with porin prepared with LDAO. In contrast to the formation of the HK-porin complex, the presence of 5 mM $MgCl_2$ suppressed the activation of creatine kinase by porin. The results suggested a different mechanism of creatine kinase porin interaction. The octamer (MW 350 kDa) was collected by centrifugation for 1 hr at 450,000xg. Control centrifugation of the porin preparation did not yield a sediment. The octamer sediment contained a dimer of porin as analyzed by SDS gel-electrophoresis and immuno decoration of the blots. Addition of the complex to artificial membranes led to incorporation of pores and stepwise increase of conductance. Contrary to the hexokinase complex, the pore bound to creatine kinase was not changed in voltage sensitivity compared to the control (Fig 3B). However, comparable with the hexokinase complex, we observed a significant increase in voltage sensitivity upon addition of 1 mM ATP and 10 mM creatine, whereas 1 mM ADP and creatine phosphate was less affective.

Conclusion

It was anticipated that the presence of macromolecules and of a transmembrane voltage physiologically restricts the permeation of adenine nucleotides through the outer membrane pore. Our results serve to explain how the exchange may be specifically regulated. ATP appeared to interact directly with the pore and increased the voltage sensitivity suggesting a further reduction of anion transport under physiological conditions. However, the pore was able to form two types of complexes with kinases which affected the permeability: i) energy consuming kinases such as hexokinase, which bind to the mitochondrial surface, seemed to interact with the N-terminus of porin ii) whereas energy transmitting kinases such as creatine kinase, located in the intermembrane space, interacted in a different way with porin. The distinct interaction of porin with the two types of kinases affected the permeability of the pore differently, according to the function of the kinases: in a complex with energy consuming kinases, the pore was less voltage sensitive and thus would allow to channel ADP, in a complex with energy transferring kinases, the pore had a normal voltage sensitivity and would thus restrict AdN permeation and separate the kinases from the external AdN pool.

Acknowledgements: This work was supported by grants of the Deutsche Forschungsgemeinschaft given to Frank Gellerich and Dieter Brdiczka.

References

Adams V, Bosch W, Hämmerle Th, Brdiczka D. (1988) Activation of low Km hexokinases in purified hepatocytes by binding to mitochondria. Biochim Biophys Acta 932:195-205.

Adams V, Bosch W, Schlegel J, Wallimann T, Brdiczka D (1989) Further characterization of contact sites from mitochondria of different tissues: toplogy of peripheral kinases. Biochim Biophys Acta 981:213-225

BeltrandelRio H, Wilson JE (1991) Hexokinase of rat brain mitochondria: relative importance of adenylatekinase and oxidative phosphorylation as sources of substrate ATP, and interaction with intramitochondrial compartments of ATP and ADP. Arch Biochem Biophys 286:138-194

Benz R, Kottke M, Brdiczka D (1990) The cationically selective state of the mitochondrial outer membrane pore: a study with intact mitochondria and reconstituted mitochondrial porin. Biochim. Biophys. Acta 1022:311-318.

Benz R, Brdiczka D (1992) The cation-selective substate of the mitochondrial outer membrane pore: singele-channel conductance and influence on intermembrane and peripheral kinases. J. Bioenergetics and Biomembranes, 24:33-39.

Bessman SP, Geiger PJ (1980) Compartmentation of hexokinase and creatine phosphokinase, cellular regulation, and insulin action in "Current Topics in Cellular Regulation", Vol 16 pp 55 ed. Horecker,B.L. and Stadtman,E.R. Academic Press NY

Brdiczka D, Riesinger I, Adams V, and Bremm G (1986) Function in metabolic regulation of contact sites between the mitochondrial boundary membranes 4th EBEC Short-Reports, p 200, Cambridge University Press, Cambridge.

Brdiczka D (1991) Review: Contact sites between mitochondrial envelope membranes. Structure and Function in energy- and protein transfer. Biochim. Biophys. Acta 1071:291-312

Bücheler K, Adams V, and Brdiczka D (1991) Localization of the ATP/ADP translocator in the inner membrane and regulation of contact sites between mitochondrial envelope membranes by ADP. A study on freeze fractured isolated liver mitochondria. Biochim Biophys Acta 1061:215-225

Colombini M (1979) A candidate for the permeability pathway of the outer mitochondrial membrane. Nature 279, 643-645.

Denis-Pouxviel C, Riesinger I, Bühler C,Brdiczka D, and Murat J-C (1987) Regulation of mitochondrial hexokinase in cultured HT 29 human cancer cells. Biochim. Biophys. Acta 902, 335-348.

DePinto V, Benz R, Palmieri F (1989) Interaction of non-classical detergents with mitochondrial porin. A new purification procedure and characterization of the pore-forming unit. Eur. J. Biochem. 183:179-187.

DePinto V, Prezioso G, Thinnes F, Link TA, and Palmieri F (1991) Peptide-specific antibodies and proteases as probes of the transmembrane topology of the bovine heart mitochondrial porin. Biochemistry 30:10191-10200.

Felgner PI, Messer JL, and Wilson JE (1979) Purification of a hexokinase binding protein from the outer mitochondrial membrane. J. Biol. Chem. 254, 4946-4949.

Fiek Ch, Benz R, Roos N, and Brdiczka D (1982) Evidence for identity between the hexokinase-binding protein and the mitochondrial porin in the outer membrane of rat liver mitochondria. Biochim. Biophys. Acta 688:429-440.

Gellerich FN, Schlame M, Bohnensack R, and Kunz W (1987) Dynamic compartmentation of adenine nucleotides in the mitochondrial intermembrane space of rat-heart mitochondria. Biochim. Biophys. Acta 890:117-126.

Gellerich FN, Wagner M, Kapischke M, Wicker U, Brdiczka D (1992) Effect of macromolecules on the regulation of the mitochondrial outer membrane pore and activity of the adenylate kinase in the inter-membrane space Biochem Biophys. Acta accepted

Gots RE, and Bessman SP (1974) The functional compartmentation of mitochondrial hexokinase. Arch. Biochem. Biophys. 163:7-14.

Holden MJ, and Colombini M (1988) The mitochondrial outer membrane channel, VADAC, is modulated by a soluble protein. FEBS Letters 241:105-109.

Kabir F, Nelson BD (1991) Hexokinase bound to rat brain mitochondria uses externally added ATP more efficiently than internally generated ATP Biochim Biophys Acta 1057:147-150.

Klug G, Krause J, Östlund AK, Knoll G, and Brdiczka D (1984) Alteration in liver mitochondrial function as a result of fasting and exhaustive exercise. Biochim. Biophys. Acta, 764, 272-282.

Knoll G, Oebel G, and Plattner H (1982) A simple sandwich-cryogen-jet procedure with high cooling rates for cryofixation of biological materials in native state Protoplasma 111, 161-176

Knoll G, and Brdiczka D (1983) Changes in freeze-fracture mitochondrial membranes correlated to their energetic state. Biochim. Biophys. Acta 733:102-110.

König T, Kocsis B, Mészaros L, Nahm K, Zoltán S, and Horváth I (1977) Interaction of a synthetic polyanion with rat liver mitochondria, Biochim.Biophys.Acta. 462:380-389

Kottke M, Adams V, Riesinger I, Bremm G, Bosch W, Brdiczka D, Sandri G, Panfili E (1988) Mitochondrial boundary membrane contact sites in brain: Biochim. Biophys. Acta 395:807-832.

Kottke M, Adams V, Wallimann T, Nalam VK, and Brdiczka D (1991) Location and regulation of octameric mitochondrial creatine kinase in contact sites. Biochim Biophys Acta 1061:215-225.

Lindén M, Gellerfors P, and Nelson BD (1982) Pore protein and the hexokinase-binding protein from the outer membrane of rat liver mitochondria are identical. FEBS-Leters 141:189-192.

Malhotra SK, and Van Harreveld A (1965) Some structural features of mitochondria in tissues prepared by freeze-substitution. J. Ultrastruct. Res. 12: 473-487.

Rojo M, Hovius R, Demel RA, Nicolay K, and Wallimann T (1991) Mitochondrial creatine kinase mediates contact formation between membranes. J. Biol Chem. 266:20290-20295.

Ohlendieck K, Riesinger I, Adams V, Krause J, and Brdiczka D (1986) Enrichment and biochemical characterization of boundary membrane contact sites in rat-liver mitochondria. Biochim. Biophys. Acta 860:672-689.

Riesinger I, Kottke M, Wallimann T, and Brdiczka D (1992) Distinct effects of digitonin on brain mitochondria and electron microscopic lacalization of creatin kinase and porin. Biochim. Biophys. Acta submitted.

Schlegel J, Zurbriggen B, Wegmann B, Wyss M, Eppenberger HM and Wallimann T (1988). Native mitochondrial creatine kinase forms octameric structures. I. Isolation of two interconver-tible mitochondrial creatine kinase forms, dimeric and octa-meric mitochondrial creatine kinase.J. Biol. Chem. 263:16942-16953

Thinnes FP, Götz H, Kayser H, Benz R, Schmidt WE, Kratzin HD und Hilschmann N (1989) I.Reinigung eines Porins aus menschlichen B-Lymphozyten (Porin 31HL) und sein topochemischer Nachweis auf dem Plasmalemm der Herkunftszelle. Biol. Chem. Hoppe-Seyler 370:1253-1264.

Weiler U, Riesinger I, Knoll G, and Brdiczka D (1985) The regulation of mitochondrial-bound hexokinases in the liver. Biochem. Medicine 33:223-235.

Wicker U, Bücheler K, Gellerich FN, Wagner M, Kapischke M, and Brdiczka D (1992) Effect of macromolecules on the structure of the inter-membrane space and the regulation of hexokinase Biochim Biophys Acta accepted.

Wilson JE (1989) Rapid purification of mitochondrial hexokinase from rat brain by a single affinity chromatography step on Affi-Gel blue Prep. Biochem. 19:13-21.

Wojtczak L, Adams V, and Brdiczka D (1988) Effect of oleate on the apparent Km of monoamine oxidase and the amount of membrane-bound hexokinase in isolated rat hepatocytes Mol. Cell. Biochem. 79:25-30.

Xie G, and Wilson JE (1990) Tetrameric structure of mitochondrially bound rat brain hexokinase: A crossliking study Arch. Biochem. Biophys. 276:285-293.

Zimmerberg J, and Parsegian VA (1986) Polymer inaccessible volume changes during opening and closing of a voltage-dependent ionic channel Nature 323:36-39

COORDINATED REGULATION OF CEREBRAL GLYCOLYTIC AND OXIDATIVE METABOLISM MEDIATED BY MITOCHONDRIALLY BOUND HEXOKINASE

John E. Wilson
Department of Biochemistry
Michigan State University
East Lansing, MI 48823 U.S.A.

Under normal conditions, oxidative metabolism of glucose represents virtually the sole pathway for generation of energy required to support cerebral function, with the rate of glucose utilization in different brain regions varying in response to altered neurophysiological activity (Sokoloff, 1981). The overall metabolic sequence can be divided into the glycolytic segment occurring in the cytoplasm and giving rise to pyruvate, and an aerobic segment occurring within the mitochondria and involving the TCA cycle as well as mitochondrial electron transport and oxidative phosphorylation. If introduction of glucose into this pathway exceeds the capacity of the mitochondrial segment, excess carbon is diverted to lactate. Since lactate is known to have neurotoxic effect (Marie and Bralet, 1991), it is evident that coordination of the glycolytic segment and the mitochondrial oxidative reactions is essential. How is this balance achieved?

The initial step in glycolysis is phosphorylation of glucose, catalyzed by hexokinase. It is evident that regulation of this reaction could play a major role in governing the rate of cerebral glucose utilization (see Wilson (1984), and references therein, for a discussion of mammalian hexokinase isozymes and their regulation). Hexokinase is inhibited by its product, glucose 6-phosphate (G6P), and this inhibition is antagonized by inorganic phosphate (P_i). Thus, hexokinase activity is sensitive to changes in cellular levels of G6P and P_i, which characteristically occur during various states associated with marked alteration in rate of glucose metabolism. There is little doubt that regulation of hexokinase activity by G6P and P_i can have a major influence on cerebral glycolytic rate.

Unlike the rest of the glycolytic enzymes, which appear to

NATO ASI Series, Vol. H 83
Molecular Biology of Mitochondrial Transport Systems
Edited by M. Forte and M. Colombini

be located within the cytoplasm, the major portion of the hexokinase in brain homogenates has been associated with the mitochondrial fraction. Parry and Pedersen (1990) have suggested that hexokinase is associated with microsomal contaminants in the mitochondrial fraction, rather than with the mitochondria themselves. However, we believe that there are alternative interpretations for their results, consistent with a mitochondrial location for the enzyme. For example, after submaximal loading of mitochondria with barium phosphate, Parry and Pedersen observed that hexokinase did not move to a density (on sucrose density gradients) comparable to that seen with succinate dehydrogenase. This was interpreted as indicating a nonmitochondrial location for hexokinase. However, it is not evident why hexokinase would be expected to move to a higher density *at all* if it were not associated with barium loaded mitochondria. We believe that a more likely interpretation is the known heterogeneity of brain mitochondria, as reported by several previous workers (see Wilson, 1972, for pertinent references), with hexokinase being associated with a subpopulation of mitochondria. The lack of correlation between various mitochondrial enzymes, including hexokinase, after sucrose density gradient centrifugation has been known for more than 20 years, and interpreted as an indication of mitochondrial heterogeneity. Thus, the lack of congruence between the distribution of glutamate dehydrogenase and mitochondrial markers such as succinate dehydrogenase has never been taken as evidence that glutamate dehydrogenase is not a mitochondrial enzyme, but rather, as an indication that a subpopulation of brain mitochondria is enriched in glutamate dehydrogenase; it is not evident that the lack of correlation between hexokinase and succinate dehydrogenase should be interpreted differently.

Most of our work has been with the hexokinase of rat brain, which we have assumed would serve as a model for brain hexokinase from other species. Later in this talk, I will tell you about some recent results which indicate that significant differences may exist in brains from various species. Nonetheless, we believe that the rat brain enzyme does provide a useful system for examining interaction of mitochondrially bound hexokinase with intramitochondrial processes, and these studies will be the primary focus of this talk.

Like hexokinase activity itself, binding of hexokinase to mitochondria is sensitive to G6P, with P_i antagonizing this effect. There is evidence to support the view that changes in the distribution of hexokinase between mitochondrially bound and dissociated forms, in response to altered cellular levels of G6P and P_i, may be an additional factor in regulation of hexokinase activity, with the bound form being more active (see Wilson, 1984).

Unlike brain mitochondria, liver mitochondria contain little if any bound hexokinase activity. Yet, like brain mitochondria, they have the ability to bind brain hexokinase in a G6P-sensitive manner. Phil Felgner was the first to isolate a protein from the outer membrane of rat liver mitochondria that conferred the ability to bind hexokinase (Felgner *et al.*, 1979); we called this protein the "hexokinase binding protein", or HBP. The HBP was quickly shown by other investigators (Fiek *et al.*, 1982; Lindén *et al.*, 1982) to be identical to porin, or VDAC.

Binding of the enzyme to the outer mitochondrial membrane is stabilized by divalent cations, with Mg^{++} likely to be the most relevant under physiological conditions; the divalent cation is thought to bridge between negatively charged groups on the enzyme and membrane, the latter presumably being phospholipids (Moller and Wilson, 1983). Binding of the enzyme is known to involve interactions with specific regions located in the N-terminal half of the enzyme (Smith and Wilson, 1991), and is critically dependent on a hydrophobic sequence at the N-terminus of the molecule (Polakis and Wilson, 1985), which is inserted into the hydrophobic core of the membrane when the enzyme is bound (Xie and Wilson, 1988).

Brain hexokinase - which consists of a single polypeptide chain with a molecular weight of about 100 kDa - is comprised of two major domains, corresponding to the N-terminal and C-terminal halves of the molecule (White and Wilson, 1989, 1990). Catalytic function resides in the *C-terminal* half; indeed, using limited proteolysis, it is possible to isolate the C-terminal half of the molecule with full retention of catalytic activity. The allosteric site, to which G6P binds with resulting inhibition, has been associated with the *N-terminal* half of the enzyme. Despite this functional division, the two halves are very similar in amino acid sequence, as deduced from the cloned cDNA (Schwab and Wilson, 1989). Moreover, both halves are similar in sequence not only to each other but also to the 50 kDa hexokinase found in yeast, consistent with evolution of the 100 kDa mammalian hexokinase by duplication and fusion of a gene for a 50 kDa ancestral hexokinase related to the present-day yeast enzyme.

Several investigators have presented results which support the view that mitochondrially bound hexokinase has "privileged" (or "preferential") access to ATP generated within the mitochondria, and that ADP generated in the reaction catalyzed by mitochondrially bound hexokinase is preferentially redirected back into the mitochondria, to be rephosphorylated (see BeltrandelRio and Wilson, 1991, and references therein). Now "privileged" access does not mean *exclusive* access, and indeed it would be catastrophic if hexokinase were to utilize every ATP generated within the mitochondrion, for surely there are many other ATP-requiring processes that need their share of the ATP. Together with Hector BeltrandelRio, we decided to quantitatively measure the fraction of ATP that was captured by hexokinase as the ATP emerged from the pore. ATP was generated intramitochondrially by oxidative phosphorylation, or by adenylate kinase or creatine kinase. *Total* ATP production was measured in the presence of excess yeast hexokinase, such that every ATP emerging from the mitochondrion was quickly scavenged by the yeast enzyme, with resulting formation of G6P. Utilization of intramitochondrially produced ATP by endogenous mitochondrially bound hexokinase was measured in the absence of added yeast hexokinase. In either case, G6P formation was followed spectrophotometrically by linking it to NADPH production, monitored at 340 nm, with addition of excess G6P dehydrogenase (Fig. 1). Thus we reasoned that the fraction of the ATP captured as it emerged from the mitochondrion could be estimated from the ratio of the rate of G6P production in the absence of added yeast hexokinase (i.e., G6P production by endogenous hexokinase) to the total rate of ATP production.

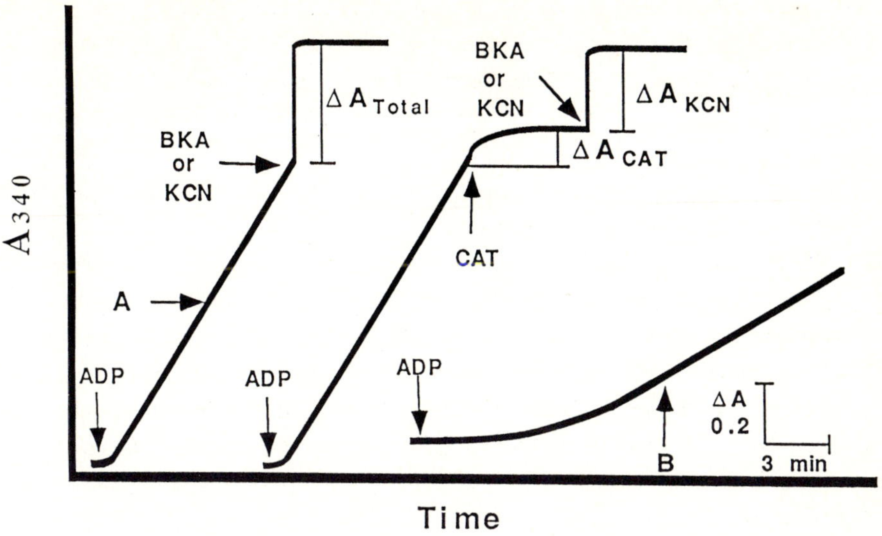

Figure 1. Glucose 6-phosphate (G6P) formation was monitored spectrophotometrically, coupling this to NADPH production ($A_{340\,nm}$) by addition of excess G6P dehydrogenase. *Total* ATP production by mitochondria was determined in the presence of excess yeast hexokinase (curves at left and center in figure), while G6P production by endogenous mitochondrially bound hexokinase was determined without addition of yeast hexokinase (curve at right). See text for interpretive comments. Reprinted from BeltrandelRio and Wilson (1992a) with permission of Academic Press.

Under our experimental conditions, ATP production by oxidative phosphorylation began virtually immediately upon addition of ADP and continued linearly for at least 30 min. In contrast, there was a transient period of 8-10 minutes during which the rate of glucose phosphorylation by the endogenous mitochondrial hexokinase increased to a steady state rate that remained constant throughout the duration of the experiments, for at least 25-30 minutes. The ratio of the steady state rate of ATP utilization by mitochondrial hexokinase to the rate of ATP production (measured with added yeast hexokinase) was about 0.4-0.5. In other words, during steady state and under these conditions, about 40-50% of the ATP being produced was being used by endogenous hexokinase. It therefore followed that the 50-60% of the ATP *not* being utilized must be accumulating in the extramitochondrial medium, and this was shown to be the case. But the levels accumulated (0.5 mM or less) were always well below saturating levels, and the steady state rate was always well below the V_{max} that could be attained by supplying the enzyme with excess exogenous ATP. Why did the rate of glucose phosphorylation by endogenous hexokinase come to this limiting steady state value, well below V_{max} and completely oblivious to the continued increase in extramitochondrial ATP?

Another initially perplexing observation was the effect of adding KCN, a potent inhibitor of electron transport and hence of ATP production by oxidative phosphorylation. Addition of KCN caused an immediate cessation of oxygen uptake, and thus ATP production must have been halted virtually immediately. Yet, addition of KCN caused a burst of glucose phosphorylation prior to complete cessation of further G6P production (Fig. 1). Addition of carboxyatractyloside (CAT), an inhibitor of the ATP/ADP translocase, had a somewhat similar effect although both the rate and extent of increase in A_{340} following addition of CAT were less than seen with KCN. If, subsequent to CAT, KCN was also added, a burst of glucose phosphorylation was again seen, with the total increase in A_{340} being equivalent with either KCN alone or CAT plus KCN. These effects were not simply related to the inhibition characteristics of these compounds since azide, like cyanide an inhibitor of electron transport, gave results equivalent to CAT, while bongkrekic acid (BKA), like CAT an inhibitor of the translocase, gave results equivalent to KCN.

With excess yeast hexokinase present, it could be shown that extramitochondrial ATP was maintained at an extremely low level. Thus the burst of glucose phosphorylation seen after addition of these inhibitors must result from release of *intra*mitochondrial ATP, and based on the differential effect of KCN and CAT, we concluded that there were (minimally) two intramitochondrial compartments involved; these were referred to as the "CAT-sensitive" and "KCN-sensitive" compartments although, as noted above, these compartments are not uniquely sensitive to CAT or KCN (BeltrandelRio and Wilson, 1991).

When ATP was generated by adenylate kinase, addition of CAT, KCN, or BKA had no effect on ATP formation nor did they evoke release of compartmented ATP (BeltrandelRio and Wilson, 1992a).

Thus, these compartments appear to be selectively associated with generation of ATP by oxidative phosphorylation.

From results of these and other experiments (BeltrandelRio and Wilson, 1991, 1992a, 1992b) , we were led to the conclusion that, *during active oxidative phosphorylation,* mitochondrially bound hexokinase comes to be *totally* dependent on ATP from intramitochondrial compartments. Thus, the transient lag in the rate of glucose phosphorylation after initiation of oxidative phosphorylation corresponded to the time required to fill the intramitochondrial compartments. Once these compartments were filled, hexokinase was seeing maximal levels of substrate and hence there was no further increase in rate of glucose phosphorylation despite continued build-up of extramitochondrial ATP, to which the mitochondrial hexokinase remained oblivious.

Additional experiments (BeltrandelRio and Wilson, 1992a, 1992b) demonstrated that both the rate of glucose phosphorylation and the levels of intramitochondrial ATP were responsive to changes in the rate of oxidative phosphorylation. For example, when ATP production by oxidative phosphorylation was manipulated by using subsaturating levels of oxidative substrates, pyruvate and malate, both the levels of intramitochondrial ATP as well as the rate of glucose phosphorylation by mitochondrially bound hexokinase were shown to vary in concert. In another series of experiments, it was demonstrated that, in the absence of oxidative phosphorylation, the rate of glucose phosphorylation by mitochondrial hexokinase exhibited Michaelis-Menten dependence on extramitochondrial ATP levels. *However*, initiation of active oxidative phosphorylation, by addition of ADP in the presence of various concentrations of extramitochondrial ATP, resulted in a transient period during which intramitochondrial compartments of ATP were generated, with the various rates of glucose phosphorylation (initially dependent of extramitochondrial ATP levels) coming to a common value that was *independent of* the levels of extramitochondrial ATP present in the system. Manipulation of the rate of oxidative phosphorylation, and thereby the level of ATP in intramitochondrial compartments, by limiting the amount of P_i present resulted in a corresponding decrease in the rate of glucose phosphorylation by mitochondrial hexokinase.

Intracellular ADP levels have been estimated to be in the range of 30 μM (Veech, 1980). ATP production by oxidative phosphorylation is responsive to ADP concentrations in this range, with the level of intramitochondrially compartmented ATP being closely correlated with oxidative phosphorylation rate (BeltrandelRio and Wilson, 1992a). Thus, the observations we have been discussing suggest a mechanism by which coordination between the glycolytic and oxidative segments of glucose metabolism can be achieved with mediation by mitochondrial hexokinase. Let us imagine the following sequence of events. Suppose that neurophysiological activity is increased in some brain region, with corresponding increase in energy demand and hence in hydrolysis of ATP. An increase in ADP levels can be anticipated, thereby increasing the rate of oxidative phosphorylation which, as we have noted, is responsive to changes in ADP concentration in the physiological range. This in turn will increase intramitochondrial ATP levels, with resulting increase in glucose phosphorylation by the mitochondrially bound hexokinase. Thus, increased energy demand has initiated a coordinated increase in both the terminal oxidative stages of glucose metabolism as well as the introduction of glucose carbon into the glycolytic pathway. Obviously, when energy demands subside, an analogous sequence of events could result in coordinated decrease in the rates of these metabolic processes.

Where are these intramitochondrial compartments of ATP? The fact that the ATP can be released in the presence of translocase inhibitors such as CAT and BKA implies that they are outside the inner mitochondrial membrane. We have used digitonin fractionation to further examine the possible location of these compartments (BeltrandelRio and Wilson, 1992a). The release of various marker enzymes in response to treatment of rat brain mitochondria with increasing concentrations of digitonin was similar to that reported by previous investigators. Of particular relevance in the present context, it should be noted that adenylate kinase - generally accepted as a marker for the intermembranal space - is totally released by relatively low digitonin concentrations (approx. 0.5 mg digitonin per mg mitochondrial protein), much lower than required to cause release of the outer membrane marker, monoamine oxidase. Digitonin is quite ineffective in causing release of hexokinase. Brdiczka (1991) and others (Dorbani *et al.*, 1987) have suggested that this is due to preferential binding of hexokinase at contact sites, thought to be exceptionally resistant to disruption by digitonin.

The KCN- and CAT-sensitive compartments were quite similar in their sensitivity to digitonin (BeltrandelRio and Wilson, 1992a). *Neither* was disrupted by digitonin levels that caused total release of adenylate kinase, leading to the conclusion that neither of these compartments is associated with the intermembrane space. Of the various marker enzymes determined, release of monoamine oxidase was most closely correlated with disruption of the compartments.

Extending previous speculations and suggestions of Dorbani *et al.* (1987) and of Brdiczka (1991) and his colleagues, we envisage something like that shown in Fig. 2. The outer mitochondrial membrane is considered to contain domains of differing cholesterol content, and hence different resistance to digitonin, which is presumed to cause membrane disruption through its ability to complex with cholesterol. If one were to look down on the surface of the membrane, one would see hexokinase bound to porin located in cholesterol-poor regions, indicated by the absence of shading and presumed to be contact sites. Within these regions are small islands of intermediate cholesterol content, which can be disrupted with moderate concentrations of digitonin, causing release of compartmented ATP underlying these regions - seen in cross-section below. Monoamine oxidase is also located in outer membrane regions of moderate cholesterol content extending between contact sites. Regions of high cholesterol content are distributed throughout these regions; these are readily disrupted by digitonin, releasing adenylate kinase from the underlying intermembranal space.

Like hexokinase, creatine kinase may be preferentially associated with contact sites (Brdiczka, 1991). It was an obvious question to ask whether creatine kinase and hexokinase might be associated with the same intramitochondrial compartments of ATP. Our conclusion (BeltrandelRio and Wilson, 1992b) is that, *if* the

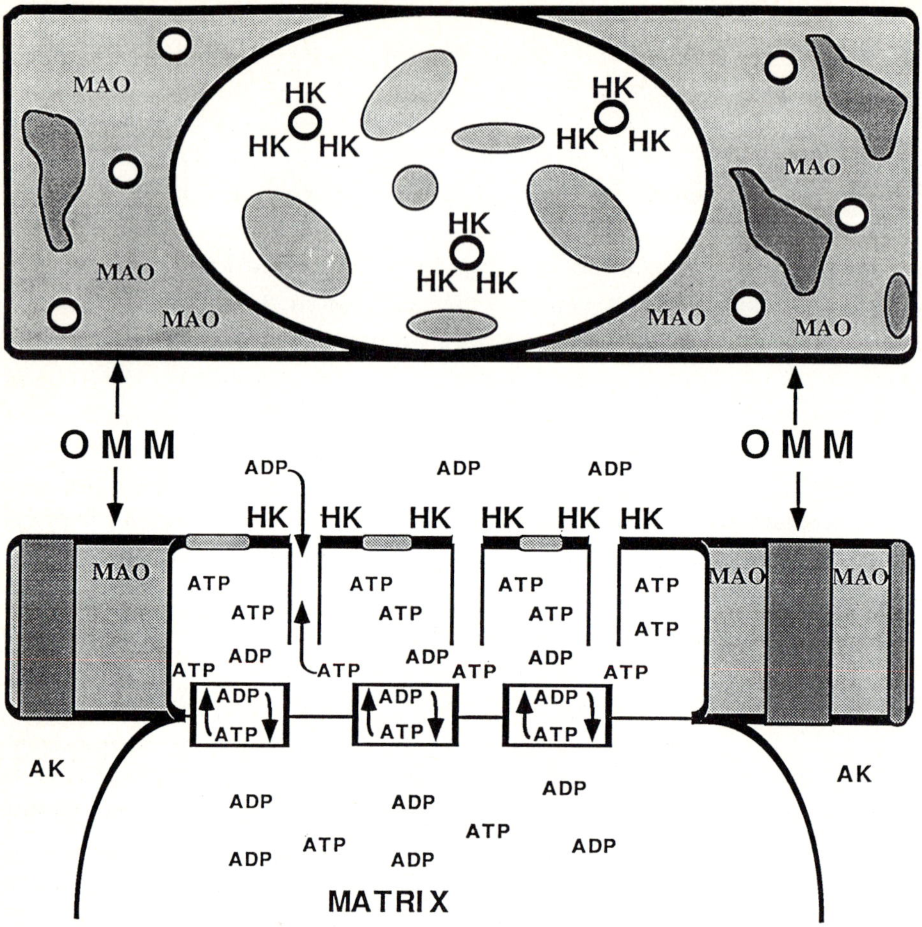

Figure 2. Schematic representation of domains in the outer mitochondrial membrane, differing in cholesterol content, and their relationship with mitochondrially bound hexokinase and intramitochondrial compartments of ATP. Reprinted from BeltrandelRio and Wilson (1992a) with permission of Academic Press.

intramitochondrial compartments providing mitochondrial hexokinase with substrate ATP are generated by active oxidative phosphorylation, then ATP generated by creatine kinase can gain access to these compartments. *However*, creatine kinase alone cannot induce formation of these compartments. Nor does supplementation of ATP production by creatine kinase increase the rate of glucose phosphorylation by mitochondrial hexokinase significantly beyond that seen with maximal rates of oxidative phosphorylation alone. In other words, our results are consistent with the view that formation of the intramitochondrial compartments providing mitochondrial hexokinase with ATP is critically dependent on active oxidative phosphorylation. While creatine kinase can provide ATP to these compartments, it cannot generate them; nor can creatine kinase significantly increase the capacity of these compartments, and hence generation of ATP by creatine kinase cannot increase the rate of glucose phosphorylation beyond that seen when the compartments are maximally filled by oxidative phosphorylation alone.

As I noted at the start of this talk, most of our work has been done with rat brain hexokinase. More than 90% of this enzyme can be released to soluble form by incubation of rat brain mitochondria with G6P, and this was initially presumed to be the case for mitochondrial hexokinases from brain of other species. However, this assumption is not valid. In work with Firoz Kabir, we have examined the sensitivity of brain mitochondrial hexokinase to solubilization with G6P. While approximately 90% of the rat brain enzyme is released by G6P, with guinea pig or rabbit brain, only about 60% can be released with G6P. With mitochondria from bovine, porcine or ovine brain, approximately 40% can be released, while only 20% of the mitochondrial hexokinase from human brain is released by similar treatment. We do not yet understand the basis for this difference in sensitivity to release by G6P. Our present results are consistent with the view that this reflects different types of binding sites rather than different forms of hexokinase *per se*. The basis for this view is that bovine brain hexokinase released with G6P has been shown to be indistinguishable from the G6P-insensitive hexokinase with respect to several physical and kinetic properties, and moreover, *both* forms of the enzyme bind to rat brain mitochondria in a G6P-sensitive manner. The most straightforward interpretation of these results is that the majority of the hexokinase binding sites on rat brain mitochondria are located in an environment that leads to release of the enzyme in the presence of G6P, whereas mitochondria from bovine (and other species') brain contain (at least) two types of binding sites, one similar to that of rat brain mitochondria while a second environment somehow renders the bound enzyme insensitive to release by G6P. It is presently unclear what physiological significance might be attached to these differences in G6P sensitivity of brain mitochondrial hexokinase - certainly they have major implications for the proposed role of G6P-dependent changes in distribution of hexokinase between mitochondrially bound and dissociated forms as a factor in regulation of activity. Questions of current interest include asking about the molecular basis for these apparently distinct mitochondrial binding environments, and whether these have any effect on interactions between mitochondrially bound hexokinase and intramitochondrial ATP generating processes and associated compartments of ATP.

REFERENCES

BeltrandelRio H and Wilson JE (1991) Hexokinase of rat brain mitochondria: Relative importance of adenylate kinase and oxidative phosphorylation as sources of substrate ATP, and interaction with intramitochondrial compartments of ATP and ADP. Arch Biochem Biophys 286:183-194

BeltrandelRio H and Wilson JE (1992a) Coordinated regulation of cerebral glycolytic and oxidative metabolism, mediated by mitochondrially bound hexokinase dependent on intramitochondrially generated ATP. Arch Biochem Biophys 296:667-677

BeltrandelRio H and Wilson JE (1992b) Interaction of mitochondrially bound rat brain hexokinase with intramitochondrial compartments of ATP generated by oxidative phosphorylation and creatine kinase. Arch Biochem Biophys in press

Brdiczka D (1991) Contact sites between mitochondrial envelope membranes. Structure and function in energy- and protein-transfer. Biochim Biophys Acta 1071:291-312

Dorbani L Jancsik V Lindén M Leterrier JF Nelson BD and Rendon A (1987) Subfractionation of the outer membrane of rat brain mitochondria: Evidence for the existence of a domain containing the porin-hexokinase complex. Arch Biochem Biophys 252:188-196

Felgner PL Messer JL and Wilson JE (1979) Purification of a hexokinase-binding protein from the outer mitochondrial membrane. J Biol Chem 254:4946-4949

Fiek C Benz R Roos N and Brdiczka D (1982) Evidence for identity between the hexokinase-binding protein and the mitochondrial porin in the outer membrane of rat liver mitochondria. Biochim Biophys Acta 688:429-440

Lindén M Gellerfors P and Nelson BD (1982) Pore protein and the hexokinase-binding protein from the outer membrane of rat liver mitochondria are identical. FEBS Letts 141:189-192

Marie C and Bralet J (1991) Blood glucose level and morphological brain damage following cerebral ischemia. Cerebrovascular and Brain Metabolism Reviews 3:29-38

Moller F and Wilson JE (1983) The influence of specific phospholipids on the interaction of hexokinase with the outer mitochondrial membrane. J Neurochem 41:1109-1118

Parry DM and Pedersen PL (1990) Glucose catabolism in brain. Intracellular localization of hexokinase. J Biol Chem 265:1059-1066.

Polakis PG and Wilson JE (1985) An intact hydrophobic N-terminal sequence is critical for binding of rat brain hexokinase to mitochondria. Arch Biochem Biophys 236:328-337

Schwab DA and Wilson JE (1989) Complete amino acid sequence of rat brain hexokinase, deduced from the cloned cDNA, and proposed structure of a mammalian hexokinase. Proc Natl Acad Sci USA 86:2563-2567.

Smith AD and Wilson JE (1991) Disposition of mitochondrially bound hexokinase at the membrane surface, deduced from reactivity with monoclonal antibodies recognizing epitopes of defined location. Arch Biochem Biophys 287:359-366

Sokoloff L (1981) Localization of functional activity in the central nervous system by measurement of glucose utilization with radioactive deoxyglucose. J Cerebral Blood Flow and Metabolism 1:7-36

Veech RL (1980) Freeze-blowing of brain and the interpretation of the meaning of certain metabolite levels. *In* Cerebrovascular Metabolism and Neural Function (Passonneau JV Hawkins RA Lust WD and Welsh FA, eds) pp. 34-41 Williams & Wilkins Baltimore London

White TK and Wilson JE (1989) Isolation and characterization of the discrete N- and C-terminal halves of rat brain hexokinase: Retention of full catalytic activity in the isolated C-terminal half. Arch Biochem Biophys 274:375-393

White TK and Wilson JE (1990) Binding of nucleoside triphosphates, inorganic phosphate, and other polyanionic ligands to the N-terminal region of rat brain hexokinase: Relationship to the regulation of hexokinase activity by antagonistic interactions between glucose 6-phosphate and inorganic phosphate. Arch Biochem Biophys 277:26-34

Wilson JE (1972) The localization of latent brain hexokinase on synaptosomal mitochondria. Arch Biochem Biophys 150:96-104

Wilson JE (1984) Regulation of mammalian hexokinase activity. *In* Regulation of Carbohydrate Metabolism (Beitner R, ed) pp. 45-85 CRC Press Boca Raton FL

Xie G and Wilson JE (1988) Rat brain hexokinase: The hydrophobic N-terminus of the mitochondrially bound enzyme is inserted in the lipid bilayer. Arch Biochem Biophys 267:803-810

Influence of Macromolecules on the Permeability of Porin Pores and Dynamic Compartmentation of Adenine Nucleotides in the Mitochondrial Intermembrane Space

F.N. Gellerich, M. Kapischke, M. Wagner and D. Brdiczka*

Institut für Biochemie, Medizinische Akademie Magdeburg, Leipziger Str. 44
O-3090 Magdeburg, Germany

In the past, the biological relevance of mitochondrially localized ATP splitting enzymes such as hexokinase or creatine kinase has been discussed as an advantage in supplying these enzymes with mitochondrially formed ATP (Saks *et al.*, 1974; Gellerich *et al.*, 1977; Bessman & Gots, 1975). A new approach to this problem became possible through experiments with reconstituted systems (Gosalvez *et al.*, 1974; Gellerich & Saks, 1982; Gellerich *et al.*, 1987; Kottke *et al.*, 1991), in which mitochondria and muscle pyruvate kinase compete for ADP produced by kinases in varied localization with respect to the mitochondrial outer membrane. For mitochondria from heart (Gellerich & Saks, 1982), liver (Gellerich, 1992), and brain (Kottke *et al.*, 1991) it was shown that the ADP supply to oxidative phosphorylation was privileged via mitochondrial creatine kinase or adenylate kinase compared to the ADP supply by extramitochondrially added enzymes such as yeast hexokinase. Therefore, channelling of the extramitochondrially formed ADP into the mitochondria seems to be the crucial problem in cellular bioenergetics (Gellerich & Saks, 1982). In the resting muscle the cytosolic ADP is in the micromolar range whereas the concentrations of ATP, creatine phosphate and creatine are in the millimolar range (Wallimann *et al.*, 1992). The low cytosolic ADP concentration is advantageous to the thermodynamic efficiency of the cell work but does not allow an optimal stimulation of oxidative phosphorylation. It is assumed that it is one of the main functions of the creatine phosphate shuttle to transport ADP into the mitochondria at low extramitochondrial ADP concentrations. In spite of extensive studies, the mechanism of the creatine phosphate

*) Fakultät für Biologie, Universität Konstanz, Germany

NATO ASI Series, Vol. H 83
Molecular Biology of Mitochondrial Transport Systems
Edited by M. Forte and M. Colombini

shuttle is not yet completely understood. It was suggested that the compartmentalized creatine kinase isoenzymes may transport ADP by means of a metabolite shuttle (For a recent review see Wallimann *et al.*, 1992). The general mechanism of metabolite shuttle is

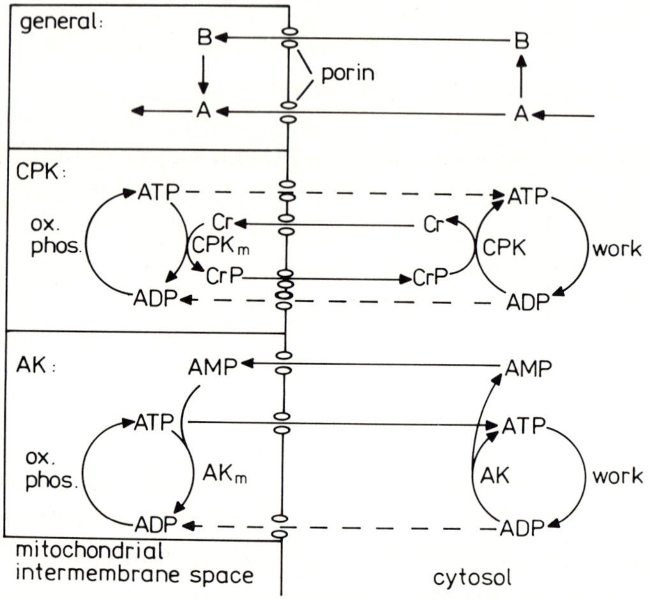

Fig.1 Scheme representing the mechanism of ATP transport into the mitochondrial intermembrane space by means of a metabolite shuttle.
General: General mechanism modified according to Meyer *et al.*, 1984. A,B, metabolites.
CPK, CrP-shuttle; CrP, creatine phosphate; Cr, creatine; Ox.Phos., oxidative phosphorylation; CPKm, CPK, mitochondrial and cytosolic creatine kinase and adenylate kinase.
AK, Adenylate kinase shuttle; AKm, AK, mitochondrial and cytosolic adenylate kinase.
Dashed arrows: transport rates of minor importance at low substrate concentrations.

shown in the upper part of Fig. 1. In contrast to other well established metabolite shuttles such as the malate/aspartate shuttle which transport metabolites (hydrogen) through the mitochondrial inner membrane into the matrix space, metabolite shuttles into the mitochondrial intermembrane space are considered here. Therefore, these shuttles do not require translocator proteins, since their metabolites can diffuse through the porin pores. The metabolite A is reversibly transformed into the metabolite B. Both or only B diffuse into the intermembrane space, thus increasing the transport rate of A into the compartment. It is a prerequisite for a metabolite shuttle that the reaction A $<===>$ B has different directions in both compartments. In the lower part of Fig. 1, two shuttles are shown which are probably special forms of the general mechanism. One is the widely accepted creatine phosphate shuttle (Wallimann *et al.*, 1992). As in the general scheme, the mitochondrial and the extramitochondrial creatine kinases act in different directions. Furthermore, this shuttle needs a sufficiently high creatine concentration since creatine instead of ADP (or both at elevated ADP concentrations) has to diffuse into the mitochondrial intermembrane space forming there ADP via mitochondrial creatine kinase. As a second ADP shuttle we propose the adenylate kinase shuttle acting in a way similar to that of the creatine phosphate shuttle. It may operate in tissues with sufficiently high adenylate kinase activities such as liver, some types of spermatozoa or muscles. This could be possible, since mitochondrial and cytosolic adenylate kinase isoenzymes are compartmentalized in a way similar to that of the creatine kinase isoenzymes. In this shuttle, AMP carries ADP equivalents (just as creatine) into the mitochondria: AMP formed in the cytosol from ADP via cytosolic adenylate kinase diffuses into the intermembrane space forming there ADP via mitochondrial adenylate kinase, and this ADP stimulates the oxidative phosphorylation.

Concentration Gradients and Dynamic Compartmentation

For thermodynamic reasons, the mitochondrial and the extramitochondrial enzymes can work in different directions only if the metabolite concentrations are different in the immediate vicinity of the compartmentalized isoenzymes. Since both pools are not separated from each other by closed membranes but by porin pores which keep open about 1% of the mitochondrial surface, these concentration differences required can exist as dynamic gradients only. The dynamic AdN compartmentation in the intermembrane space of mitochondria from heart (Gellerich & Saks, 1982; Gellerich *et al.*, 1987), liver (Gellerich, 1992), and brain (Kottke *et al.*, 1991) has been demonstrated previously. To obtain quantitative data concerning the diffusion barrier we analyzed heart mitochondrial respiration *vs.* addition of increasing pyruvate kinase activity. Stationary respiratory rates of rat heart mitochondria were adjusted by addition of either creatine or glucose plus yeast hexokinase (Fig. 2). Then, increasing amounts of pyruvate kinase were added additionally. It is evident that the minimal respiratory rate adjustable by excessive pyruvate kinase in the creatine kinase system was substantially elevated when compared to the hexokinase system. At the same rate of respiration (50 nmol O_2/mg/min), the extramitochondrial ATP/ADP ratio was 46 and 150 (arrows) in the HK and in the CPK system, respectively. Since mitochondria phosphorylate in both systems at the same velocity, it should be expected that in the immediate vicinity of the adenine nucleotide translocator the same ATP/ADP ratio occur. Unfortunately, there is no possibility to detect directly the concentrations of adenine nucleotides in the intermembrane space. We, therefore, measured extramitochondrial AMP

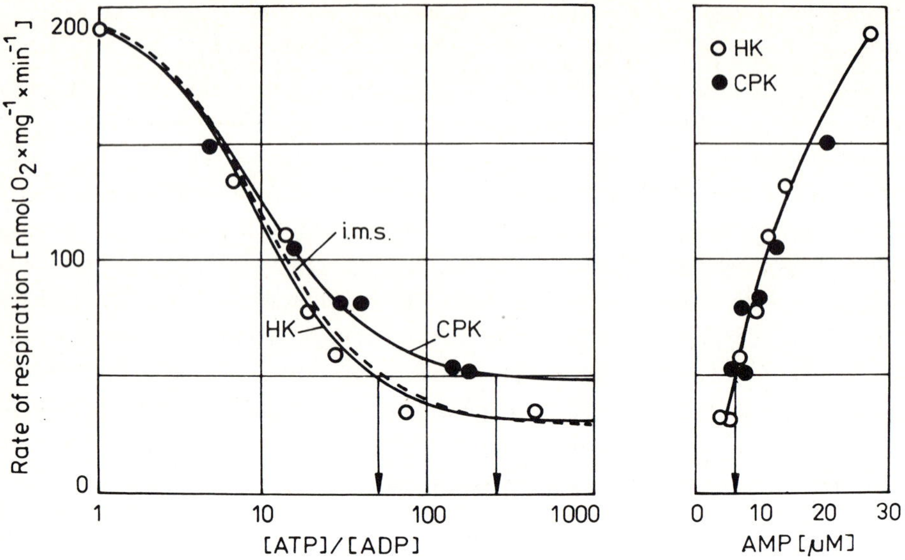

Fig.2 Effect of a different localization of ADP-regenerating enzymes on the dependence of mitochondrial phosphorylation on extramitochondrial ATP/ADP ratios and AMP concentrations in reconstituted systems. Rat heart mitochondria (0.14 mg/ml) were incubated with 5 mM PEP, 4 mM glutamate, 2 mM malate and either 2.2 U/mg yeast HK or 25 mM creatine and increasing amounts of PK. Line i.m.s. represents the computed dependence of respiration on the [ATP]/[ADP] ratio in the intermembrane space (Gellerich *et al.*, 1987).

which is capable of indicating the ATP/ADP ratio in the intermembrane space where mitochondrial adenylate kinase metabolically connects AMP with ATP and ADP. AMP diffuses into the extramitochondrial space like ADP and ATP do, but unlike the latter two, AMP cannot be metabolized there. As shown in Fig. 2, the AMP concentration near the rate of 50 nmol O_2/mg/min is virtually the same in both systems (approx. 7 μM). Hence, it is concluded that ATP/ADP ratios in the intermembrane space are in effect identical in both systems. As described previously (Gellerich *et al.*, 1987) a diffusion rate constant of $(8.7 \pm 4.7)\ 10^4$ μl/mg/min for ADP and ATP was estimated, resulting in rate-dependent concentration gradients of up to 13.7 μM AdN between the extramitochondrial space and the AdN-translocator at the maximum rate of oxidative phosphorylation of rat heart mitochondria. Since the cellular ADP and ATP concentrations are much different from each other we simulated the effect of the metabolite concentration on the extent of its dynamic compartmentation in the intermembrane space and used for that a mathematical model of the complete reconstituted system. As shown in Fig.3, we varied the diffusion rate for ADP

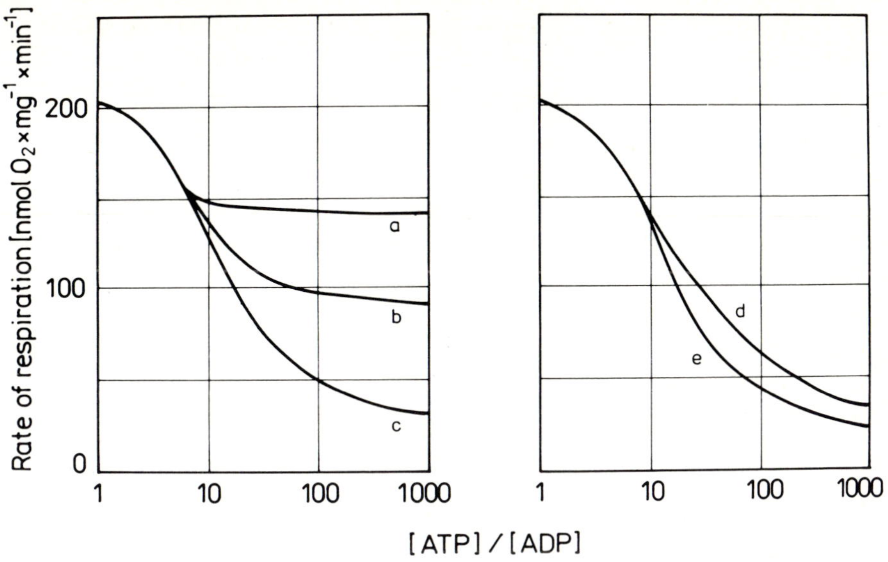

Fig.3 Simulated effects of separated ATP or ADP diffusion limitations on the mitochondrial respiration *vs.* extramitochondrial ATP/ADP ratios in the presence of pyruvate kinase. Curves (a)-(e) were computed as follows: (Left) with unlimited ATP diffusion ($k_T = 10^6$) but limited ADP diffusion $k_D = 10^3$ (a), 10^4 (b) and 10^5 (c) µl/mg/ml respectively. (Right) Unlimited ADP diffusion ($k_D = 10^6$) but limited ATP diffusion ($k_T = 10^3$ (d), 10^4, 10^5 (e) µl/mg/ml respectively (Schlame, 1983).

but maintained the rate of ATP diffusion (left part of Fig.3) and vice versa (right). When the diffusion rate constant (k_D) was decreased the ADP compartmentation in the intermembrane space increased which was indicated by increasing rates of respiration. In contrast, however, the decreasing ATP diffusion rate had virtually no influence on the shape of the curves (right part of Fig. 3). This different behaviour can be explained by the different concentrations of adenine nucleotides. At an AdN concentration of 1 mM and an ATP/ADP ratio of about 100 the ADP and the ATP concentrations were 10 and 1,000 µM respectively. Therefore, a concentration gradient of 10 µM is equivalent to 100% of the total ADP concentration but to only 1% of the total ATP concentration. These results support the assumption that, despite the possible existence of concentration gradients for all metabolites which are exchanged across the mitochondrial outer membrane, the ADP compartmentation only is important to regulation of the cellular energy metabolism. These ADP gradients probably contribute to the concentration differences which are required for the ADP transport by a metabolite shuttle. In the dog heart, the measured concentrations of ATP, CrP and Cr were 5, 9.3 and 11 mM, respectively (Balaban, 1989). From the CPK equilibrium an ADP concentration of 56 µM ADP was calculated. At a maximum transport rate an AdN gradient of at least 14 μM should exist between the intermembrane space and the cytosol; then, the ADP concentration is 49 μM and 63 μM in both pools. Under these conditions the ratio of the equilibrium and the actual metabolite concentrations is 1.11 in the intermembrane space and 0.90 in the cytosol, indicating that indeed, from the thermodynamic viewpoint, the creatine kinases could work in different directions. This calculation was done with that concentration gradient determined for rat heart mitochondria incubated in a normal isotonic medium (Gellerich *et al.*, 1987). We assume, however, that these gradients are even more pronounced in the intact cell.

EFFECT OF ONCOTIC PRESSURE ON FREQUENCY OF BOUNDARY MEMBRANE CONTACT SITES

In the intact cell, mitochondria were embedded in a 30% protein solution (Cameron & Fullerton, 1990), while isotonic media were used without macromolecules to isolate and investigate mitochondria. Because of the missing oncotic pressure, mitochondria exhibit a large intermembrane space in these media compared to mitochondria (Bakeeva *et al.*, 1972) *in vivo*. The artificially increased mitochondrial intermembrane space can be reduced by adding macromolecules as polyvinyl pyrrolidone, ficol or albumin to isolation and incubation media (Bakeeva *et al.*, 1972; Wrogemann *et al.*, 1985). Contact sites between the two mitochondrial envelope membranes have been described as specific structures which are important to the uptake of mitochondrial precursor proteins (Schweiger *et al.*, 1987) and the mitochondrial energy transfer (Brdiczka, 1990; Brdiczka, 1991). To investigate whether or not the volume reduction of mitochondrial intermembrane space is accompanied by an increase in contact sites between the two envelope membranes we performed freeze fracture experiments with rat liver mitochondria incubated in 10% dextran 70 under conditions of active and resting respiration. The presence of dextran induced the frequency of contact sites to 156% (active state) and 253% (resting state), while the volume of the intermembrane space strongly decreased (Wicker *et al.*, 1992). These findings led to the question whether or not the exchange of extra- and intramitochondrial adenine nucleotides through the porin pores might be influenced by macromolecules.

EFFECT OF ONCOTIC PRESSURE ON FUNCTIONAL PROPERTIES OF MITOCHONDRIA AND KINETIC PROPERTIES OF SOLUBLE ENZYMES

Before analyzing the effect of macromolecules on the exchange of adenine nucleotides across the mitochondrial outer membrane we studied their influence on basic properties of heart and liver mitochondria. As previously shown (Gellerich *et al.*, 1992) the addition of 10% (w/v) BSA or dextrans did not affect basic mitochondrial parameters. We further investigated the influence of 10% dextran 70 on kinetic properties of soluble hexokinase from yeast, pyruvate kinase from muscle, as well as of released mitochondrial adenylate kinase and creatine kinase. As shown in Table 1, addition of 10% dextran 70 to the assay had no effect on the kinetic properties of the soluble enzymes. Neither the Michaelis constants nor the maximal velocities of the investigated enzymes were altered significantly.

INFLUENCE OF DEXTRAN ON THE EXCHANGE OF ADENINE NUCLEOTIDES ACROSS THE OUTER MITOCHONDRIAL MEMBRANE

Several aspects must be considered to explain the observed reduction of the metabolite exchange between the intermembrane space and the extramitochondrial compartment (Fig. 4): i) due to the increased number of contact sites and the reduced volume of the intermembrane space the resistance of lateral diffusion within the intermembrane space could have been reinforced, ii) the permeability of the outer membrane pore might have been reduced by oncotic stress (Zimmerberg & Parsegian, 1986), iii) the diffusion of metabolites through the homogeneous bulk phase to the pores

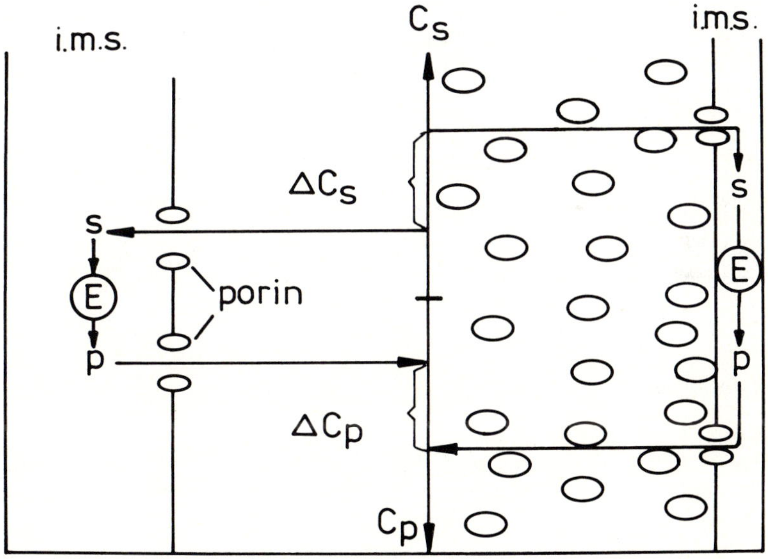

Fig.4 Scheme representing the possible mechanisms of the action of macromolecules on the exchange of adenine nucleotides across the outer mitochondrial membrane. Explanations in the text. C_P, C_S, concentrations of P and S in the bulk phase. Isotonic medium without (left) and with (right) macromolecules. ΔC_s, ΔC_p, concentration gradients increased by the action of dextran, between the bulk phase and the intermembrane space.

In general, these effects increase the diffusion resistance of the exchanging metabolites. The apparent Michaelis constant of an enzyme located in the intermembrane space has two terms: one is the K_M of the homogeneous reaction and the second term reflects the concentration gradient between the compartment where the enzyme is located and the bulk phase (diffusion term). It was clearly shown that dextran did not influence the kinetic constants of homogeneous enzymes (Table 1). Therefore, an increase in K_M reflects changes in the diffusion term into the compartment where the reaction takes place. Additionally, depending on the type of inhibition, the increase in product concentration in the intermembrane space could cause an inhibition of the enzyme rate. To investigate the influence of macromolecules on the flux of adenine nucleotides across the outer mitochondrial membrane we determined the apparent Michaelis constants of adenylate

Table 1. Effect of 10% dextran 70 on kinetic constants of homogeneous and compartmentalized reactions. Kinetic constants of enzymes were determined by means of the optical test, those of the mitochondrial respiration by oxygraphic measurements in the presence of exceeding yeast hexokinase. MP, Mitoplasts from rat heart mitochondria; O.P., oxidative phosphorylation; PK, soluble pyruvate kinase from muscle; HK, soluble yeast hexokinase; AK, adenylate kinase compartmentalized in liver mitochondria and released; CPK, creatine kinase compartmentalized in heart mitochondria, in mitoplasts and released.

REACTION		CONTROL		10% DEXTRAN	
		K_{AdN}	V_{max}	K_{AdN}	V_{max}
		[μM] ADP	[U/mg]	[μM] ADP	[U/mg]
PK	Soluble	372±29	223±33	378±20	216±29
AK	RLM	118±10	0.85±0.02	192±20	0.76±0.03
AK	Released	168± 9	0.90±0.02	172± 9	0.90±0.02
		[μM]ATP	[U/mg]	[μM] ATP	[U/mg]
HK	Soluble	306±15	280±4	312±13	282±3
CPK	RHM	341±50	3.1±0.1	553±65	3.1±0.1
CPK	MP	248±30	1.9±0.2	237±35	1.8±0.10
CPK	Released	181±19	1.8±0.0	180±16	1.8±0.03
		[μM]ADP	[nmolO_2/mg /min]	[μM]ADP	[nmolO_2/mg /min]
O.P.	RLM	46±5	67± 3	60± 7	66± 4
O.P.	RHM	40±5	132±10	85±11	140±10

kinase, creatine kinase or of the oxidative phosphorylation. Addition of dextran had significant effects exclusively on the compartmentalized enzymes compared to the soluble enzymes (Table 1). The K_{ADP} of mitochondrial adenylate kinase increased from 107 to 165 μM ADP and V_{max} decreased from 0.92 to 0.82 U/mg in the presence of dextran. The effect of dextran on creatine kinase of heart mitochondria was even more pronounced (341 $\pm$ 50 μM ATP without and 553 $\pm$ 65 μM ATP with dextran). Oxygraphic measurements were conducted to determine the apparent Michaelis constant for the oxidative phosphorylation of heart and liver mitochondria. In the absence of dextran, both types of mitochondria exhibited a similar K_{ADP} (46 $\pm$ 5 μM ADP for liver and 40 $\pm$ 5 μM ADP for heart mitochondria). In the presence of dextran, the K_{ADP} showed a more marked increase for heart mitochondria (85 $\pm$ 11 μM) than for liver mitochondria (60 $\pm$ 7 μM). In contrast to the increasing Michaelis constants the maximal rates were not changed by the action of dextran.

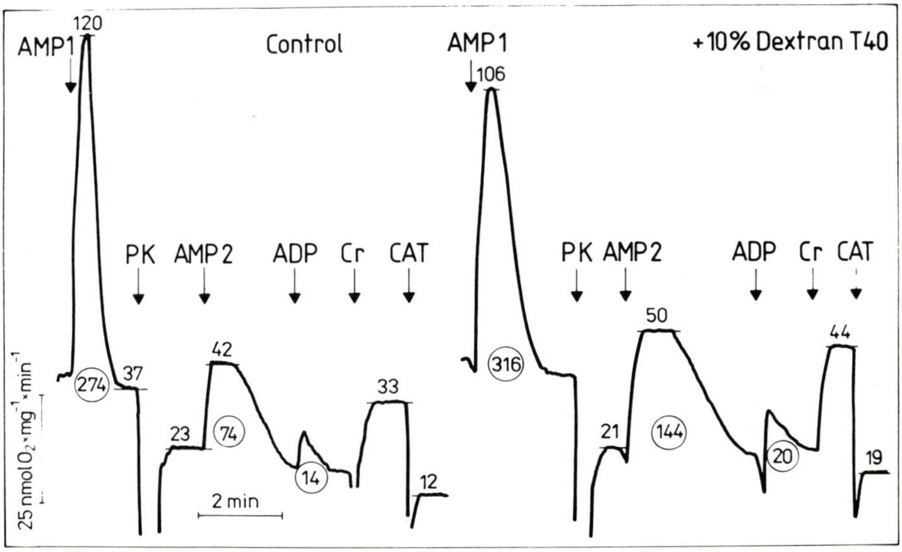

Fig. 5 Effect of 10% dextran on the dynamic compartmentation in the mitochondrial intermembrane space. Incubation of rat heart mitochondria (0.64 mg/ml) as in Fig. 2. Additions: AMP1 and AMP2, 84 and 167 μM AMP respectively; PK, 115 U/mg pyruvate kinase, ADP, 500 μM ADP and 59 μM AMP, Cr, 25 mM creatine; CAT, 1 μmol CAT/mg. The numbers in the circles represent the oxygen consumption after ADP or AMP addition (nmol O_2/mg). The numbers on the curves (first deviation of the oxygen-time signal) indicate the rate of respiration (nmol O_2/min).

The liberation of ADP from the intermembrane space was investigated with functionally intact mitochondria and soluble pyruvate kinase both competing for the ADP produced in the intermembrane space by adenylate kinase or creatine kinase. As can be seen from Fig. 5 the presence of 10% dextran increased the AMP-induced oxygen consumption (relative oxygen consumption) from 74 nmol O_2/mg/min (14% of that without pyruvate kinase) to 144 nmol O_2/mg/min (23%). Also, the creatine-induced rate of respiration increased from 33 to 44 nmol O_2/mg/min. Obviously, the elevated oncotic pressure reduces the permeability of pores which, in turn, requires higher ADP concentration gradients. The higher ADP gradients cause greater ADP concentrations in the intermembrane space, thus favouring oxidative phosphorylation of ADP.

Experiments similar to those shown in Fig. 5 were performed with rat liver mitochondria. In the presence of increasing amounts of dextrans with differing molecular weights the relative AMP-induced oxygen consumption was determined. The AMP-induced oxygen consumption was plotted versus concentration [mM] and content [mg/ml] of the various dextrans (Fig. 6). The amount of ADP consumed by oxidative phosphorylation in the presence of pyruvate kinase was 10% of that in the absence of the enzyme. However, correlated to gradual additions of dextran, the ADP increased which was retained in the intermembrane space and utilized. At 10% dextran it was almost three times higher than in the absence of the macromolecules. Plotting the data *versus* the molar dextran concentration [mM] (Fig. 6) revealed that ADP compartmentation increased with the molecular weight of dextrans examined.

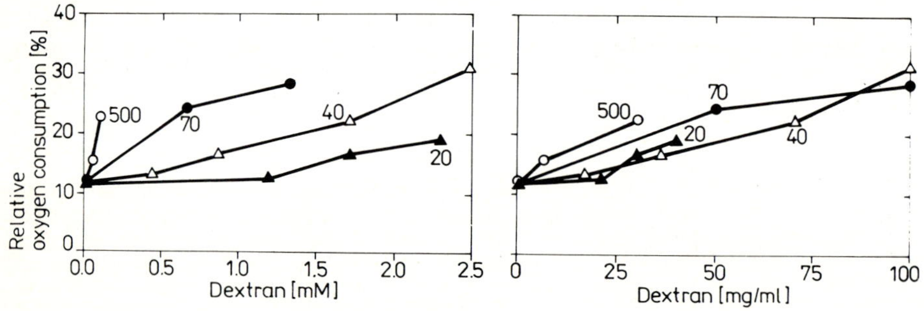

Fig. 6 Effect of different dextrans on the ADP compartmentation in the intermembrane space. Rat liver mitochondria (0.9-1.2 mg/ml) were incubated as in Fig. 5 but with different concentrations of different dextrans as indicated. The AMP-induced oxygen consumption in the absence and in the presence of pyruvate kinase of the same incubation was taken as a measure of the ADP compartmentation in the intermembrane space. 500, 70, 40, 20, dextrans with molecular weights of 500, 70, 40, and 20 kDa, respectively.

Assuming that the mitochondrial outer membrane was essential to the dextran effects on dynamic AdN compartmentation, we investigated the kinetic properties of mitochondrial creatine kinase in heart mitochondria with disrupted outer membrane in the absence and in the presence of dextran. Since in this system the mitochondrial creatine kinase remains bound to the inner membrane a net transport of adenine nucleotides occurs between the mitochondrial membrane and the bulk phase. This transport is comparable to that occurring in intact mitochondria except for the transport through the pores and concentration gradients in the intermembrane space which are absent in digitonin-treated mitochondria. As shown in Table 1, after disruption of the mitochondrial outer membrane the dextran effect on the K_{ATP} disappeared completely indicating that the net transport through the homogeneous phase was not influenced by dextran. In view of these results, there are only two possibilities which may account for the diffusion resistance observed: i) the decreased intermembrane space and ii) the oncotic stress on the porin pores. To discriminate between these both possibilities experiments were performed with isolated porin pores reconstituted in black membranes.

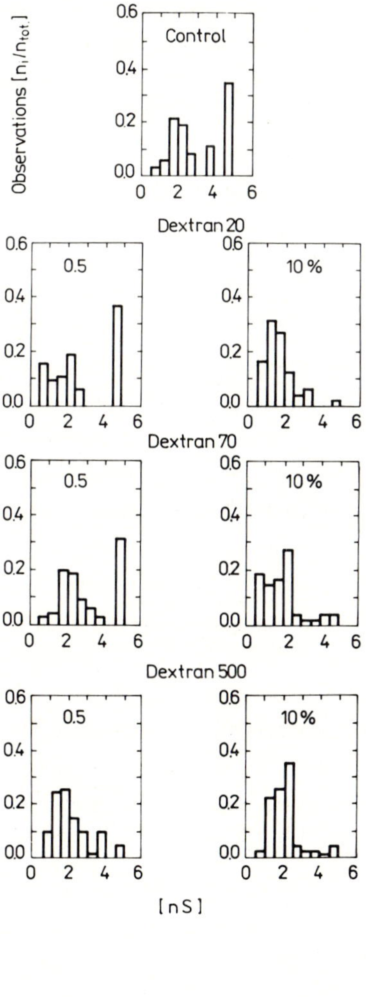

Fig. 7 Effect of different dextrans on the conductivity of porin pores inserted into "black" membranes. Histograms of conductance fluctuations of pores in membranes formed of diphytaloyl phosphatidylcholine/n-decane in 1M KCl solution obtained as described previously (Benz, 1990). Two ng/ml rat liver porin were used and 10 mV were applied. The histograms were obtained from 50 to 110 single insertions into the membrane.

THE ROLE OF PORIN PORES

Experiments were performed with isolated and purified pore protein added to both sides of the diphytanoyl lecithin membrane. Dextrans of different molecular weights (20, 70 or 500 kDa) were present at a concentration of either 10% or 0.5 mM in the medium. The current at 10 mV membrane potential increased stepwise due to the insertions of pores into the membrane.

Histograms of 40 to 100 single current steps at 10 mV are shown in Fig. 7. In the absence of dextran (control) 35% of the incorporated pores had a conductance of about 5 nS. In the presence of 10% dextran virtually all pores inserted into the membrane had a lower conductance. When 0.5 mM of the various dextrans were added, the frequency of smaller current steps increased correlated to the increase of molecular weight of the dextrans. It was shown that addition of macromolecules to reconstituted porin pores (Zimmerberg & Parsegian, 1986) caused a change in the pore structure to a low conductance state. Our results are completely consistent with these observations. In addition, we observed that these effects did not correlate with the concentration but were dependent on the molecular weight of the added macromolecules.

Since it was assumed that the oncotic stress on the porin is caused by the oncotic pressure (Zimmerberg & Parsegian, 1986) we measured the dependence of the oncotic pressure on the molar concentration [mM] and the content [mg/ml] of different dextrans. Plotting the data *versus* the molar concentration exhibited a correlation with the molecular weight (Gellerich *et al.*, 1992) which was very similar to the effect on the ADP compartmentation (Fig. 6). Obviously, the oncotic pressure was, under these conditions, not a colligative property, but dependent on the molecular weight of the dextran.

CONCLUSIONS

Addition of 10% (w/v) dextran to reconstituted systems consisting of isolated mitochondria and soluble enzymes allowed to investigate the influence of oncotic pressure on the exchange of adenine nucleotides between the intermembrane space and the cytosol.

Dextrans significantly increased the Michaelis constants of enzymes localized in the mitochondrial intermembrane space but had no influence on their maximal rates. Dextrans decrease the outflow of ADP formed by mitochondrial creatine kinase and adenylate kinase through the porin pores, thus increasing the rate of oxidative phosphorylation in a pyruvate kinase containing system. From these results we concluded that extramitochondrial macromolecules remarkably increase the dynamic ADP compartmentation within the intermembrane space.

The mitochondrial porin pores obviously occur in the intact cell at a low permeability state for ADP due to the oncotic pressure exerted by the cytosolic proteins. Therefore, it is an important problem in cellular energy transport to shuttle ADP across the porin pores into the intermembrane space. This can be done either by direct binding of ADP-generating enzymes such as hexokinase or glycerol kinase to porin (Wicker *et al.*, 1992, and accompanying lecture) or by metabolite shuttles such as the creatine phosphate shuttle or the newly proposed adenylate kinase shuttle. The mitochondrial and the cytosolic isoenzymes involved in these metabolite shuttles have to work in different directions which is only possible at sufficiently high concentration gradients across the mitochondrial outer membrane. The experimentally determined dynamic ADP concentration gradients of up to 13 μM are adequate to account for different directions of mitochondrial and extramitochondrial creatine kinases in the presence of virtually unchanged concentrations of ATP, Cr and CrP.

ACKNOWLEDGEMENTS: This work was supported by grants of Deutsche Forschungsgemeinschaft given to F.N. Gellerich. (Ge 663/1-1) and D. Brdiczka (773/3-2). The authors thank Prof.Dr. W. Kunz, Prof.Dr. R. Bohnensack and Dr. W. Kunz for stimulating discussions, as well as Dr. M. Schlame and Prof.Dr. R. Bohnensack for computer simulation of the results shown in Figs. 2 and 3.

REFERENCES

Bakeeva LE, Chentsov YS, Jasaitis AA Skulachev VP (1972) The effect of oncotic pressure on heart muscle mitochondria. Biochim Biophys Acta **275**:319-332

Balaban R, (1989) Control of mitochondrial respiration in the heart *in vivo*. Mol Cell Biochem **89**:191-197

Benz R, Kottke M, Brdiczka D (1990) The cationically selective state of the mitochondrial outer membrane pores: A study with intact mitochondria and reconstituted mitochondrial porin. Biochim Biophys Acta **1022**:311-318

Bessman SP, Gots RE (1975) The hexokinase acceptor theory of insulin action - Hormone control of functional compartmentation. Life Sci **16**:1215-1225

Brdiczka D (1990) Interaction of mitochondrial porin with cytosolic proteins. Experienta **46**:161-167

Brdiczka D (1991) Review: Contact sites between mitochondrial envelope membranes. Structure and function in energy- and protein transfer. Biochim Biophys Acta **1071**: 291-312

Cameron IL, Fullerton GD (1990) A model to explain the osmotic behavior of hemoglobin and serum albumin. Biochem Cell Biol **68**:894-898

Gellerich FN, Augustin HW (1977) Studies on the functional significance of mitochondrial bound hexokinase in rabbit reticulocytes. Acta Biol Med Germ **36**:571-577

Gellerich FN, Saks VA (1982) Control of heart mitochondrial oxygen consumption by creatine kinase: The importance of enzyme localization. Biochem Biophys Res Commun **105**:1473-1481

Gellerich FN, Schlame M, Bohnensack R, Kunz W (1987) Dynamic compartmentation of adenine nucleotides in the mitochondrial intermembrane space of rat heart mitochondria. Biochim Biophys Acta **722**:381-391

Gellerich FN (1992) The role of adenylate kinase in dynamic compartmentation of adenine nucleotides in the mitochondrial intermembrane space. FEBS Lett **297**:55-58

Gellerich FN, Wagner M, Kapischke M, Wicker U, Brdiczka D (1992) Effect of macromolecules on the regulation of the mitochondrial outer membrane pore and the activity of adenylate kinase in the intermembrane space Biochim Biophys Acta; submitted

Gosalvez M, Perez-Garcia J, Weinhouse S (1974) Competition for ADP between pyruvate kinase and mitochondrial oxidative phosphorylation as a control mechanism in glycolysis. Eur J Biochem **46**:133-140

Kottke M, Adams V, Wallimann T, Nalam VK, Brdiczka D (1991) Location and regulation of octameric mitochondrial creatine kinase in the contact sites. Biochim Biophys Acta **1061**:215-225

Meyer RA, Sweeny HL, Kushmerick MJ (1984) A simple analysis of the "phosphocreatine shuttle". Am J Physiol **246**:C365-C348

Saks VA, Chernousova GB, Gukovski DE, Smirnov VN, Chazov EI (1974) Studies of energy transport in heart cells. Mitochondrial isoenzyme of creatine phosphokinase, kinetic properties and regulatory action of Mg ions. Eur J Biochem **57**:273-290

Schlame M (1985) Einfluß der Membranbindung mitochondrialer Phosphokinasen auf den Umsatz von Adeninnucleotiden. Thesis, Medizinische Akademie Magdeburg.

Schwaiger M, Herzog V, Neupert W (1987) Characterization of translocation contact sites involved in the import of mitochondrial proteins. J Cell Biol **105**:235-246

Wallimann T, Wyss M, Brdiczka D, Nicolay K, Eppenberger HM (1992) Significance of intracellular compartmentation, structure and function of creatine kinase isoenzymes for cellular energy homeostasis: The Phospho-Creatine Circuit. Biochem J **281**: 215-225

Wicker U, Bücheler K, Gellerich FN, Wagner M, Kapischke M, Brdiczka D (1992) Effect of macromolecules on the structure of mitochondrial intermembrane space and the regulation of hexokinase. Biochim Biophys Acta; submitted.

Wittenberg JB (1970) Myoglobin-facilitated oxygen diffusion: Role of myoglobin in oxygen entry into muscle. Physiol Rev **50**:559-636

Wrogemann K, Nylen EG, Adamson I, Pande SV (1985) Functional studies on *in situ*-like mitochondria isolated in the presence of polyvinyl pyrrolidon. Biochem Biophys Acta **806**:1-8

Zimmerberg J, Parsegian VA (1986) Polymer inaccessible volume changes during opening and closing of a voltage-dependent ionic channel. Nature **323**:36-39

RESTRICTIONS OF METABOLITE PERMEATION THROUGH THE OUTER MITOCHONDRIAL MEMBRANE OF PORIN-DEFICIENT YEAST MUTANT.

J.Michejda, H.Kmita, O.Stobienia, M.Budzińska and G.J.-M.Lauquin[+]
Department of Bioenergetics, Poznań University, Fredry 10, 61701 Poznań Poland
[+]Laboratoire de Physiologie Moleculaire et Cellulaire, IBCN-CNRS, 1 rue Camille St Saens, 33077 Bordeaux Cedex, France

Abstract

Several limitations in transport properties of the outer mitochondrial membrane were found in intact mitochondria of porin-deficient mutants of S. cerevisiae, which grows as efficient as the isogenic wild strain. Restrictions concerned the effectiveness of ADP and CATR in triggering and terminating the phosphorylating state and of the external NADH in supporting the electron flow, with all three types of molecules apparently competing for the entry. These restrictions disappeared in mitoplasts obtained from mutant mitochondria. The existence of another pore is postulated. The novel pore seems to be controlled by cations which might act either directly on the pore or indirectly by screening the negative charge of the transported species. The novel pore is similarly sensitive to the synthetic polyanion as is VDAC; in the absence of VDAC the outer mitochondrial membrane is easily damaged by this inhibitor.

Keywords

Yeast mitochondria / Mitochondrial outer membrane / VDAC / Porin-free mutant / Novel pore / Transport restrictions / Bioenergetics / Synthetic polyanion

Introduction

Mitochondrial porin, encoded in the nucleus, is the major integral protein present in thousands of copies in the outer mitochondrial membrane. Porin has been isolated, sequenced, manipulated genetically, and crystallized in the plane of the outer membrane (for recent review, see Mannella, 1992). Incorporated into planar phospholipid bilayers, porin forms anion channels - known as VDAC - which reveal a voltage dependent gating (Colombini, 1979) and allow for a non-specific passage of low-molecular weight hydrophilic solutes. The functional

NATO ASI Series, Vol. H 83
Molecular Biology of Mitochondrial Transport Systems
Edited by M. Forte and M. Colombini

properties of VDAC estimated in reconstituted system, i.e. a diameter of approximately 2 nm, a defined exclusion limit of about 2500 Da and a slight anion selectivity, should make VDAC in its open state a crucial device for the passage of substrates and metabolites through the outer membrane (Benz, 1985). Therefore, the physiological control of VDAC conductance would be vital for the energetic and metabolic functions of mitochondria.

In the reconstituted system such control is exerted by the moderate increase in transmembrane potential which switches VDAC to a lower conductivity state (closed substate) with reduced effective pore size and reverses the anion selectivity of the channel towards cation selectivity (Ludwig et al., 1989). The voltage sensor of VDAC is a possible target for substances capable of reducing the conductance of VDAC in the reconstituted system, like the synthetic polyanion (Colombini et al., 1987; Benz et al., 1990) and the soluble protein extracted from various types of mitochondria (Holden and Colombini, 1988; Liu and Colombini, 1991, 1992).

In intact mitochondria, Konig's synthetic polyanion was found to decrease the permeability of the outer mitochondrial membrane for adenine nucleotides (Benz et al., 1988; Benz and Brdiczka, 1992), the substrates for adenilate kinase and creatine kinase, enzymes located in the intermembrane space. More recently, the soluble mitochondrial protein was found to similarly reduce the adenilate kinase activity in intact mitochondria and, moreover, to decrease the ADP-dependent phosphorylating state respiration, an intrinsic function of the inner mitochondrial membrane (Liu and Colombini, 1992). These findings support the view that VDAC is the device by which the outer membrane participates in the overall regulation of mitochondrial functions while the soluble mitochondrial protein might play a modulatory role by lowering the threshold of the membrane potential needed for the closure of VDAC.

Therefore, the viability of yeast mutants depleted of porin, grown on a nonfermentable carbon source, evokes understandable interest. One such mutant obtained by Dihanich et al. (1987) revealed almost a 4 time decrease in cytochrome content and a strong retardation of growth including a long adaptation time to the glycerol medium. However, other mutants constructed by Guo and Lauquin (1986, 1988) were growing quite similarly to the wild isogenic strain and required no adaptation time, if pH 5.5 was controlled in the growth medium (Michejda et al., 1990). Moreover, the cells of the latter mutants grown on glycerol revealed a tight coupling in situ and their mitochondria were coupled in vitro (Michejda et al., 1988, 1989, 1990).

Reconstituted in a planar bilayer the bulk proteins isolated from the outer membrane of porin depleted mutant obtained by Dihanich et al. (1987), revealed the presence of transient pores. These pores were found to be slightly cation selective, of a smaller conductance at 0.21 nS in 0.1 M KCl (wild strain 0.4 nS) and of an estimated diameter of 1.2 nm, compared to that of 1.7 nm in the wild type yeast mitochondria (Dihanich et al., 1989; Benz et al., 1989).

These data are in agreement with the results obtained in our laboratories with intact mitochondria of other porin-depleted mutants. The studies were performed with physiological methods implying bioenergetic tests for measuring the restrictions in metabolite access to the inner mitochondrial membrane. It was shown that in mutant mitochondria, compared to wild type mitochondria, much higher concentrations of ADP and the presence of Mg^{2+} are necessary for effective triggering of the phosphorylating state supported by ethanol (Michejda et al., 1988, 1990). Therefore, the existence of a novel pore was postulated (Michejda et al., 1989).

In the present paper we report further characteristics of the limitations in metabolite and inhibitor transport in mitochondria of porin-depleted mutant B5. Because all restrictions in the access of NADH, adenine nucleotides and carboxyatractylate to the inner membrane (or intermembrane space) disappeared with the removal of the outer membrane, we conclude that the restrictions truly reflect the transport properties of a novel pore in intact mutant mitochondria. Moreover, it was found that, like VDAC, the novel pore is sensitive to the synthetic polyanion which, at higher concentrations, easily induces the rupture of the outer membrane lacking porin.

Materials and Methods

The wild strain of yeast Saccharomyces cerevisiae (DBY747, Mata, ura3-52, leu2-3, leu2-112, his3- Δ1, trp1-289) and porin-free mutant B5, derived from the above parent strain (Guo 1988; Michejda et al., 1989) were the source of mitochondria. Cultures were grown at 28°C on YP medium, at pH 5.5. Glycerol 3% was used as a non-fermentable carbon source. Mitochondria were isolated essentially as described by Daum et al. (1982) from the cells grown to the late logarithmic phase. The recovery of mitochondrial protein in the mutant strain was not lower than in the wild one. Mitoplasts were obtained by "swelling-shrinking" procedure, essentially as described by Daum et al. (1982).

The standard incubation medium (2.5 ml) contained: 0.65 M mannitol, 10 mM Hepes pH 6.9, 10 mM KPi, 5 mM KCl, 0.2% BSA (defatted), 0.8-1.5 mg mitochondrial protein. Other additions are specified in the legend of figures.

The membrane potential of mitochondria, $\Delta\psi$, was measured at 25°C (simultaneously with oxygen uptake) according to Kamo et al. (1979) with TPP^+-specific electrode, inserted into a Rank oxygen electrode vessel. The degree of integrity of the outer mitochondrial membrane was measured with exogenous cytochrome c essentially according to Douce et al. (1984).

Adenylate kinase activity was determined spectrophotometrically (Brdiczka et al., 1968) in 2 ml medium containing: 0.6 M mannitol, 0.1 mM EGTA, 10 mM Tris-HCl pH 7.4 and 5 mM $MgCl_2$ and 1 mM KCN. The presence of KCN was necessary to inhibit the oxidation of the external NADH by the respiratory chain of yeast mitochondria. Treatment of isolated mitochondria with synthetic polyanion (synthesized by Dr T.Konig and kindly given by Dr L.Wojtczak) was performed as described by Benz et al. (1988).

Results

With ethanol as a freely permeating respiratory substrate, the parallel recordings of membrane potential ($\Delta\psi$) and oxygen uptake performed in the same incubation revealed that in mitochondria of several mutants depleted of VDAC the rate of electron flow and the level of generated membrane potential were closely comparable to those observed in mitochondria of isogenic wild strains (Michejda et al., 1989). This finding proved that destroying the porin gene had no major impact on the respiratory chain of these mutants. However, two main changes were found in mutant mitochondria as compared to wild type mitochondria. The effective triggering of the phosphorylating state in mutant mitochondria required much higher concentrations of ADP ($K_{0.5}$ 200 µM) than in wild type mitochondria (40 µM) and was strongly dependent on the presence of Mg^{2+} or Ca^{2+} ions in the incubation.

In B-5 mutant mitochondria in the presence of 400 µM ADP (twice $K_{0.5}$), allowing for 90 percent of the maximal state 3 respiration, the respiratory control ratio (RC) rised sharply from 1.4 in the absence of Mg^{2+} to 3.4 in the presence of 6 - 10 mM Mg^{2+} (Fig. 1).

On the contrary, in isogenic wild strain mitochondria at the same ADP concentration (ten times the respective $K_{0.5}$) the respiratory control slightly decreased with the increase of Mg^{2+} concentration. The shaded field in Fig. 1 illustrates how differently the effectiveness of ADP in triggering the phosphorylating state is dependent on Mg^{2+} concentration in wild type and mutant mitochondria. This dramatic effect of Mg^{2+} was clearly distinct from a

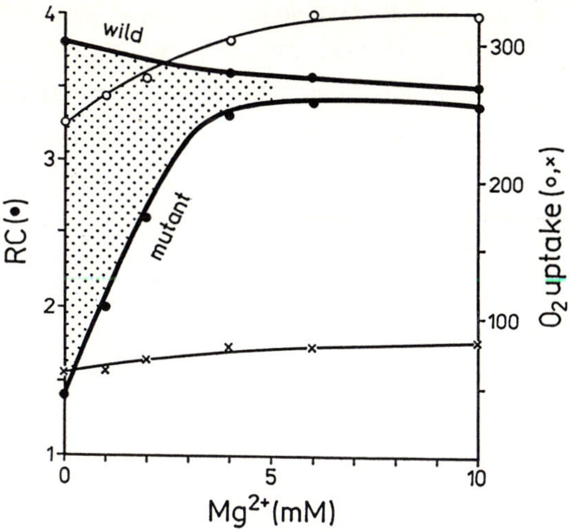

Figure 1. Dependence of the respiratory control ratio, RC (•) and of the oxygen uptake (o , x) on magnesium ion concentration in mutant and wild type mitochondria. Conditions of incubation as in Methods. Additions: 20 mM ethanol, 400 µM ADP (•). Respiratory rate in mutant mitochondria: (x) - state 4, (o) - uncoupled state, in the presence of valinomycin. Shaded field: Mg^{2+}-dependent difference in the RC ratio between mutant and wild type mitochondria. At 0, 1 and 2 mM Mg^{2+} the RC values for mutant mitochondria may be too high, as they were calculated against state 4 previous to ADP addition; at these concentrations of Mg^{2+} it was technically impossible to follow the return of a sluggish state 3 respiration to that of the consecutive state 4.

minor stimulatory influence of Mg^{2+} ions on the electron flux in state 4 and in the uncoupled state observed equally in both types of mitochondria.

Ca^{2+} ions were similarly efficient as Mg^{2+} ions in restoration of a high value of respiratory control ratio in mutant mitochondria (see Fig. 1) and in the faciliation of terminating state 3 by CATR in mutant mitochondria (see Fig. 2). It has to be emphasized that Ca^{2+} ions are not electrogenically accumulated in yeast mitochondria and therefore do not interfere with the level of $\Delta\psi$.

In view of the apparent restrictions in triggering state 3 by ADP in mutant mitochondria, the effectiveness of external NADH as the respiratory substrate was examined. In yeast mitochondria the electrons from external NADH are accepted by the rotenone-insensitive dehydrogenase at the outer surface of the inner mitochondrial membrane. Though the molecular size of NADH is almost double that of ADP, the rate of oxygen uptake and the level of membrane potential in state 4 respiration supported by 1.5 mM NADH were quite similar in mutant and wild type mitochondria (Fig. 2) and closely comparable with those supported by ethanol.

In wild type mitochondria NADH-supported respiration was tightly coupled regardless of the presence or absence of Mg^{2+} with $K_{0.5}$ 40 µM ADP for the respiratory control ratio as with ethanol. Therefore, the small initial prepuls of 50 µM ADP (needed for "conditioning" the energetic transitions in yeast, plant

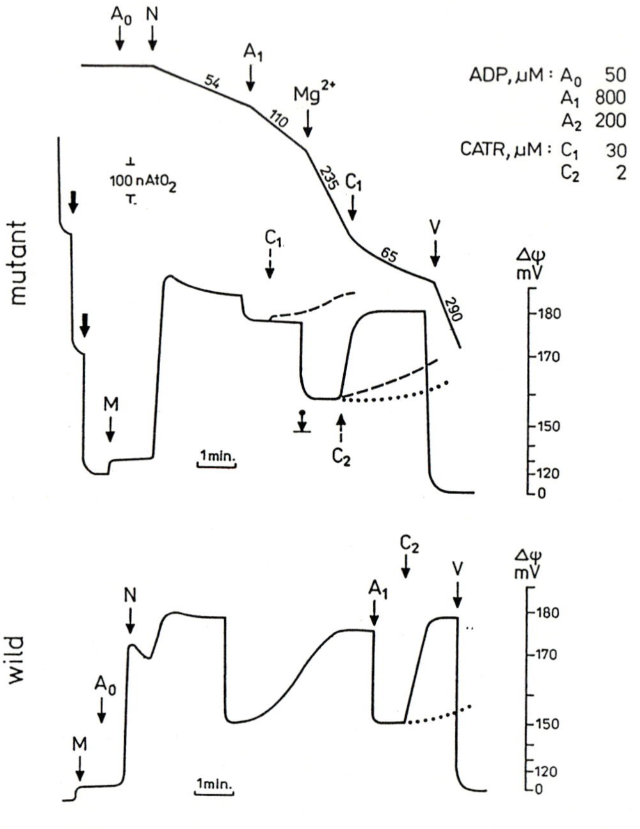

Figure 2. Triggering of the phosphorylating state and its termination by CATR in mutant mitochondria and mitoplasts. Incubation conditions as in Methods. $MgCl_2$ originally omitted in the incubation medium. 0.95 mg of mitochondrial protein (M). Additions: NADH (N) 1.5 mM; ADP (A), various concentrations: $MgCl_2$ (Mg^{2+}) 4 mM; CATR (C), various concentrations (dashed lines); valinomycin (V) 0.15 µg. Dotted lines - in the absence of CATR. Last two additions of TPP^+ (calibration) marked by short arrows; final concentration - 8 µM TPP^+. Scale of $\Delta\psi$(mV) for mitochondria corrected for an unspecific shift in TPP^+ tracing due to the addition of Mg^{2+} (marked with a dotted arrow). Oxygen uptake in nAt . min^{-1} . mg^{-1} mitochondrial protein marked on the oxygraphic tracing. In mitochondria at 400 µM ADP and in the presence of 4 mM Mg^{2+} (not saturating conditions), the values of RC and ADP ratios were 3.3 and 1.6, respectively (not shown).

and protozoan mitochondria) applied prior to the respiratory substrate resulted in a two-step early course of $\Delta\psi$ generation. With the later addition of 200 μM ADP the respiratory control ratio of about 4.0 and ADP:O ratio of 1.6 were obtained. Termination of the phosphorylating state by carboxyatractylate (CATR), an inhibitor of adenine nucleotide translocator, could be easily achieved with 1 μM CATR within 0.5 min as indicated by a return of $\Delta\psi$ level to a value even higher than that in previous state 4 (Fig. 2, lower tracing).

On the contrary, in mutant mitochondria the response of the resting state to a relatively high concentration of 800 μM ADP was only a slight one. (This explains why the prepulse of only 50 μM ADP did not interfere with a fast one-step generation of state 4 membrane potential). However, the fully active phosphorylating state, reflected in a large depression of $\Delta\psi$ and a high respiratory rate, could be initiated by a later addition of Mg^{2+} or Ca^{2+} ions. Termination of this phosphorylating state within 0.5 min required as much as 30 μM CATR in the presence of 4 mM Mg^{2+}; lower concentrations of CATR were less effective. In the absence of Mg^{2+} the effect of 30 μM CATR was much delayed and could only be accelerated by further increasing of CATR concentrations. The dependence of respiratory control ratio and of electron flow on Mg^{2+} concentration in wild type and mutant mitochondria was quite similar to that with ethanol as respiratory substrate (see Fig. 1).

In mutant mitochondria the decrease in pH of the incubation medium from 6.9 to 6.2 apparently facilitated the triggering of state 3 by ADP and its later termination by CATR, largely decreasing the requirement for Mg^{2+} ions for these actions.

The bioenergetic tests applied for quantitation of the constraints imposed on permeation of ADP and CATR in VDAC-depleted mutant mitochondria base on rather complex events involving the efficiency of adenine nucleotide translocator and ATP synthetase. The ability of yeast mitochondria to oxidize external NADH offers a more simple approach of estimation of the substrate concentration half-saturating the respiration. The value of $K_{0.5}^{NADH}$ supporting state 4 respiration (high $\Delta\psi$) in the absence of Mg^{2+} was found to be ten times higher in mutant than in wild type mitochondria (Fig. 3, left). The presence of 4 mM Mg^{2+} reduced this difference by only 30 percent, while the same concentration of Mg^{2+} was able to reduce by 80 percent the value of $K_{0.5}^{ADP}$ for triggering the phosphorylating state supported by ethanol in mutant mitochondria (Fig. 3, right).

The values of $K_{0.5}^{ADP}$ in Figure 3 are given for ethanol as a substrate to avoid interference in a simultaneous transport of the two molecules, ADP and

NADH through the pore in the outer membrane. The existence of such interference in mutant mitochondria was suggested by two observations. (i) $K_{0.5}^{ADP}$ for triggering the phosphorylating state was significantly increased when 1.5 mM NADH replaced ethanol as the respiratory substrate. (ii) In mitochondria, preincubated with CATR, $K_{0.5}^{NADH}$ supporting state 4 oxygen uptake was significantly increased by the addition of 1 mM ADP which could not trigger the phosphorylating state in these conditions. This apparent interference between NADH and ADP passage across the outer membrane was intensified in the absence of Mg^{2+} and was absent from wild type mitochondria.

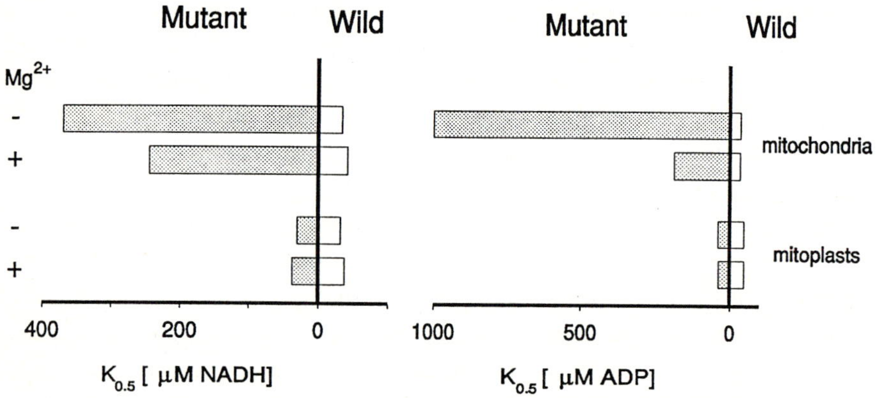

Figure 3. Effect of Mg^{2+} and the removal of outer mitochondrial membrane (mitoplasts) on $K_{0.5}$ values for external NADH-supported state 4 respiration and for ADP-triggered phosphorylating state. $K_{0.5}^{ADP}$ - the concentration of ADP yielding half-maximal stimulation of oxygen uptake (state 4 - state 3 transition) with 20 mM ethanol as the respiration substrate. $K_{0.5}^{NADH}$ - the concentration of external NADH supporting half-maximal oxygen uptake in state 4, calculated from oxygraphic and spectrophotometric assays. Incubation conditions as in Methods; 4 mM $MgCl_2$, if present.

All restrictions on accessibility of NADH, ADP and CATR to intact mutant mitochondria including the competition between these molecules for entry, found with bioenergetic tests disappeared following the removal of the outer membrane. In particular, in mitoplasts obtained from mutant mitochondria, $K_{0.5}^{NADH}$ for state 4 respiration and $K_{0.5}^{ADP}$ for triggering the phosphorylating state decreased to the values found in intact wild type mitochondria (Fig. 3) and became independent on Mg^{2+} concentration. As a consequence, the kinetics of energetic transitions (state 4 - state 3, driven by ADP and state 3 - state 4, triggered by CATR) monitored on tracings of oxygen uptake and of $\Delta\psi$ examined in mitoplasts prepared from mutant mitochondria closely resembled those recorded for wild type mitochondria (Fig. 2, lower tracing). In mitoplasts the stimulation of the electron flow by Mg^{2+} or Ca^{2+} was still apparent to the same degree as in intact mitochondria of both types, but the dramatic dependence of the respiratory control ratio on Mg^{2+} (Fig. 1, shaded field) disappeared. It has to be emphasized that mitoplasts of both types were coupled, though they revealed significantly reduced oxygen uptake. In mitoplasts obtained from wild type mitochondria the values of $K_{0.5}$ for NADH and ADP remained the same as in intact mitochondria (Fig. 3).

Intact mitochondria and mitoplasts were examined for the presence of an effective outer membrane permeability barrier with the test of oxidation of the externally added cytochrome c in the presence of ascorbate (see Methods). Full accessibility of cytochrome c was proved in mitoplasts prepared from both types of mitochondria while intact mitochondria of mutant and wild strain revealed only 3-8 percent outer membrane permeability for cytochrome c.

Therefore, one can conclude that the restrictions observed in bioenergetic effects of metabolites and CATR in mutant mitochondria were connected to outer membrane permeability. The apparent competition for entry between NADH and ADP was mentioned. More spectacular competition of that kind, strongly controlled by Mg^{2+} ions, was observed in mutant mitochondria between ADP and CATR for triggering and terminating of the phosphorylating state, respectively (Fig. 4). It has to be emphasized that CATR is a noncompetitive inhibitor of adenine nucleotide translocase and any delay in its action would be due to its slowed down entry to the intermembrane space.

When high concentration of ADP and CATR were added simultaneously to intact mutant mitochondria, the difference between the rapid action of ADP immediately triggering state 3 and the delayed action of CATR terminating this state was strongly emphasized in the absence of Mg^{2+} ions (Fig. 4, left). Ethanol was used as a respiratory substrate to avoid the presence of another molecule competing for the entry - NADH. In Figure 4 decrease in $\Delta\psi$ indicates initiation of the phosphorylating state and therefore, is a measure of a rapid ADP entry. The shaded field illustrates how dramatically the action of 30 μM CATR (increase in $\Delta\psi$) was delayed in the absence of Mg^{2+}.

Even in the presence of Mg^{2+}, lower concentrations of CATR (e.g. 10 μM) acted in terminating state 3 with much delay (for 2-4 min) while the initial depression of $\Delta\psi$ due to unchanged concentration of ADP was significantly

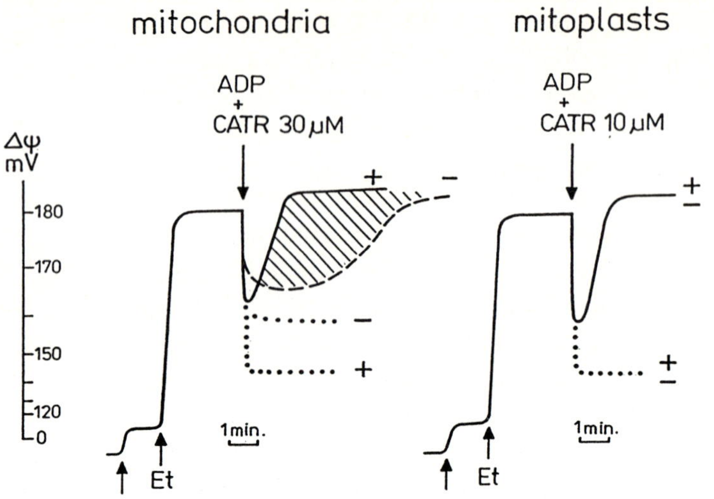

Figure 4. Competition for respiratory transitions between ADP and CATR apparent in mutant mitochondria is strongly decreased by Mg^{2+} ions or by the removal of the outer membrane (mitoplasts). Incubation conditions as in Methods. Et - ethanol 20 mM as a freely permeating respiratory substrate. $MgCl_2$ 4 mM, if present. 1.5 mM ADP was added simultaneously with CATR. Shaded field illustrates the restrictions imposed on CATR permeation in mutant mitochondria in the absence of Mg^{2+}. Dotted lines illustrate the depression in $\Delta\psi$ level triggered by ADP in the absence of CATR (phosphorylating state).

deeper (not shown). In the absence of Mg^{2+}, 10 μM CATR was not effective at all and $\Delta\psi$ remained for many minutes at the low level induced by ADP (Fig. 4 left, dotted line).

On the contrary, in mitoplasts, the presence of Mg^{2+} ions was no longer necessary for the fast action of CATR (Fig. 4 right). Much lower concentrations of CATR were effective in terminating state 3, though $\Delta\psi$ depression due to simultaneously added ADP was larger. This again points to the outer membrane being the barrier responsible for the apparent competition for entry between ADP and CATR observed in energy state transitions in intact mitochondria of the mutant. The barrier was obviously more restricting for CATR than for ADP. In general, the recordings of changes in $\Delta\psi$ in mutant mitoplasts due to simultaneous addition of ADP and CATR were similar to those obtained with intact wild type mitochondria.

VDAC was found to be sensitive to a synthetic polyanion (copolymer of methacrylate, maleate and styrene; Konig's polyanion). This was proved in the reconstituted system (Colombini et al., 1987; Benz et al., 1990) in which polyanion increased the sensitivity of VDAC towards $\Delta\psi$, thus facilitated rendering the channel into a closed substate at a much smaller $\Delta\psi$ applied than that needed in the absence of polyanion. The decrease of VDAC conductance by synthetic polyanion was also inferred from the inhibitory action of this polyanion on the apparent activity of adenilate kinase in intact rat liver and brain mitochondria (Benz et al., 1988, 1990). The activity of this enzyme, localized in the intermembrane space, depends on fluxes of adenine nucleotides across the outer mitochondrial membrane.

Similar inhibition of the apparent activity of adenilate kinase was observed in intact wild type mitochondria of yeast (Fig. 5). In mutant mitochondria lower concentrations of the synthetic polyanion also decreased kinase activity. However, higher concentrations of polyanion (> 40 µg/mg protein) resulted in an apparent increase in kinase activity which was paralleled by a rupture of the outer

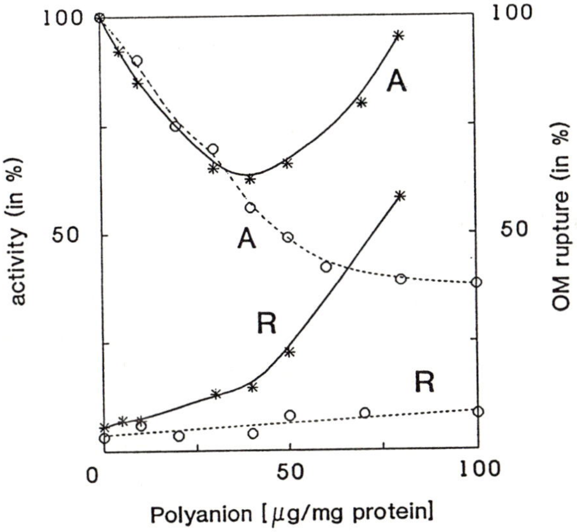

Figure 5. Inhibition of adenilate kinase activity (A) in intact mitochondria by the synthetic polyanion and the effect of the latter on the integrity of outer mitochondrial membrane (OM) expressed in percent of OM rupture (R). Determination of adenilate kinase activity, incubation with the synthetic polyanion and the test of outer mitochondrial membrane integrity performed as in Methods. (o) wild and (*) mutant mitochondria.

membrane as proved with the cytochrome c test (Fig. 5). Two conclusions can be drawn from these results: (i) the putative novel channel is also sensitive to the synthetic polyanion, and (ii) the absence of VDAC renders the outer membrane more fragile and/or more suspectible to polyanion.

In the control experiments with rat liver mitochondria the inhibition of adenilate kinase activity by polyanion was reversed by 0.1% triton X-100 to exactly 100 percent of the initial value as found previously by Benz et al. (1988, 1990). However, yeast mitochondria treated with triton revealed about two times the activity of kinase than that measured in intact mitochondria. Taking the triton stimulated activity as 100 percent, the activity measured in intact yeast mitochondria was latent by 45 and 65 percent in wild type and mutant mitochondria, respectively. The latency in wild type mitochondria is difficult to explain. Incidently, it was similar to the one observed in creatine kinase activity in rat brain mitochondria (Benz et al., 1990). In spite of the absence of VDAC, the expected total latency was not observed in intact mutant mitochondria. This should not be surprising as the assay of adenilate kinase activity implied both a high concentration of AMP and ATP (1.5 mM) and the presence of 5 mM $MgCl_2$. In these conditions the novel pore was quite effective in transporting nucleotides (see Fig. 1-3).

Discussion

All the data reported in this and previous papers from our laboratories prove, that in the absence of VDAC some traffic of large size anionic molecules, including adenine nucleotides takes place in intact mitochondria of porin depleted mutants. The conditions required for such traffic to be effective, i.e. high concentrations of adenine nucleotides and the presence of divalent cations, are obviously met in cytoplasm, as proved by an efficient respiratory coupling in mutants cells in situ (Michejda et al., 1990). The existence of a novel pore was postulated (Dihanich et al., 1989; Michejda et al., 1989).

The electric characteristics of such a pore was obtained in reconstituted system with proteins extracted from the outer mitochondrial membrane of porin depleted mutant (Benz et al., 1989). These characteristics corroborate closely with the restrictions in the access of metabolites and CATR to the inner mitochondrial membrane, observed in intact mutant mitochondria with biochemical approach (this paper). Electrically proven reduction in channel size is reflected in a requirement of higher concentrations of a single transported species for an efficient mitochondrial function. The cationic selectivity could explain the necessity of a Mg^{2+} or Ca^{2+} presence for the transport of metabolic anions in intact mitochondria. This protective action of Mg^{2+} is visualized in decreasing $K_{0.5}$ value less for NADH (M.W. 709, slightly anionic) than for ADP (M.W. 460, three negative charges) while it is most visible in permeation of CATR (M.W. 760, four negative charges).

The stimulatory effect of bivalent cations on transport seems to be exerted by screening the negative charges of the transported species or of the channel (Fig. 6, b). However, the direct action of bivalent cations on regulating the pore size and conductivity (in analogy to the $\Delta\psi$ dependence of VDAC) in intact mitochondria can not be excluded (Fig. 6, a). It has to be emphasized that the amount of total magnesium in isolated mutant mitochondria was not less than that in wild type mitochondria (unpublished results).

The validity of bioenergetic tests applied in these studies for characterization of transport restrictions in the outer membrane of VDAC-depleted mitochondria

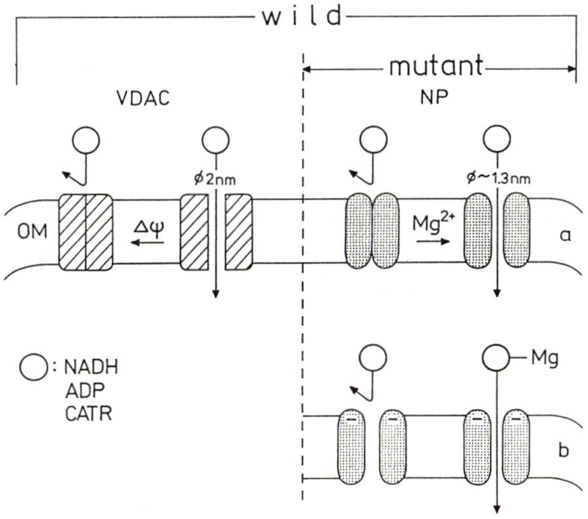

Figure 6. A scheme illustrating transport properties of the porin channel (VDAC) in wild type mitochondria and compared with those of novel pore (NP) apparently unmasked in mutant mitochondria. The passage of anions across the outer membrane (OM) via novel pore seems to be controlled by cations (e.g. Mg^{2+}) which might act either (a) directly on the channel towards its opening, or (b) indirectly, by screening the negative charge of either transported species or of the pore at constant opening state of the channel. In the latter case (b) the lining of the channel with negative charges would play a crucial role. $\Delta\psi$ symbolizes $\Delta\psi$-dependent closing of VDAC.

is justified by a finding that all these restrictions disappear in mitoplasts obtained from these mitochondria with swelling-shrinking procedure (and in digitonin-treated mitochondria - unpublished results). There was no difference in the substrate / inhibitor concentration efficiency in bioenergetic tests between mutant and wild type mitoplasts.

Porin depleted yeast mutant mitochondria are of particular interest in the solving of several problems in which the presence of normal porin might interfere. For example, the mutant obtained by Dihanich et al. (1987) was used to exclude the possible involvement of porin in the activity of the isolated pyruvate carrier (Nałęcz et al., 1991). Mutant B5 (Guo, 1987), the object of present studies, was used to prove that the specific electrical signals obtained in mitochondrial membranes with the "tip-dip" were not related to the porin activity and ascribed to a peptide sensitive channel, PSC (Fevre et al., 1990). These signals were found to be the same in yeast mitochondria containing VDAC and in those depleted of it (Fevre et al., 1990). From this unexpected result, these authors concluded that VDAC activity is lost during the experimental procedure of the "tip-dip" method.

The identity of PSC signals in wild strain and mutant mitochondria would explain why no additional or intensified band was visible in the protein pattern of the mutant when compared with that of the parent strain (Dihanich et al., 1989). The absence of any specific electrophoretic band in mutant mitochondria also suggests, that even if the novel pore were not identical to PSC, it probably coexist with VDAC in wild strain mitochondria and is not overexpressed in the absence of porin. The electrical signals of two different channels, though both anion selective, were also found in protein fraction with porin activity obtained from fish mitochondria (De Pinto et al., 1991).

One could also conclude, that the PSC channel, being different from VDAC, may be identical to the postulated novel pore. The electric signals of PSC were shown to characterize a cationically selective channel of rather large conductance. It was found to be permeable to large cations like tetraethylamonium. In contrast to VDAC, PSC is reversibly blocked by a basic peptide with a mitochondrial targeting sequence and its voltage-dependence is asymetrical with respect to 0 mV (Fevre et al., 1990). The electrical characteristics of PSC was similar to that of channels from another porin depleted mutant, incorporated into planar bilayer, except the latter channels were devoid of voltage-dependence (Dihanich et al., 1989). This difference might be due to different experimental protocols or different mutants examined.

If identical with the postulated novel pore, the PSC channel, probably primarily involved in protein import to mitochondria, could also, due to its large conductance, serve as an alternative (rescue) channel for metabolites as expected for the postulated novel pore. The latter activity would be of physiological importance in the absence of VDAC or during its closure. However, in the absence of VDAC, the double function of the PSC channel would probably result in some interference between its activity in protein import and that in transport of metabolites to mitochondria. This could, in due course, lead to restrictions in growth rate of cultures, in the functional assembly of mitochondria, and in the mitochondrial number, measured as the yield of mitochondrial protein. However, this was not the case with mutant B5, used in the present studies. Further studies are required to justify the assumption that boths channels are identical.

Acknowledgements

This study was supported by a grant from the Polish Cmte. Sci. Rsrch. PB 2223-4-91.

References

Benz R. (1985) Porin from bacterial and mitochondrial outer membranes. CRC Crit Rev in Biochem. 19, 145-190.

Benz R., Brdiczka D, (1992) The Cation-Selective Substrate of the Mitochondrial Outer Membrane Pore: Single-Channel Conductance and Influence on Intermembrane and Peripheral Kinases. J.Bioenerg.Biomembr. 24, 33-39.

Benz R., Kottke M., Brdiczka D. (1990) The cationically selective state of the mitochondrial outer membrane pore: a study with intact mitochondria and reconstituted mitochondrial porin. Biochim.Biophys.Acta 1022, 311-318.

Benz R., Schmid A., Dihanich M. (1989) Pores from mitochondrial outer membrane of yeast and a porin-deficient yeast mutant: A comparison. J.Bioenerg. Biomembr. 21, 439-450.

Benz R., Wojtczak L., Bosch W., Brdiczka D. (1988) Inhibition of adenine nucleotide transport through the mitochondrial porin by a synthetic polyanion. FEBS Lett. 231, 75-80.

Colombini M. (1979) A candidate for the permeability pathway of the outer mitochondrial membrane. Nature 279, 643-645.

Colombini M., Yeung C.L., Tung J., Konig T. (1987) The mitochondrial outer membrane channel, VDAC, is regulated by a synthetic polyanion. Biochim. Biophys.Acta 905, 279-286.

Daum G., Bohni P.C., Schatz G. (1982) Import of proteins into mitochondria. Cytochrome b_2 and cytochrome c peroxidase are located in the intermembrane space of yeast mitochondria. J.Biol.Chem. 257, 13028-13033.

De Pinto V., Zara V., Benz R., Gnoni G.V., Palmieri F. (1991) Characterization of pore-forming activity in liver mitochondria from Anguilla anguilla. Two porin in mitochondria? Biochim.Biophys.Acta 1061, 279-286.

Dihanich M., Schmid A., Oppeliger W., Benz R. (1989) Identification of a new pore in the mitochondrial outer membrane of a porin-deficent yeast mutant. Eur.J.Biochem. 181, 703-708.

Dihanich M., Suda K., Schatz G. (1987) A yeast mutant lacking mitochondrial porin is respiratory-deficient, but can recover respiration with simultaneous accumulation of an 86 kD extramitochondrial protein. EMBO J. 6, 723-728.

Douce R., Bourguignon J., Brouquisse R., Neuberger M. (1984) Isolation of plant mitochondria. General principles and criteria of integrity. Methods Enzymol. 148, 403-415.

Fevre F., Chich J-F., Lauquin G.J-M., Henry J-P., Thieffry M. (1990) Comparision of mitochondrial cationic channels in wild-type and porin-deficient mutant yeast. FEBS Lett. 262, 201-204.

Guo X.Y. (1988) Isolement et caracterisation du gene de la porine mitochondriale de S. cerevisiae. Etude de l'expression du gene et construction de mutants deficients en porine. D.Sc. thesis, University of Aix - Marseille II.

Guo X.J., Lauquin G.J-M. (1986) Mitochondrial porin-deficient mutant of S.cerevisiae: in vitro construction and preliminary properties. EBEC Reports 4, 292

Holden M.J., Colombini M. (1988) The mitochondrial outer membrane channel, VDAC, is modulated by a soluble protein. FEBS Lett. 241, 105-109.

Kamo N., Maratsugu M., Hongoh R., Kobatake Y. (1979) Membrane potential of mitochondria measured with an electron sensitive to tetraphenyl phosphonium and relationship between proton electrochemical potential and phosphorylation potential in steady state. J.Membr.Biol. 49, 105-121.

Liu M., Colombini M. (1991) Voltage gating of the mitochondrial outer membrane channel VDAC is regulated by a very conserved protein. Am.J.Physiol. 260, C371-C374.

Liu M.Y., Colombini M. (1992) Regulation of mitochondrial respiration by controlling the permeability of the outer membrane through the mitochondrial channel, VDAC, Biochim.Biophys.Acta 1098, 255-260.

Ludwig O., Benz R., Schultz J.E. (1989) Porin of Paramecium mitochondria isolation characterization and ion selectivity of the closed state. Biochim.Biophys. Acta 978, 319-327.

Mannella C. (1992) The "ins" and "outs" of mitochondrial membrane channels. TIBS 17, 315-320.

Michejda J., Guo X.J., Lauquin G.J-M. (1988) Energy coupling in cells and mitochondria of porin yeast mutants. EBEC Reports 5, 132

Michejda J., Guo X.J., Lauquin G.J-M. (1990) The respiration of cells and mitochondria of porin deficient yeast mutants is coupled. Biochem. Biophys. Res. Commun. 171, 354-361.

Nałęcz M., Nałęcz K., Azzi A. (1991) Purification and functional characterization of the pyruvate (monocarboxylate) carrier from bakers yeast mitochondria (Saccharomyces cerevisiae). Biochim.Biophys.Acta 1079, 87-95.

ROLE OF PORIN-KINASE INTERACTIONS IN DISEASE

V. Adams and E.R.B. McCabe
Department of Pediatrics and Institute for Molecular Genetics
Baylor College of Medicine
One Baylor Plaza
Houston, Texas 77030
U.S.A.

Beyond the catalytic activities of these assemblies and the factors that fine-tune their operations, we need to be aware of their social relations in the cell -- their attachments to other protein units, to membranes, and to the skeletal framework...

A. Kornberg in For the Love of Enzymes *(1989)*

Hexokinase (ATP: D-hexose 6-phosphotransferase, EC2.7.1.1) and glycerol kinase (ATP: glycerol phosphotransferase, EC2.7.1.30) catalyze the phosphorylation of their respective hexose and triose substrates in order to initiate mammalian metabolism of these compounds. The priority of these enzymes in metabolism alone would make them worthy subjects of investigation. For many of us, however, their interest also lies in their "social relations in the cell" as Kornberg (1989) so nicely stated. There is abundant evidence that hexokinase and glycerol kinase interact with porin, the voltage-dependent anion-selective channel (VDAC) of the outer mitochondrial membrane, in a developmental stage and tissue specific manner in the normal individual (McCabe 1983; Adams et al 1991a). There is also evidence that binding is altered in certain disease states. We will discuss the role of the

NATO ASI Series, Vol. H 83
Molecular Biology of Mitochondrial Transport Systems
Edited by M. Forte and M. Colombini

porin-kinase interactions in a variety of diseases and the metabolic impact of these interactions, or, in some cases, their disruption.

CANCER

The most prominent example of the involvement of a porin-kinase interaction in disease involves tumor cells. Increased mitochondrial binding of hexokinase has been observed in a number of cell lines, including hepatoma cells (Parr & Pedersen 1983; Nelson & Kabir 1986; Arora & Pedersen 1988; Arora et al 1990), Ehrlich ascites tumor cells (Bustamante et al 1981), and colon cancer cells (Denis-Pouxviel et al 1987). Arora and Pedersen (1988) showed that the specific activity of mitochondrial bound hexokinase exceeded that of all normal tissues examined, and was approximately 3-fold the highest normal tissue, brain. They concluded that bound hexokinase had preferential access to mitochondrial ATP, and suggested that this binding, and the associated enhanced production of glucose-6-phosphate, was important, not only for the glycolytic energy production characteristic of tumor cells, but also for the biosynthesis of the nucleic acids and phospholipids required for rapid tumor growth. While the concept of achieving a metabolic advantage by increased binding of hexokinase to the mitochondria most definitely has intuitive appeal, it must be noted that not all investigators feel that the bound hexokinase in tumor cells is functionally coupled to the mitochondrial matrix (Nelson & Kabir 1985; Denis-Pouxviel et al 1987; Gauthier et al 1989; Gauthier et al 1990). Therefore, while there is clear evidence for increased mitochondrial binding of hexokinase in tumor cells, the metabolic role of this phenomenon remains to be demonstrated experimentally to the satisfaction of all.

Important to this demonstration will be the availability of cloned reagents from tumor cells in order to evaluate expression in model systems. A hexokinase cDNA from the c37 mouse hepatoma cell line has been cloned (Arora et al 1990) and appears to have striking sequence similarity with hexokinase 1 (HK1) from human kidney (Nishi et al 1988), rat brain (Schwab & Wilson 1989), and bovine brain (Griffin et al 1991). This similarity includes absolute conservation of the N-terminal 15 amino acid mitochondrial targeting sequence (Gelb et al 1992) among these four HK1s. The mouse tumor hexokinase cDNA was overexpressed in active form in *E. coli* and partially purified, and was capable of binding to rat liver mitochondria (Arora et al 1990). A partial cDNA from Novikoff ascites tumor cells was isolated and contained the coding information for the entire C-terminal half of the protein (Thelen & Wilson 1991). This partial tumor cDNA was identical in sequence to the rat muscle HK2 cDNA and both coded a 5.2kb mRNA. The rat cDNAs for HK1-3 cloned by Wilson's group (Schwab & Wilson 1989; Thelen & Wilson 1991; Schwab & Wilson 1991) have also been used for hybridization with Southern blots of DNA from the AS-30D rat hepatoma cell line (Wigley & Nakashima 1992). Characterization of the expression and function of these individual genes will be a fertile area for investigations of tumorigenesis.

The underlying mechanism responsible for enhanced hexokinase binding in tumor cells remains to be elucidated. Building on Bessman's acceptor theory for the mechanism of insulin action (Bessman 1954; Bessman 1966; Bessman 1972), Golshani (1992) recently suggested that certain growth factors and oncogenes might act through stimulation of hexokinase binding to mitochondria. The evolutionary relationships and structural similarities between insulin and the two insulin-like growth factors, IGF-I and IGF-II, offer intriguing possibilities regarding mitochondrial binding of HK2 in tumors, since HK2 is the hexokinase isoenzyme found in insulin sensitive tissues (Katzen & Schimke 1965).

BRAIN HYPOXIA AND ISCHEMIA

Oxygen and glucose are critical substrates for energy metabolism in the brain. Wilson (1986) has reviewed the evidence supporting the concept that the distribution of hexokinase between the cytoplasm and the mitochondria is influenced by the concentration of metabolites, with the proportion bound increasing with increased glycolytic activity. Wilson's group and others showed an increase in bound hexokinase with ischemia, attributed to increased glucose utilization associated with metabolite changes during ischemia, including decreased glucose 6-phosphate and glucose 1,6-diphosphate, and increased inorganic phosphate. The redistribution of hexokinase was quite rapid with increased mitochondrial activity noted within 0.5 min of ischemia induced by decapitation (Wilson 1986).

The clinical situation is frequently associated with periods of hypoxia or ischemia followed by improved oxygenation. We have recently investigated the influence of ischemia and reperfusion on hexokinase binding in the brain using a piglet model, in order to determine the potential for recovery after an ischemic insult (Adams et al 1992a). These results confirmed the previous observations that bound hexokinase increased during the ischemic phase. Although there was a tendency toward recovery to baseline levels of bound enzyme during the reperfusion phase, this was incomplete after 2 hrs. The limitations of this model have not permitted evaluation of more chronic residual effects, but one would speculate that the effects on hexokinase binding contribute to the cellular damage associated with hypoxia and ischemia.

DIABETES MELLITUS

Due to the prominence of glucose and its altered metabolism in diabetes mellitus there has been considerable interest in the role of hexokinase in this disease. While characterizing the tissue distribution of hexokinase isoenzymes, Katzen and Schimke (1965) observed that HK2 was found in insulin sensitive tissues. Walters and McLean (1968) found that particulate hexokinase fell from 46% of the total in mammary gland tissue from lactating control rats to 11% in this tissue one hour after injection with anti-insulin antibodies. Insulin has been shown to increase mitochondrial bound hexokinase in rat epididymal fat pad (Borrebaek & Spydevold 1969) and muscle (Bessman 1972). In rat muscle it was shown that the bound hexokinase was HK2. These and other reports indicate a role for the hexokinase-porin interaction in the pathogenesis of diabetes mellitus at the end organ level.

Recent information implicates glucokinase (HK4) even more directly in the insulin release mechanism of the pancreatic islets and indicates linkage of the HK4 gene to this disorder in some families with type 2 or non-insulin dependent diabetes mellitus (NIDDM). Islet cell glucokinase has been implicated as a glucose sensor by generating a metabolic signal, that, when integrated with the intermediary metabolism of the ß-cell, results in signal transduction and insulin release (Matschinsky 1990). Since it appears that a portion of the islet cell glucokinase is mitochondrial bound, in contrast to hepatic glucokinase (Sener et al 1986; Malaisse-Lagae & Malaisse 1988), it is possible that this interaction may be involved in the insulin release mechanism. Genetic linkage was established between the HK4 gene on chromosome 7p and diabetes mellitus in several pedigrees with an autosomal dominant form of NIDDM, known as maturity onset diabetes of the young (MODY), characterized by a younger age of onset than typical NIDDM (Froguel et al 1992; Hattersley et al 1992). The combined lod score for the linked pedigrees in these two studies exceeded 20 at a recombination fraction (Θ) of zero. There were three families out of a total of 18 in which linkage between NIDDM and HK4 clearly was excluded, indicating genetic heterogeneity. One family (F8) contained an individual who did not inherit the HK4 allele segregating with NIDDM in that family and who appeared to have a different phenotype from her relatives (Froguel et al 1992). The F8 pedigree was subsequently used for mutation analysis and this individual did not inherit the HK4 nonsense mutation associated with MODY in that family (Vionnet et al 1992). The identification of this mutation in the HK4 locus, which is so tightly linked to diabetes mellitus in these families, is strong support for a role of glucokinase in the etiology of this disorder, although a role for mitochondrial bound glucokinase remains speculative.

INBORN ERRORS OF METABOLISM

Inherited deficiencies of glycerol kinase and hexokinase have been reported and the reader is referred to recent reviews (McCabe 1989; Adams et al 1991a) for more complete discussions of these. These disorders represent very different clinical conditions, and the role of porin-kinase interactions in their pathogenesis has yet to be determined.

Glycerol Kinase Deficiency (GKD). This is an X-linked disorder which maps to Xp21, immediately telomeric to the gene for Duchenne muscular dystrophy (DMD) (McCabe 1989; Adams et al 1991a). The majority of patients with GKD have a contiguous gene syndrome with DMD, GKD and adrenal hypoplasia congenita (AHC) due to deletions of all three of these loci. There are also patients with isolated GKD, presumably due to point mutations or small rearrangements within the GK locus. One group with isolated GKD presents in adulthood when they are incidentally ascertained with pseudohypertriglyceridemia due to elevated circulating free glycerol concentrations (McCabe 1989). The other group presents in childhood with episodes of vomiting and acidemia, which may be associated with central nervous system depression, hypoglycemia and/or hypothermia (Eriksson et al 1983; Ginns et al 1984; Howell et al 1989; Bonham & Crawford 1990). These patients respond well to reduced fat, and hence glycerol intake, and intravenous glucose support during intercurrent illnesses, to prevent lipolysis and release of glycerol. It has become clear that patients with GKD as part of the contiguous gene syndrome have acute episodes similar to those with symptomatic isolated GKD and it is thought that these metabolic and central nervous system disturbances may be a consequence of the deficiency of this enzyme. Interestingly, there are extremely rare patients with very similar acute presentations associated with glycerol ingestion or lipolytic stress who have normal glycerol kinase activity (McCabe 1989). This disorder has been referred to as the glycerol intolerance syndrome and one of the proposed etiologies is disrupted compartmentation of glycerol kinase with normal enzyme activity (McCabe 1983; McCabe & Seltzer 1986).

A physiological role for glycerol kinase beyond phosphorylation of glycerol is a matter of question. It binds to porin in proportions somewhat similar to hexokinase, with 85% bound in brain (McCabe 1989). The observations of acute, but reversible, central nervous dysfunction in these patients suggest a critical role for this enzyme in brain metabolism, particularly during periods of metabolic stress.

Hexokinase Deficiency. Patients with hexokinase deficiency most typically present with problems referable to red cell metabolism and non-spherocytic hemolytic anemia (McCabe 1989). One individual is reported with a skeletal myopathy, apparently without hematologic problems (Poulton & Nightingale 1988). HK1 and HK3 are both present in red cells, but HK1 is the primary isoform (Stocchi et al 1982). During erythrocyte development the hexokinase exists in both soluble and mitochondrial bound forms (Stocchi et al 1988; Magnani et al 1984a; Magnani et al 1984b). Only the soluble fraction is subject to degradation in a process that is ATP- and ubiquitin-dependent (Magnani et al 1984b; Magnani et al 1986). With loss of the mitochondria during development, the hexokinase is released and the degradation rate is reduced. The result is that hexokinase activity is also reduced in mature erythrocytes. With the first report of red cell hexokinase deficiency (Valentine et al 1967), it was recognized that the hexokinase deficiency might be a major problem for the younger reticulocytes, accustomed to a much higher activity of this enzyme in the normal situation. Since a significant portion of the hexokinase is mitochondrial bound in the younger cells and aerobic metabolism is more important to them, the non-spherocytic hemolytic anemia may result from disruption of the normal porin-kinase interaction in the deficient cells. The sparing of other organs in these patients, and particularly the central nervous system, confirms that our understanding of this disorder and this enzyme is incomplete, but suggests the possibility that the isolated red cell phenotype is due to a tissue-specific reduction in expression. It is also possible that generalized HK1 deficiency will be associated with severe mental retardation, possibly with involvement of other organ systems, and may be an embryonic lethal.

FUNCTION DEDUCED FROM STRUCTURE

In order to more fully understand the role of these porin-kinase interactions in disease, we must gain a more complete appreciation of their function in the normal cell. Molecular genetic investigations of porin and hexokinase are making major contributions in this area. In this section we will describe recent studies in our laboratory that derive directly from a precise knowledge of the deduced amino acid sequence for HK1.

HK1 Mitochondrial Targeting Sequence. When human HK1 was first cloned it was noted that there was remarkable similarity between the two halves of the protein with an N-terminal "overhang" of 15 amino acids which did not match up with sequence in the carboxy-half of the protein (Nishi et al 1988). Comparison of additional sequences showed that these 15 amino acids were absolutely conserved in human, rat, mouse and bovine HK1s without a single substitution (Griffin et al 1991). Previous investigations had shown that this N-terminal domain participated in the mitochondrial binding of HK1, since binding was eliminated by removal of nine of the 15 N-terminal amino acids by limited chymotryptic digestion (Polakis & Wilson 1985). However, it was not known whether this domain at the N-terminus was sufficient for binding.

Availability of sequence information allowed us to generate a reporter gene construct to directly test the hypothesis that this N-terminal domain was both necessary and sufficient for binding of HK1 to its receptor on the outer mitochondrial membrane (Gelb et al 1992). A polymerase chain reaction (PCR) amplification strategy was used to generate the chimeric reporter construct encoding the fifteen amino acid N-terminal domain of HK1 coupled to the chloramphenicol acetyltransferase (CAT) gene, inserted into the mammalian expression vector pCMVB (Tuttle & Wilson 1970), and referred to as pCMVHKCAT. The construct was electroporated into mouse Hepa 1-6 hepatoma cells, which, like other hepatoma cells (Arora & Pedersen 1988; Arora et al 1990), have a relatively high HK activity with a significant proportion mitochondrial bound. After digitonin treatment at a concentration that caused cytosolic leak, HKCAT, like HK, remained predominantly in the pellet fraction, whereas CAT (expressed from the control pCMVCAT construct) remained primarily unbound. To determine if the HKCAT was binding to mitochondria, membrane free extracts of electroporated hepatoma cells were incubated with rat liver mitochondria, and 91% of HKCAT bound to the mitochondria, compared with only 12% of CAT. Specificity of HKCAT binding to mitochondrial HK1 receptor sites was demonstrated by showing that isolated rat brain HK1 competed with HKCAT for these binding sites, and that N,N'-dicyclohexylcarbodiimide (DCCD), in doses which prevent HK1 binding to porin, blocked the binding of HKCAT. Deletional mutant constructs showed decreased binding with increasing size of the deletion in the HK domain of HKCAT, the largest effects observed with loss of amino acids two through nine of the HK domain. We concluded that the N-terminal 15 amino acids were capable of conferring mitochondrial binding to CAT, and that there was specificity of this binding to the HK1 receptor.

A number of questions remain regarding the role of these N-terminal 15 amino acids in HK1 binding. This α-helical domain is extremely hydrophobic. It is unclear whether it is the hydrophobicity or the specific amino acid sequence that is critical for binding, but the absolute identity of this sequence among the four mammalian HK1s characterized to date suggests that the specific amino acid sequence is important. Rat HK2 also binds to mitochondria and its N-terminal domain shows 53% sequence identity with HK1. Does HK2 compete with HK1 for the same receptor site and what is the role of HK2 in this interaction in model systems? The occurrence of multiple porin isoforms had been proposed (Ludwig et al 1986), and it now appears that there are at least two, only one of which binds HK1 (Blachly-Dyson et al 1992). Is it possible that individual hexokinase isozymes interact with specific porin isoforms? These questions and many others can be addressed experimentally with molecular genetic approaches.

The importance of these issues goes beyond merely understanding the basic characteristics of porin-kinase interactions. If the mitochondrial interaction of porin with hexokinase gives tumor cells a metabolic advantage, then it follows that disruption of this interaction might reduce this advantage.

Gene therapy possibilities include specific tumor cell delivery of an appropriate antisense construct or the coding sequence of a protein that would compete effectively to block the binding of hexokinase to porin. In 1986, we proposed a genetic engineering approach to some of these issues (McCabe & Seltzer 1986), and we have now entered the era when these possibilities may become reality.

Protein Kinase Activity of HK1. Comparison of the sequences of mammalian HK1s showed that the putative ATP binding domains in both halves of HK1 were quite similar to these regions in several protein kinases, including the cAMP- and cGMP-dependent protein kinases and the oncogenes c-myc and c-src (Arora et al 1990; Griffin et al 1991). It was known that protein kinase A phosphorylated HK4 at serine residues (Ekman & Nilsson 1988), and that yeast HK-PII showed protein kinase activity, including autophosphorylation, as well as phosphorylation of an artificial substrate, casein (Herrero et al 1989). Therefore, we undertook investigations to determine if HK1 had protein kinase activity.

Initial experiments using rat brain HK1 showed that HK1 did possess protein kinase activity (Adams et al 1991b). HK1 was incubated with [γ-^{32}P]ATP for 1 hr and electrophoresed on a denaturing polyacrylamide gel. A single band, approximately 100kDa in size, was observed with silver staining and autoradiography. This was consistent with autophosphorylation of HK1, though a minor contaminating band with protein kinase activity could not be ruled out. When purified HK1 was incubated with histone H2A as a model substrate, phosphorylation of both proteins was observed, the histone to a greater extent than HK1.

Subsequent investigations were undertaken to determine the amino acid specificity of rat brain HK1 autophosphorylation (Adams et al 1992b). An anti-phosphotyrosine monoclonal antibody gave a weakly positive signal with freshly prepared HK1 before incubation with ATP, suggesting that HK1 was phosphorylated on tyrosine residues in its native state. After incubation with ATP the signal observed with the anti-phosphotyrosine antibody was much stronger. In order to confirm the tyrosine phosphorylation, HK1 was incubated with [γ-^{32}P]ATP, and phosphoamino acid analysis performed. To our surprise serine, threonine and tyrosine were all labeled. The results suggest that HK1 is a dual specificity protein kinase, capable of autophosphorylation at serine/threonine and tyrosine residues (Lindberg et al 1992).

At this time, the specific positions of the phosphorylated residues in the sequence have yet to be determined. However, computer analysis of the HK1 sequence, searching for protein kinase substrate phosphorylation consensus sequences, indicates that only one of the 20 tyrosines in the molecule meets the criteria, and this is the tyrosine at position 27. One must be cautious with such predictions, since the literature contains reports demonstrating tyrosine phosphorylation at sites which do not satisfy the criteria for a consensus sequence. However, tyrosine 27 would be quite interesting, lying just beyond the receptor binding domain, within the regulatory region of HK1. Phosphorylation in this region might induce a conformational change in the N-terminal domain involved in mitochondrial binding and effect an altered binding. Altered conformational change in this region has been observed with the binding of glucose 6-phosphate to HK1 (Wilson 1973). It is also possible that HK1 phosphorylation of other proteins in the contact sites could be involved in regulating the porin-kinase interaction, since it has been observed that HK1 does not randomly bind to porin, but shows a preference for porin at the contacts between the inner and outer mitochondrial membranes (Kottke et al 1988).

The observation that porin is present in plasma membrane (Thinnes et al 1989; Kayser et al 1989; Jurgens et al 1991) argues for regulation of the pore to prevent indiscriminate passage of the relatively large molecules which can pass through isolated porin (Benz 1985). We would propose that hexokinase, interacting with porin at its cytoplasmic interface, is a candidate for gating the pore, and that its protein kinase activity may play a role in this process (Figure). The mechanism we propose is analogous to the "ball and chain" model that has been developed for certain ion channels (Jan & Jan 1989), including the cystic fibrosis transmembrane conductance regulator (CFTR) (Kerem et al 1989) and the *Shaker* potassium channel (Barinaga 1990; Hoshi et al 1990; Zagotta et al 1990). The prominence of ATP binding domains in the CFTR protein is interesting in relation to HK1. Whereas these other models incorporate the "ball" and the "chain" within the same protein, ours proposes that hexokinase is the "ball," with four in a tetrameric structure (Gouchon & Wilson 1990) moving from the open to the closed configuration. We have drawn the "chain" as part of the hexokinase molecule and would suggest that this could be the N-terminal α-helical domain. However, porin could also contribute to the "chain," and also contains an N-terminal α-helical domain. Consistent with this proposal is the major conformational change observed in yeast hexokinase with substrate binding (Anderson et al 1979).

Although the physiological relevance of the HK1 protein kinase activity remains unknown, it is possible to speculate that this may afford an entirely new role for hexokinase in metabolism. Protein kinases are known to be involved in membrane receptors and signal transmission (Ralph et al 1990). Communication between the mitochondrion and the nucleus is required for coordinate expression of proteins with subunits encoded by each of these genomes. A membrane associated protein kinase, situated at the cytoplasmic outlet for ATP transit, would provide a mechanism for monitoring a key

Figure. "Ball and chain" model for regulation of porin conductance by hexokinase. Upper panel shows the "ball," a single hexokinase molecule, binding to porin, represented as a simple cylinder, by way of its "chain," perhaps the N-terminal α-helical domain. Conformational changes induced by ligand binding, protein phosphorylation or other mechanisms are proposed to move the hexokinase molecule from an open or non-occlusive position to one in which it closes or occludes one quadrant of the pore, with the movement indicated by the double headed arrow. The lower figures represent overhead views of this model. The lower left panel shows the four hexokinase molecules in the open position. The lower right panel shows the four hexokinase molecules in the closed position. According to this model, hexokinase would modulate conductance through porin, serving to gate the pore.

mitochondrial process, ATP generation, and communicating the status of ATP production by the mitochondria to the nucleus or other organelles. Porin bound glucokinase could be involved in signal transduction for insulin release through its protein kinase activity.

SUMMARY

We have reviewed the possible roles of porin-kinase interactions in a variety of disorders including cancer, brain hypoxia and ischemia, diabetes mellitus and two inborn errors of metabolism. This discussion has focused on hexokinase because of the more extensive knowledge of this protein, but has included glycerol kinase in the context of its inherited deficiency. We also have considered the deduction of functional activities from structural information acquired by cloning and sequencing the genes for these proteins. These deduced functions were then tested experimentally. Detailed knowledge of these structure-function relationships permits one to speculate about possible therapeutic implications and an expanded role for hexokinase in intracellular communication.

ACKNOWLEDGEMENTS

The investigations from the authors' laboratory and the preparation of this manuscript have been supported in part by a research grant (1 R01 HD22563), the Baylor Mental Retardation Research Center (1 P30 HD24064), the Baylor Child Health Research Center (1 P30 HD2723), and the BCM Human Genome Program Center (1 P50 HG00210).

REFERENCES

Adams V, Griffin L, Towbin J, Gelb B, Worley K, McCabe ERB (1991a) Porin interaction with hexokinase and glycerol kinase: Metabolic microcompartmentation at the outer mitochondrial membrane. Biochem Med Metab Biol 45:271-291

Adams V, Griffin LD, Gelb BD, McCabe ERB (1991b) Protein kinase activity of rat brain hexokinase. Biochem Biophys Res Commun 177:1101-1106

Adams V, Gray S, McCabe L, Yamashita Y, Le SP, McCabe ERB, Goddard-Finegold J (1992a) Effects of ischemia and reperfusion on mitochondrial bound hexokinase in piglet brain. In preparation.

Adams V, Schieber A, McCabe ERB (1992b) Autophosphorylation of hexokinase: Evidence for a new "dual specificity" protein kinase. In preparation.

Anderson CM, Zucker FH, Steitz TA (1979) Space-filling models of kinase clefts and conformation changes: comparison of the surface structures of kinase enzymes implicates closing clefts in their mechanism. Science 204:375-380

Arora KK, Fanciulli M, Pedersen PL (1990) Glucose phosphorylation in tumor cells: Cloning, sequencing, and overexpression in active form of a full length cDNA encoding a mitochondrial bindable form of tumor hexokinase. J Biol Chem 265:6481-6488

Arora KK, Pedersen PL (1988) Functional significance of mitochondrial bound hexokinase in tumor cell metabolism -- evidence for preferential phosphorylation of glucose by intramitochondrially generated ATP. J Biol Chem 263:17422-17428

Barinaga M (1990) Playing tetherball in the nervous system. Science 250:506-507

Benz R (1985) Porin from bacterial and mitochondrial outer membranes. CRC Crit Rev Biochem 19:145-190

Bessman S P (1954) A contribution to the mechanism of diabetes mellitus. In Najjar VA (ed) Fat Metabolism. Johns Hopkins Press Baltimore, pp 133-137

Bessman SP (1966) A molecular basis for the mechanism of insulin action. Amer J Med 40:740-749

Bessman SP (1972) Hexokinase acceptor theory of insulin action, new evidence. Isr J Med Sci 8:344-351

Blachly-Dyson E, Zambronicz E B, Yu W H, Adams V, McCabe E R B, Adelman J, Colombini M, Forte M (1992) Cloning, functional expression in yeast, and mapping of two human isoforms of the outer mitochondrial membrane channel, VDAC. In Proceedings of the Molecular Biology of Mitochondrial Transport Systems (This volume). Springer Verlag Berlin

Bonham JR, Crawford M (1990) The effect of isolated glycerol kinase (EC2.7.1.30) deficiency on glucose homeostasis -- a cause for concern in affected patients. Abstracts Vth International Congress Inborn Errors of Metabolism OC8.5:(Abstract)

Borrebaek B, Spydevold O (1969) The effects of insulin and glucose on mitochondrial-bound hexokinase activity of rat epididymal adipose tissue. Diabetologia 5:42-43

Bustamante E, Morris HP, Pedersen PL (1981) Energy metabolism of tumor cells: Requirement for a form of hexokinase with a propensity for mitochondrial binding. J Biol Chem 256:8699-8704

Denis-Pouxviel C, Riesinger I, Buehler C, Brdiczka D, Murat JC (1987) Regulation of mitochondrial hexokinase in cultured HT 29 human cancer cells, an ultrastructural and biochemical study. Biochim Biophys Acta 902:335-348

Ekman P, Nilsson E (1988) Phosphorylation of glucokinase from rat liver in vitro by protein kinase A with a concomitant decrease of its activity. Arch Biochem Biophys 261:275-282

Eriksson A, Lindstedt S, Ransnas L, von Wendt L (1983) Deficiency of glycerol kinase (EC2.7.1.30). Clin Chem 29:718-722

Froguel P, Vaxillaire M, Sun F, Velho G, Zouali H, Butel MO, Lesage S, Vionnet N, Clément K, Fougerousse F, Tanizawa Y, Weissenbach J, Beckmann JS, Lathrop GM, Pass P, Permutt MA, Cohen D (1992) Close linkage of glucokinase locus on chromosome 7p to early-onset non-insulin-dependent diabetes mellitus. Nature 356:162-164

Gauthier T, Denis-Pouxviel C, Murat JC (1989) Study on ATP-generating system and related hexokinase activity in mitochondria isolated from undifferentiated or differentiated HT29 adenocarcinoma cell. Biochim Biophys Acta 975:231-238

Gauthier T, Denis-Pouxviel C, Murat JC (1990) Respiration of mitochondria isolated from differentiated and undifferentiated HT29 colon cancer cells in the presence of various substrates and ATP generating systems. Int J Biochem 22:411-417

Gelb BD, Adams V, Jones SN, Griffin LD, MacGregor GR, McCabe ERB (1992) Targeting of hexokinase 1 to liver and hepatoma mitochondria. Proc Natl Acad Sci USA 89:202-206

Ginns EI, Barranger JA, McClean SW, Schaefer E, Brady RO, Young R, Goodman SI, McCabe ERB (1984) Juvenile form of glycerol kinase deficiency with episodic vomiting, acidemia and stupor. J Pediatr 104:736-739

Golshani S (1992) Insulin, growth factors, and cancer cell energy metabolism: An hypothesis on oncogene action. Biochem Med Metab Biol 47:108-115

Gouchon X, Wilson JE (1990) Tetrameric structure of mitochondrially bound rat brain hexokinase: A crosslinking study. Arch Biochem Biophys 276:285-293

Griffin LD, Gelb BD, Wheeler DA, Davison D, Adams V, McCabe ERB (1991) Mammalian hexokinase 1: Evolutionary conservation and structure to function analysis. Genomics 11:1014-1024

Hattersley AT, Turner RC, Permutt MA, Patel P, Tanizawa Y, Chiu KC, O'Rahilly S, Watkins PJ, Wainscoat JS (1992) Linkage of type 2 diabetes to the glucokinase gene. Lancet 339:1307-1310

Herrero P, Fernandez R, Moreno F (1989) The hexokinase isozyme PII of Saccharomyces cerevisiae is a protein kinase. J Gen Microbiol 135:1209-1216

Hoshi T, Zagotta WN, Aldrich RW (1990) Biophysical and molecular mechanisms of *Shaker* potassium channel inactivation. Science 250:533-538

Howell RR, Grier R, Dominguez B, Draehn DK (1989) Neurological manifestations of juvenile glycerol kinase deficiency: Improvement with nutritional therapy. Brain Dysfunct 2:126-130

Jan LY, Jan YN (1989) Voltage-sensitive ion channels. Cell 56:13-25

Jurgens L, Ilsemann P, Kratzin HD, Hesse D, Eckart K, Thinnes FP, Hilschmann N (1991) Studies on human porin. IV. The primary structures of "porin 31HM" purified from human skeletal muscle membranes and of "porin 31HL" derived from human B lymphocyte membranes are identical. Biol Chem Hoppe-Seyler 372:455-463

Katzen HM, Schimke RT (1965) Multiple forms of hexokinase in the rat: tissue distribution, age dependency, and properties. Proc Natl Acad Sci USA 54:1218-1225

Kayser H, Kratzin HD, Thinnes FP, Gotz H, Schmidt WE, Eckart K, Hilschmann N (1989) Zur Kenntnis der Porine des Menschen. II. Charakterisierung und Primarstruktur eines 31-kDa-Porins aus menschlichen B-Lymphozyten (Porin 31HL). Biol Chem Hoppe-Seyler 370:1265-1278

Kerem B, Rommens JM, Buchanan JA, Markiewicz D, Cox TK, Chakravarti A, Buchwald M, Tsui L-C (1989) Identification of the cystic fibrosis gene: Genetic analysis. Science 245:1073-1080

Kornberg A (1989) For the Love of Enzymes, Harvard University Press Cambridge, MA

Kottke M, Adams V, Riesinger I, Bremm G, Bosch W, Brdiczka D, Sandri G, Panfili E (1988) Mitochondrial boundary membrane contact sites in brain: Points of hexokinase and creatine kinase location, and control of Ca^{2+} transport. Biochim Biophys Acta 935:87-102

Lindberg RA, Quinn AM, Hunter T (1992) Dual-specificity protein kinases: Will any hydroxyl do? Trends Biochem Sci 17:114-119

Ludwig O, DePinto V, Palmieri F, Benz R (1986) Pore formation by the mitochondrial porin of rat brain in lipid bilayer membranes. Biochim Biophys Acta 860:268-276

Magnani M, Stocchi V, Dacha V, Fornaini G (1984a) Rabbit red blood cell hexokinase: Intracellular distribution during reticulocyte maturation. Mol Cell Biochem 63:59-65

Magnani M, Stocchi V, Dacha V, Fornaini G (1984b) Rabbit red blood cell hexokinase: Evidences for an ATP-dependent decay during cell maturation. Mol Cell Biochem 61:83-92

Magnani M, Stocchi V, Chiarantini L, Serafini G, Dacha V, Fornaini G (1986) Rabbit red blood cell hexokinase: Decay mechanism during reticulocyte maturation. J Biol Chem 261:8327-8333

Malaisse-Lagae F, Malaisse WJ (1988) Hexose metabolism in pancreatic islets: Regulation of mitochondrial hexokinase binding. Biochem Med Metab Biol 39:80-89

Matschinsky FM (1990) Glucokinase as glucose sensor and metabolic signal generator in pancreatic ß-cells and hepatocytes. Diabetes 39:647-652

McCabe ERB (1983) Glycerol kinase deficiency: An inborn error of compartmental metabolism. Biochem Med 30:215-230

McCabe E R B (1989) Disorders of glycerol metabolism. In Scriver CR, Beaudet AL, Sly WS, Valle D (eds) The Metabolic Basis of Inherited Disease. McGraw-Hill New York, pp 945-965

McCabe E R B, Seltzer W K (1986) Glycerol kinase deficiency: Compartmental considerations regarding pathogenesis and clinical heterogeneity. In Brautbar N (ed) Myocardial and Skeletal Muscle Bioenergetics: Proceedings of 2nd International Congress on Myocardial and Cellular Bioenergetics and Compartmentation. Plenum Publishing Corp. New York, pp 481-494

Nelson BD, Kabir F (1985) Adenylate kinase as a source of ATP for tumor mitochondrial hexokinase. Biochim Biophys Acta 841:195-200

Nelson BD, Kabir F (1986) The role of the mitochondrial outer membrane in energy metabolism of tumor cells. Biochemie 68:407-415

Nishi S, Seino S, Bell GI (1988) Human hexokinase: Sequences of amino- and carboxy-terminal halves are homologous. Biochem Biophys Res Commun 157:937-943

Parry DM, Pedersen PL (1983) Intracellular localization and properties of particulate hexokinase in the Novikoff ascites tumor. J Biol Chem 258:10904-10912

Polakis PG, Wilson JE (1985) An intact N-terminal sequence is critical for binding of rat brain hexokinase to mitochondria. Arch Biochem Biophys 236:328-337

Poulton KR, Nightingale S (1988) A new metabolic muscle disease due to abnormal hexokinase activity. J Neurol Neurosurg Psychiatry 51:250-255

Ralph RR, Darkin-Rattray S, Schofield P (1990) Growth-related protein kinases. BioEssays 12:121-124

Schwab DA, Wilson JE (1989) Complete amino acid sequence of rat brain hexokinase, deduced from the cloned cDNA, and proposed structure of a mammalian hexokinase. Proc Natl Acad Sci USA 86:2563-2567

Schwab DA, Wilson JE (1991) Complete amino acid sequence of the type III isozyme of rat hexokinase, deduced from cloned cDNA. Arch Biochem Biophys 285:365-370

Sener A, Malaisse-Lagae F, Giroix MH, Malaisse W (1986) Hexose metabolism in pancreatic islets: Compartmentation of hexokinase in islet cells. Arch Biochem Biophys 251:61-67

Stocchi V, Magnani M, Canestrari F, Dacha M, Fornaini G (1982) Multiple forms of human red blood cell hexokinase: Preparation, characterization, and age dependence. J Biol Chem 257:2357-2364

Stocchi V, Magnani M, Piccoli G, Fornaini G (1988) Hexokinase microheterogeneity in rabbit blood cells and its behaviors during reticulocyte maturation. Mol Cell Biochem 79:133-136

Thelen AP, Wilson JE (1991) Complete amino acid sequence of the type II isozyme of rat hexokinase, deduced from a cloned cDNA: Comparison with a hexokinase from Novikoff ascites tumor. Arch Biochim Biophys 286:645-651

Thinnes FP, Gotz H, Kayser H, Benz R, Schmidt WE, Kratzin HD, Hilschmann N (1989) Zur Kenntnis der Porine des Menschen. I. Reinigung eines Porins aus menschlichen B-Lymphozyten (Porin 31HL) und sein topochemischer Nachweis auf dem Plasmalemm der Herkunftszelle. Biol Chem Hoppe-Seyler 370:1253-1264

Tuttle JP, Wilson JE (1970) Brain hexokinase: A kinetic comparison of soluble and particulate forms. Biochim Biophys Acta 212:185-188

Valentine WN, Oski FA, Paglia DE, Baughan MA, Schneider AS, Naiman JL (1967) Hereditary hemolytic anemia with hexokinase deficiency -- role of hexokinase in erythrocyte aging. N Engl J Med 276:1-11

Vionnet N, Stoffel M, Takeda J, Yasuda K, Bell GI, Zouali H, Lesage S, Velho G, Iris F, Pass P, Froguel P, Cohen D (1992) Nonsense mutation in the glucokinase gene causes early-onset non-insulin-dependent diabetes mellitus. Nature 356:721-722

Walters E, McClean P (1968) The effect of anti-insulin serum and alloxan-diabetes on the distribution and multiple forms of hexokinase in lactating rat mammary gland. Biochem J 109:737-741

Wigley WC, Nakashima RA (1992) Evidence for multiple genes coding for the isozymes of hexokinase in the highly glycolytic AS-30D rat hepatoma. FEBS Lett 300:153-156

Wilson JE (1973) Ligand induced conformations of rat brain hexokinase: Effects of glucose 6-phosphate and inorganic phosphate. Arch Biochem Biophys 159:543-549

Wilson J E (1986) Regulation of mammalian hexokinase activity. In Beitner R (ed) Regulation of carbohydrate metabolism, Vol. I. CRC Press, Cleveland, pp 45-85

Zagotta WN, Hoshi T, Aldrich RW (1990) Restoration of inactivation in mutants of *Shaker* potassium channels by a peptide derived from ShB. Science 250:568-571

GENES ENCODING HUMAN VDAC PROTEINS: IDENTIFICATION OF MULTIPLE ISOFORMS, EXPRESSION IN YEAST AND CHROMOSOMAL LOCALIZATION

Elizabeth Blachly-Dyson and
Michael Forte
Vollum Institute for Advanced Biomedical Research
Oregon Health Sciences University
3181 S.W. Sam Jackson Park Rd.
Portland, OR 97201

Introduction

The voltage-dependent anion channel (VDAC, also known as mitochondrial porin) of the outer mitochondrial membrane is a small (~30 kDa) abundant protein found in cells from all eukaryotic kingdoms. VDAC forms large (~3nm), voltage-gated pores when incorporated into planar lipid bilayers (Colombini, 1989; Sorgato and Moran, 1993). Physiologically, VDAC is thought to be the primary pathway for the movement of adenine nucleotides through the mitochondrial outer membrane, controlling the traffic of these essential compounds to and from the mitochondria, as well as the passage of the substrates and products of a variety of metabolic pathways. VDAC has also been shown to be the binding site for hexokinase (HEX) and glycerol kinase (GK) on the outer surface of the mitochondrion (Adams et al.,1991). Binding of these enzymes to the mitochondria may allow them preferential access to mitochondrial ATP (Brdiczka, 1991) although recently this view has been questioned (Kabir and Nelson, 1991). The association of these enzymes with the outer membrane may constitute a significant metabolic regulatory interaction, since association of HEX and GK with mitochondria is dynamic and varies among different tissues, during development and depending on the metabolic state of the cell (Adams et al, 1991). For example, cells in highly malignant tumors have an increased percentage of

NATO ASI Series, Vol. H 83
Molecular Biology of Mitochondrial Transport Systems
Edited by M. Forte and M. Colombini

mitochondrially bound hexokinase when compared to normal cells (Nakashima et al, 1988). Furthermore, binding of HEX and GK to VDAC may occur specifically at contact sites between the inner and outer mitochondrial membranes, thus linking cytoplasmic metabolism and ADP production as regulated by these enzymes with the regulation of mitochondrial respiration and oxidative phosphorylation in the mitochondrial matrix (Kottke et al, 1988; Brdiczka, 1990). In addition, it appears that VDAC is part of a complex forming the mitochondrial benzodiazepine receptor (McEnery et al, 1992).

Since HEX and GK may be bound to mitochondria through VDAC only at contact sites, while VDAC is distributed throughout the outer membrane, it has been suggested that there may be multiple forms of VDAC in mammals that differ in their localization within the outer mitochondrial membrane and in physiological function (Brdiczka, 1990). We have used a molecular cloning approach to look for multiple forms of VDAC in humans, and have identified cDNAs representing the transcripts of two different human VDAC genes, HVDAC1 and HVDAC2 (Blachly-Dyson et al, 1993). When these genes were expressed in yeast, they resulted in the production of proteins which formed channels in bilayers with the characteristics expected of VDAC. Mitochondria isolated from yeast expressing HVDAC1, but not HVDAC2 demonstrated specific binding of mammalian hexokinase. In the course of mapping the chromosomal locations of these genes, we have identified two additional sequences highly homologous to the HVDAC1 gene (Blachly-Dyson et al, submitted).

Identification of HVDAC genes

The HVDAC1 gene was cloned by using two nested pairs of degenerate oligonucleotides to prime PCR amplification of first strand cDNA from a human B-lymphocyte cell line (Blachly-Dyson et al, 1993). The primers were based on the published protein sequence of VDAC from human B-lymphocytes (Kayser et al., 1989). The amplified fragment, which encoded a peptide identical to a portion of the published protein sequence, was used to screen a human pituitary cDNA library, and a cDNA clone containing the entire protein coding sequence was obtained, and named HVDAC1. This DNA was subsequently used to screen a human liver cDNA library at low stringency, and five overlapping clones were obtained that corresponded to a different, related sequence designated HVDAC2. The proteins encoded by the two sequences are identical at 211/283 positions (75%), with conservative replacements at many of the remaining positions. The most notable differences between the two sequences are the presence of an 11 amino acid amino-terminal extension in HVDAC2 relative to HVDAC1, and the subtitution of two acidic residues (Glu 211 and Asp 212) in HVDAC2 for two basic residues (Lys 200 and Lys 201) in HVDAC1.

During PCR analysis aimed at determining the location of the HVDAC1 gene (see below), it became clear that primers specific for HVDAC1 amplified at least two additional, related sequences located at distinct positions in the human genome (Blachly-Dyson et al, submitted). Amplification of these sequences from specific cell lines missing individual human chromosomes and restriction analysis allowed the identification of two additional genes encoding proteins highly similar to HVDAC1. These genes have been designated HVDAC3 and HVDAC4. At this time, the derived amino acid sequence of the HVDAC proteins encoded by these genes is known only between the primer sites used in the amplification. The sequence of the PCR fragment from HVDAC3 is identical to HVDAC1 in 256 of 272 nucleotides (94%) (Fig. 1). When the sequence is translated in a reading frame equivalent to that of HVDAC1, the predicted polypeptide is identical to HVDAC1 up to codon 247, which is a stop codon in HVDAC3, whereas the HVDAC1 open reading frame continues to codon 283. This observation raises the possibility that HVDAC3 represents a pseudo gene rather than a functional gene encoding a HVDAC protein. The HVDAC4 PCR fragment is identical to HVDAC1 in 257 of 272 nucleotides (94.5%). It contains an open reading frame that terminates at the same position as the HVDAC1 termination site. Within this open reading frame are 11 nucleotide sequence differences between HVDAC1 and HVDAC4, resulting in 3 amino acid differences between the encoded polypeptides. Nucleotide sequence comparisons indicate that HVDAC3 and HVDAC4 are each more similar to HVDAC1 than they are to each other. Only 245 of 272 nucleotides (90.1%) are identical between HVDAC3 and HVDAC4.

```
HVDAC1    GCCTGCTTCTCGGCTAAAGTGAACAACTCCAGC
HVDAC3    GCTTGCTTCTTGGCTAAAGTGAACAACTCCAGC
HVDAC4    AGCTGCTTCTCGGCTAAAGTGAATAACTCCAGC
HVDAC2    GCTTCCATTTCTGCAAAAGTCAACAACTCTAGC
          231          235               240

HVDAC1    CTGATAGGTTTAGGATACACTCAGACTCTAAAG
HVDAC3    CTGATAGGTTTAGGATAAACTCAGACCCTAAAG
HVDAC4    CTGATGGGTTTAGGATACACTCAGACCCTAAAA
HVDAC2    TTAATTGGAGTAGGCTATACTCAGACTCTGAGG
                   245              250

HVDAC1    CCAGGTATTAAACTGACACTGTCAGCTCTTCTG
HVDAC3    CCAGGTATCAAACTGACACTGTCAGCTCTTCTG
HVDAC4    CCAGGTATCAAATTGACACTGTCAGCTCTTCTG
HVDAC2    CCTGGTGTGAAGCTTACACTCTCTGCTCTGGTA
                255              260

HVDAC1    GATGGCAAGAACGTCAATGCTGGTGGCCACAAG
HVDAC3    GATCTCAAGAAGGTCAGTGCTGGTGGCCCCATG
HVDAC4    GATGGCAAGAACGTCAATGGTGGTGGCCACAAG
HVDAC2    GATGGGAAGAGCATTAATGCTGGAGGCCACAAG
             265              270

HVDAC1    CTTGGTCTAGGACTGGAATTTCAAGCATAAATG
HVDAC3    CTTGGTCTAGGACTGGAATTTCAAGCATAAATG
HVDAC4    CTTGGTTTAGGCCTGGAATTTCAAGCATAAATG
HVDAC2    GTTGGGCTCGCCCTGGAGTTGGAGGCTTAATCC
          275              280

HVDAC1    AATACTGTACAATTGTTTAATTTTAAACTATTT
HVDAC3    AATACTGTACAATTGTTCAATTTTAAACTATTT
HVDAC4    AATACTGTACAATTGTTTAATTTTAAACTATTT
HVDAC2    AGCTGAAAGAAACCTTTGGGAATGGATATCAGA

HVDAC1    TGCAGCATAGCTACC-TTCAGAATTTAGTGTAT
HVDAC3    TGCAGCATAGCTACCCTTCAGAATTTGGTGTGT
HVDAC4    TGCAGCATAGCTACC-TTCAGAATTTACTGTAT
HVDAC2    AGATTTGGCCTTAATATATTTCCATTGTGACCA

HVDAC1    CTTTTAATGTTGTATGTCTGGGATGCAAGTATT
HVDAC3    CCTTTAATGTTGTATGTCTGGGATGCAAGTATT
HVDAC4    CGTTTAATGTTGTACGTCTGGAATGCAAGTATT
HVDAC2    GCAGCAGGCTTTTTCCCCCAAGAAGATGATCA

HVDAC1    GCTAAATAT
HVDAC3    GCTAAATAT
HVDAC4    GCTAAATAT
HVDAC2    AAACAAAGG
```

Figure 1 - Sequence comparison of HVDAC sequences. Sequences of cloned PCR fragments from HVDAC3 (line2) and HVDAC4 (line3) are aligned with the homologous portion of the HVDAC1 gene (top) and HVDAC2 (bottom). Underlining indicates nucleotides that differ from the HVDAC1 sequence. The termination codons are boxed. Numbering indicates codon number based on HVDAC1.

Expression of Human VDAC cDNAs in yeast cells

In order to confirm that HVDAC1 and HVDAC2 cDNAs encode a protein with the characteristics expected of VDAC, constructs were generated containing the promoter, 5' untranslated region, and first 9 codons of the yeast VDAC gene fused to HVDAC1 at codon 11 and HVDAC2 at codon 22 (Blachly-Dyson et al, 1993). These constructs were introduced into yeast cells lacking the endogenous yeast VDAC gene, mitochondria from the resulting transformants were isolated, and VDAC was purified from the mitochondrial membranes. Purified samples contained a single protein band of appropriate molecular weight as assessed by silver-stained SDS polyacrylamide gels (data not shown). These proteins were then introduced into synthetic phospholipid bilayers and their electrophysiological properties tested. HVDAC1 expressed in yeast formed channels with a single-channel conductance (4.1 nS in 1M HCl) and selectivity ($P(Cl^-)/P(K^+) = 1.8$) essentially identical to those of yeast VDAC. HVDAC2 expressed in yeast formed channels that initially appeared very similar to HVDAC1 channels, but appeared to lose their voltage dependence with time.

Yeast cells lacking the endogenous yeast VDAC gene are temperature-sensitive for growth on glycerol-based media. Introduction of HVDAC1 or HVDAC2 into these cells restored the ability of the cells to grow on glycerol at elevated temperature as well as did reintroduction of the yeast VDAC gene. Thus,the proteins encoded by these cDNAs are capable both of forming VDAC-like channels and of performing the essential physiological function of VDAC in yeast cells. To determine whether the two human isoforms differ in their ability to bind hexokinase, mitochondria prepared from yeast strains expressing HVDAC1, HVDAC2 or lacking any VDAC were compared for their ability to bind rat brain hexokinase. Only HVDAC1 was observed to bind the enzyme, while mitochondria containing HVDAC2 exhibited the same background level of binding as mitochondria lacking VDAC. This background level of hexokinase binding is similar to that seen for mitochondria containing wild-type yeast VDAC. (Presumably, this reflects the species dependence of binding, since yeast VDAC has been demonstrated to bind yeast hexokinase (Forte, et al. 1987a)). This suggests that HVDAC1 but not HVDAC2 may be responsible for respiratory control involving the binding of this enzyme to the outer membrane, giving it preferential access to mitochondrially generated ATP (Adams et al, 1991).

Expression of human VDAC cDNAs

Since essentially all cell types contain mitochondria, they presumably all contain some form of VDAC. With at least 4 different human VDAC genes, it is possible that each HVDAC gene is uniquely expressed at distinct subcellular locations, differentially in specific tissues or in unique spatial and temporal patterns during development. This view would be consistent with existing data. For example, antibodies generated to HVDAC1 purified from mitochondria appear to cross-react immunocytochemically with the plasma membrane (Babel et al, 1991). Biochemical and physiological studies have also suggested that VDAC-like proteins may be present in the plasma membrane (Blatz and Magleby, 1983; Bureau et al, 1991). In addition, it is known that the distribution of hexokinase binding sites is dynamic and varies between different cell types, during development, and between cells in different metabolic states (Adams et al, 1991). To examine whether HVDAC1 and HVDAC2 may be differentially expressed in individual tissues, Northern blot analysis and specific PCR amplification were performed on RNA from a variety of human tissues and cell lines (Blachly-Dyson et al, 1993). Except for one cell line (NB5) which contained only HVDAC2, all samples tested contained both HVDAC1 and HVDAC2 RNAs. However, since this analysis was performed, we have discovered that two additional HVDAC sequences, HVDAC3 and HVDAC4, exist within the human genome. Although we do not yet know whether these encode functional VDAC proteins, the existence of these additional, highly related genes leaves open the possibilbility that individual HVDAC genes may be differentially expressed.

Mapping the chromosomal locations of human VDAC genes

To examine whether mutations in any of the HVDAC genes may result in an identified human genetic syndrome, we determined the chromosomal location of the human VDAC genes. Pairs of oligonucleotide primers specific for the HVDAC1 and HVDAC2 sequences were used to amplify DNA from a panel of human-rodent hybrid cell lines (NIGMS human/rodent somatic cell mapping panel #1 and #2; Drwinga et al, 1993). The results with HVDAC2 primers indicated that this gene is located on chromosome 21. In contrast, as mentioned above, HVDAC1 primers amplified products of the expected size from a large number of the cell lines. Restriction analysis of the PCR products and sequencing of PCR products from human genomic DNA and one of the cell lines indicated that HVDAC1 was most likely located on the X chromosome and demonstrated the existence of two additional sequences highly homologous to HVDAC1 in the human genome, designated HVDAC3 and HVDAC4. Subsequent PCR and restriction analysis of individual cell lines in the mapping panel demonstrated that HVDAC3 is located on chromosome 12 and HVDAC4 on chromosome 1.

The location of HVDAC 1 on the X chromosome was further defined by PCR analysis of an X-chromosome-specific hybrid cell mapping panel (Luty et al,1990) and fluorescence in situ hybridization (FISH) (Fig. 2). Results of these techniques indicate that HVDAC1 is located in the interval Xq13-Xq21. The FISH analysis also identified a band on chromosome 1 in the interval 1q24 1q25. This probably represents cross-hybridization of the probe to HVDAC4.

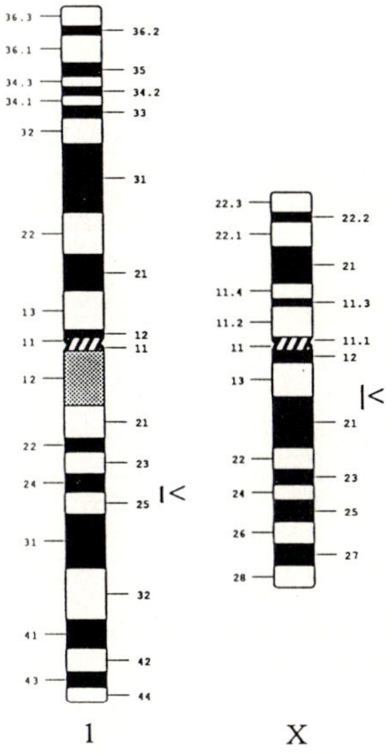

Figure 2- The position of FISH hybridization signals with HVDAC1 cDNA probe is schematically indicated by bars on a standard chromsome ideogram.

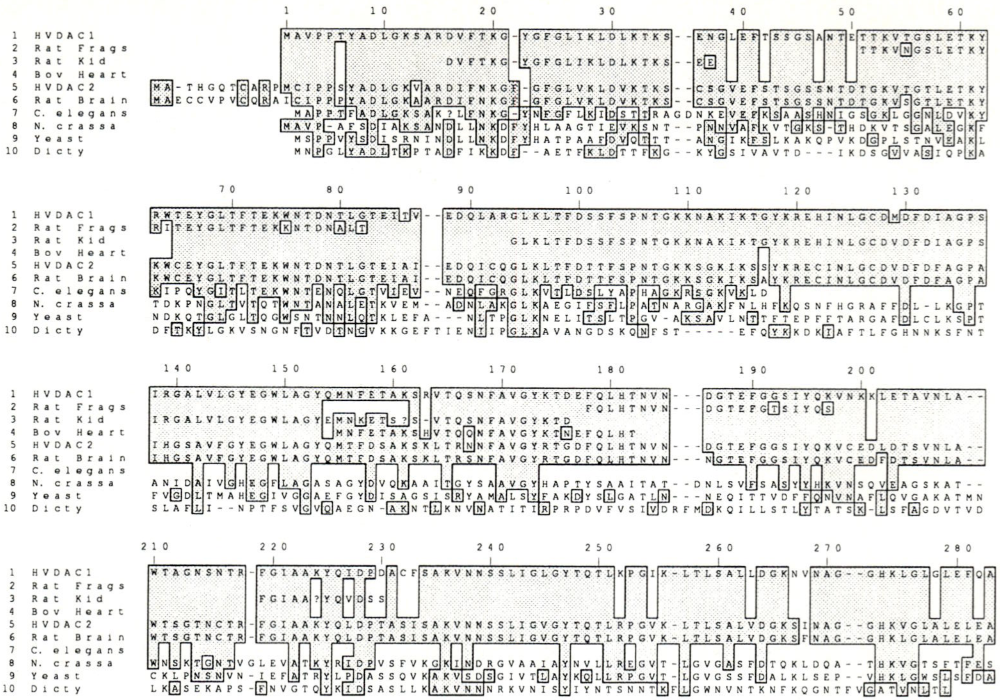

Figure 3. Alignment of published VDAC sequences. The available sequences of VDAC proteins and cDNA translations were aligned by eye. Boxes indicate residues that are identical to HVDAC1 and/or HVDAC2. Dashes indicate gaps introduced to improve alignment. Sources of sequences are HVDAC1 and HVDAC2, Blachly-Dyson et al., 1993; Rat frags (rat brain VDAC protein fragments), Bureau et al., 1992; Rat Kid (rat kidney) and Bov. heart (bovine heart), protein fragments Kayser et al., 1989; Rat brain, Bureau et al., 1992; C. elegans, Waterston et al., 1992; N. crassa, Kleene et al., 1987; Yeast (Saccharomyces cerevisiae), Forte et al., 1987b; Dictyosteleum, Troll et al., 1991.

Relationship to other VDAC genes

Since this work was begun, the sequences of additional mammalian and invertebrate VDAC genes and proteins have become available. Fig.3 summarizes the sequences that are present in the literature and databases. The mammalian sequences all appear to be similar either to HVDAC1 or HVDAC2. The *C. elegans* fragment is, not surprisingly, more closely related to the mammalian sequences than to the fungal sequences. Surprisingly, the Dictyostelium sequence (Troll et al., 1992) is distinct from all other described VDAC proteins. Despite these wide variations in sequence, in cases where the properties of the individual VDAC channels has been investigated, the functional similarity between each VDAC protein is striking. When reconstituted in planar lipid bilayers, each forms channels with unit conductance of 4.0-4.5nS in 1M KCl that are open in the absence of an applied transmembrane voltage but close with characteristic voltage dependence in response to application of either positive or negative potentials. Finally, each has a roughly 2:1 preference for monovalent anions over cations. These observations indicate that the proposed β barrel structure of these proteins tolerates extensive amino acid variations without substantial alteration in the structural characteristics underlying the functional properties of this channel.

References

Adams V, Griffin L, Towbin J, Gelb B, Worley K, McCabe E. (1991).Porin interaction with hexokinase and glycerol kinase: metabolic microcompartmentation at the outer mitochondrial membrane. *Biochem Med Metab Biol* **45**: 271-291

Babel D, Walter G, Gotz H, Thinner F, Jurgens L, Konig U, Hilschmann N (1991) Studies on human porin VI: Production and characterization of eight monoclonal mouse antibodies against the human VDAC "porin 31HL" and their application for histological studies in human skeletal muscle. *Biol Chem Hoppe-Seyler* **372**: 1027-1034

Blachly-Dyson E, Zambronicz E, Yu W-H, Adams V, McCabe E, Adelman J, Colombini M, Forte M (1993) Cloning and functional expression in yeast of two human isoforms of the outer mitochondrial membrane channel, the voltage-dependent anion channel.*J Biol Chem* **268**: 1835-1841

Blatz A, Magleby K (1983) Single voltage-dependent chloride-selective channels of large conductance in cultured rat muscle. *Biophys J* **43**: 237-241

Brdiczka D (1990) Interaction of mitochondrial porin with cytosolic pro-teins. *Experientia* **46**: 161-167

Bureau M, Khrestchatisky M, Heeren M, Zambrowwicz E, Kim H, Grisar T, Colombini M, Tobin A, Olsen R (1992) Isolation and cloning of a voltage-dependent anion channel-like Mr 36,000 polypeptide from mammalian brain. *J Biol Chem* **267**: 8679-8684

Colombini M (1989) Voltage gating in the mitochondrial channel, VDAC. *J Membr Biol* **111**: 103-111

Drwinga H L, Toji L H, Kim C H, Green AE, Mulivor R A (1993) NIGMS human/rodent somatic cell hybrid mapping panels 1 and 2. *Genomics* **16**: 311-314

Forte M, Adelsberger-Mangan D, Colombini M (1987a) Purification and characterization of the voltage-dependent anion channel from the outer mitochondrial membrane of yeast.*J Membr Biol* **99**: 65-72

Forte M, Guy H, Mannella C (1987b) Molecular genetics of the VDAC ion channel: structural model and sequence analysis. *J Bioener Biomembr* **19**:341-350

Kabir F, Nelson D (1991) Hexokinase bound to rat brain mitochondria uses externally added ATP more efficiently than internally generated ATP. *Biochim Biophys Acta* **1057**: 147-150

Kayser H, Kratzin H, Thinnes F, Gotz H, Schmidt W, Eckart K, Hilschmann N (1989) Zur kenntnis der porine des menschen II. Charakterisierung und primarstruktur eines 31-kDa-porins aus menschlichen B-lymphozyten (Porin 31HL) *Biol Chem Hoppe-Seyler* **370**:1265-1278

"PORIN 31HL" IN THE PLASMALEMMA OF HUMAN CELLS: A VDAC DISCUSSED AS PART OF A CHLORIDE CHANNEL COMPLEX IN NORMAL AND CYSTIC FIBROSIS B-LYMPHOCYTE CELL LINES

F.P. Thinnes, D. Babel, M. Heiden, A. Hein, L. Jürgens, U. König and N. Hilschmann

Max-Planck-Institut für Experimentelle Medizin
Abteilung Immunchemie
D-3400 Goettingen
Germany

Key terms: Porin; VDAC; Cytoplasma Membrane; Chloride Channel; Sodium Channel; Cystic Fibrosis; $GABA_A$-Receptor

Summary: In 1989, showing the expression of "Porin 31HL" in the plasmalemma of human B lymphocytes, our laboratory gave first evidence on the multitopological localization of VDAC. Looking for a function of the channel in the outer cell membrane, we referred to five corresponding traits shared by "Porin 31HL" and plasmalemma integrated mammalian chloride channels. We furthermore proposed VDAC to be part of a chloride channel complex. VDAC on outer cell surfaces needs biochemical regulation. In apical membranes of epithelia it might be regulated by the Cystic Fibrosis Transmembrane Conductance Regulator (CFTR). Our model on the interaction of VDAC and CFTR may reconcile data on the outwardly rectifying depolarization-induced chloride channel (ORDIC) and the ohmic low-conductance channel formed by the CFTR. Both channels are affected in Cystic Fibrosis and very recent data support that CFTR might regulate ORDIC. As basis for immunotopological work our laboratory had purified and sequenced "Porin 31HL" on the

NATO ASI Series, Vol. H 83
Molecular Biology of Mitochondrial Transport Systems
Edited by M. Forte and M. Colombini

protein level, thus giving the first primary structure analysis of a VDAC from the animal kingdom. Comparing the topochemistry, the channel characteristics and the primary structure of VDAC from normal and Cystic Fibrosis B-lymphocyte cell lines, we were looking for indications of the involvement of VDAC in the CF syndrome.

"PORIN 31HL": A VDAC IN THE PLASMALEMMA OF VARIOUS HUMAN CELLS

In 1989, our laboratory gave first evidence on the localization of the VDAC "Porin 31HL" in another eukaryotic cell compartment than outer mitochondrial membranes. Then, we used rabbit antisera against purified "Porin 31HL" or synthetic peptides of the N-terminal part of this channel. In cytotoxicity and indirect immunofluorescence experiments we demonstrated the expression of the VDAC in the plasmalemma of the transformed human B-lymphocyte cell line H2LCL. Furthermore,

normal	B lymphocytes
normal	T lymphocytes
normal	skeletal muscle cells
H2LCL	EBV transformed normal B lymphoblasts
SB	EBV transformed normal B lymphoblasts
CF/F	EBV transformed Cystic Fibrosis B lymphoblasts
CF/W	EBV transformed Cystic Fibrosis B lymphoblasts
HT29	epithelial cells of a colon carcinoma
DAUDI	Burkitt tumor cells
KM3	acute lymphoblastic leukemia cells
CEM	acute lymphoblastic leukemia cells
HSB2	acute lymphoblastic leukemia cells
K562	erythro leukemia cells

Table 1: Human cell types giving surface labelling in indirect immunofluorescence tests with anti "Porin 31HL" antibodies. According to König et al. (1991), supplemented.

those experiments showed the accessibility of the acetylated N-terminal part of the channel molecule on the outer cell surface (Thinnes et al. 1989).

Meanwhile, we confirmed both topological aspects of VDAC in outer cell membranes by the application of eight mouse monoclonal anti-"Porin 31HL" antibodies in immunotopological studies on various human cell types. In these experiments, comparing transformed human normal and Cystic Fibrosis B-lymphocyte cell lines, no differences in surface labelling by anti-"Porin 31HL" antibodies could be seen. Furthermore, the outer membranes of the epithelial colon carcinoma cell line HT29 clearly reacted with our porin antibodies (Tab.1).

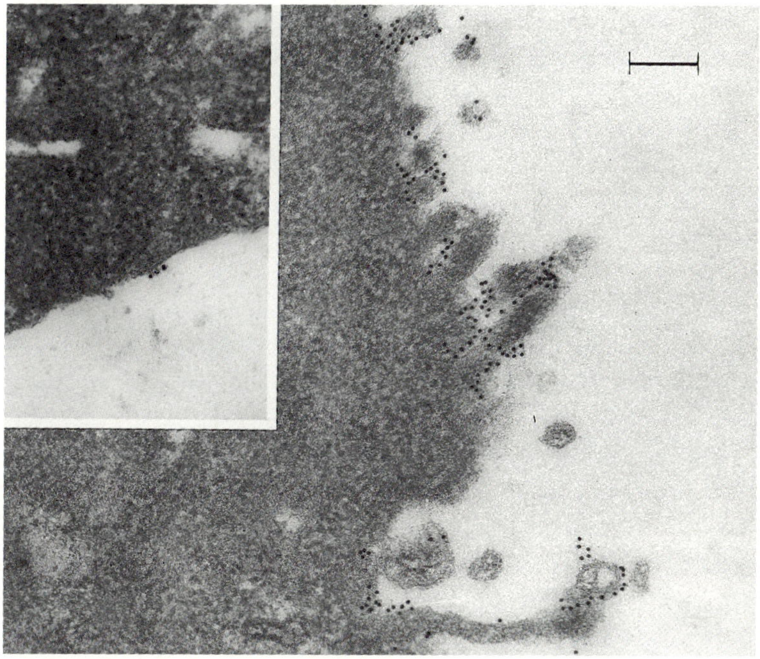

Fig.1: Indirect immunogold labelling of the plasmalemma of the B-lymphocyte type acute lymphoblastic leukemia cell line KM3 by anti-"Porin 31HL" antibodies (from Cole et al. 1992).

The demonstration of the VDAC in the cytoplasmic membrane of normal human T and B lymphocytes showed this phenomenon not to be caused by cell transformation (König et al. 1991). More recently, in a histotopological study on cryosections of human skeletal muscle we obtained simultaneous labelling of porin channels in the plasmalemma and outer mitochondrial membranes of cells (Babel et al. 1991). We furthermore confirmed our early light microscopy results on the expression of VDAC in the plasmalemma of human cells by giving first electron microscopy data on an immunogold labelling of surfaces (Fig. 1) of the B-lymphocyte type acute lymphoblastic leukemia cell line KM3 (Cole et al. 1992).

There are supporting immunochemical results on the expression of VDAC in the cytoplasmic membrane from other laboratories. First, applying an antiserum against porin purified from rat mitochondria (provided by Dieter Brdiczka, Universität Konstanz), labelling of several sites in ultrathin cryosections of rat liver parenchymal cells was found: predominantly on the plasma membrane facing the bile canaliculi and on membranes of putative autophagic vacuoles (Norbert Roos, University of Oslo; personal communication, in Babel et al. 1991).

Second, the laboratory of Michel Bureau (Université de Liège; personal communication, in Babel et al. 1991) has immunotopological results indicating the expression of porin channels in the plasmalemma and other cell compartments of mammalian cells. Furthermore, in a combined protein and cDNA level study, Michel Bureau and collegues recently presented data indicating a tight in vivo association of VDAC with the plasmalemma-integrated $GABA_A$-receptor complex. In benzodiazepine affinity chromatography purified $GABA_A$-receptor protein complex preparations the authors observed a 36 kDa protein in mammalian central nervous system cells, which by sequencing proved to be a member of the VDAC family (see below). The tight association of channel active VDAC and $GABA_A$-receptor molecules has furthermore been demonstrated by coprecipitation of some $GABA_A$-receptor by an anti-36 kDa VDAC antiserum (Bureau et al. 1992). The copurification of the plasmalemma-associated $GABA_A$-receptor and VDAC directly supports our data on the localization of VDAC in the cytoplasmic membrane.

PRIMARY STRUCTURE ANALYSES OF HUMAN AND MAMMALIAN PORINS

Our laboratory completely sequenced, on the protein level, "Porin 31HL" (Kayser et al. 1989) and "Porin 31HM" (Jürgens et al. 1991), purified from crude membrane fractions of human B lymphocytes and skeletal muscle, respectively, thus giving the first primary structure analyses of porins from the animal kingdom. The sequences of both molecules proved to be identical. In both studies, we did not find any indication of variability of porins on the phenotypic level, neither between the two cell types studied, nor interindividually in muscle probes of different dead bodies (Jürgens et al. 1991).

Recently, Michael Forte and coworkers presented data on cDNA cloning and sequencing of human porin isoforms: HVDAC1 and HVDAC2 (Blachly-Dyson et al. 1991, 1992). HVDAC1 is identical to "Porin 31HL" except for an additional N-terminal methionine. HVDAC2 is 75% identical to HVDAC1 with a N-terminal extention of eleven amino acids. Most substitutions are conservative, some going from charged to uncharged amino acids and vice versa. There is one outstanding difference: two Lys in positions 199 and 200 of "Porin 31HL" are replaced in HVDAC2 at corresponding positions by Glu and Asp respectively. For "Porin 31HL" this part of the channel was proposed to be expressed outside the bilayer of outer mitochondrial membranes (De Pinto et al. 1991). In rat VDAC cDNA (Bureau et al. 1992), positions corresponding to amino acids 199 and 200 of "Porin 31HL" show identical building blocks as in HVDAC2, indicating that both molecules belong to mammalian VDAC of type 2. For a comparison of known primary structure data on mammalian VDACs see the Appendix.

To look for indications of the involvement of porin channels in the Cystic Fibrosis syndrome, we meanwhile purified "Porin 31HL/CF" from the total membrane fraction of a transformed human B-Lymphocyte cell line of a CF patient. Part of the material was sequenced after reduction and carboxymethylation, tryptic cleavage and separation of peptides by reversed phase HPLC. Up to now, we clearly identified peptides representing 97% of a molecule identical to "Porin 31HL" from normal B lymphocytes (Hein 1992). We are on the way to complete and reexamine these results.

"PORIN 31HL" IN THE PLASMALEMMA: PUTATIVE BIOCHEMICAL BASIS OF THE ANION CHANNEL OF LARGE CONDUCTANCE (BLATZ & MAGLEBY 1983)

In our first paper on "Porin 31HL", purified from total membrane preparations of the transformed human B-lymphocyte cell line H2LCL, we already presented data proving this molecule to belong to the VDAC family (Thinnes et al. 1989). We meanwhile confirmed these early results by broadly testing its channel characteristics (Benz et al. 1992). Here, artificial lipid bilayer membrane experiments clearly defined the function of "Porin 31HL" as a channel-forming protein, showing two maxima in distribution of single-channel conductances: 2.4 or 4.3 nS in 1M KCl. "Porin 31HL" was found to be voltage-dependent, switching to ion-permeable substates at membrane potentials larger than 20 mV. In the open state the human porin exhibited the characteristics of a general diffusion pore, showing an estimated effective diameter of about 1.7 nm. These data are in agreement with a sixteen strands ß-barrel channel model, recently proposed for a transmembrane arrangement of "Porin 31HL" (De Pinto et al. 1991). The open state of "Porin 31HL" proved to be slightly anion-selective, while in the lower conductance state the channel favored the passage of cations. All these results are in agreement with those of eukaryotic porins purified from the outer mitochondrial membranes of various eucaryotic cells. More recently, the VDAC function of "Porin 31HL" was confirmed by the expression of identical or homologous human cDNAs, HVDAC1 or HVDAC2, in yeast, lacking the endogenous VDAC gene (Blachly-Dyson et al. 1992).

Monoclonal mouse antibodies against the N-terminal part of "Porin 31HL" blocked its reconstitution but had otherwise no influence on its channel properties. These results suggest the N-terminal region of plasmalemma standing VDAC probably not to be involved in channel gating (Benz et al. 1992).

We discussed plasmalemma-associated "Porin 31HL" as the biochemical basis of chloride-selective anion channels of large conductance (maxi channels), first observed by Blatz & Magleby (1983) in patch-clamping studies on outer cell membranes of cultured rat myotubes (Thinnes et al. 1989). Meanwhile, Babel et al. (1991) demonstrated the expression of VDAC in the sarcolemma of human skeletal muscle cells. Maxi channels in excised plasmalemma patches of various cell types show large conductances and voltage-dependent gating kinetics similar to those of VDAC (Blatz & Magleby 1983, Thinnes 1992, Benz et al. 1992). Multiple-conductance chloride channels of human T lymphocytes manifested different characteristics under cell-attached patch-clamping conditions as compared to those of excised patches (Pahapill et al. 1992), indicating that the channel complexes may undergo structural changes, depending on experimental conditions and membrane potential (Benz et al. 1992). Studying a mouse B-lymphocyte cell line, multiple conductance levels of a voltage-dependent stilbene-sensitive anion channel were observed, furthermore showing several differences between cell-attached and excised-patch recordings. The conductance steps observed were interpreted on the basis of different subconductance states of a single channel

(Bosma 1989). Thus, these studies may reconcile data on maxi chloride channels and outwardly rectifying medium conductance chloride channels (Gögelein 1988, Gallin 1991, Wine 1991; see below).

Anion channels of large conductance were identified in apical membranes of a Cl^--secretory epithelium (Kolb et al. 1985), a tissue as affected in CF. There are patch-clamping data, showing that our transformed human B-lymphocyte cell line H2LCL also expresses this channel in the plasmalemma (H.-A. Kolb, Universität Konstanz, personal communication). We compared the channel characteristics of porins from corresponding preparation steps, starting with total membrane fractions of transformed normal ("Porin 31HL") or Cystic Fibrosis ("Porin 31HL/CF") B-lymphocyte cell lines. First indications of differences, seen in our "Porin 31HL/CF" preparation by artificial diphytanoyl phosphatidylcholine/n-decane membrane experiments, were reexamined by precipitation of both porin preparations, their reconstitution by cholesterol and final incorporation into artificial planar bilayers. Using this approach, no difference between normal and CF porin could be validated (A. Schmid & R. Benz, Universität Würzburg, personal communication). These results probably point towards the influence of lipids in the regulation of membrane integrated porin channels.

PORIN IN THE PLASMALEMMA AS PART OF A CHLORIDE CHANNEL COMPLEX

Cystic fibrosis (CF) is a polyexocrinopathy affecting mainly the respiratory, digestive and reproductive tracts of man. The basic defect of the syndrome is a severe reduction in chloride permeability of several epithelia (Quinton 1990), this in consequence of defective structure or regulation of cAMP-dependent chloride channels - or chloride channel complexes -

in the apical membranes of epithelial cells (Welsh 1990). In CF, besides the decrease in Cl^- efflux, the Na^+ influx into epithelial cells is raised (Boucher et al. 1988), part of which is mediated by amiloride-insensitive cation channels not yet defined on the molecular level (Jorissen et al. 1991). Furthermore, corresponding chloride channel disturbances were found in human T- and B-lymphocyte cell lines (Chen et al. 1989, Garber 1992).

In the 1980's, studies of cystic fibrosis focussed, on the one hand, on an outwardly rectifying depolarization-induced Cl^- channel (ORDIC) showing intermediate conductance in apical membranes of epithelial cells and the plasmalemma of T- and B-lymphocyte cell lines. In cell attached patches, this channel was observed in normal but not in CF cells. By patch excision, it was activated in healthy and in CF cells, observations leading to the hypothesis that in CF its regulation is defective (Kunzelmann et al. 1989, Welsch 1990, Chen et al. 1989). There are meanwhile reservations about ORDIC as the major component of the pathogenesis of the CF syndrome (for discussion see: Wine 1991, Ward et al. 1991, Cabantchik et al. 1992).

On the other hand, at that time the Cystic Fibrosis Transmembrane Conductance Regulator (CFTR) molecule was discussed as a modulator of an associated chloride channel molecule in apical membranes of epithelial cells (Riordan et al. 1989, Hyde et al. 1990). VDAC in the plasmalemma of cells needs biochemical regulation: porin channels, ubiquitous or at least widespread in this cell compartment, have to be kept closed in the resting state of cells by interactions with other molecules. These modulators may vary with cell types (Thinnes et al. 1991). From here recent data, demonstrating that ORDIC appears to be ubiquitous while CFTR mRNA levels vary even in epithelial cells (Ward et al. 1991), may nevertheless be in agreement with our suggestion on the regulation of plasmalemma-associated porin channels by CFTR in cells affected in Cystic Fibrosis (see below).

In 1990, we first referred to five traits, shared by "Porin 31HL" and chloride channels putatively affected in CF, furthermore proposing that VDAC might be part of a chloride channel complex (Thinnes et al. 1990, 1991). These are:

1) Expression in the plasmalemma of different cells (Porin 31HL: Thinnes et al. 1989, König et al. 1991, Babel et al. 1991, Cole et al. 1992 - Chloride channels: for reviews see Gögelein 1988, Gallin 1991, Wine 1991).

2) Molecular masses of about 30 kDa (Porin 31HL: Thinnes et al. 1989, Kayser et al. 1989 - Chloride channels: Jessen et al. 1986, Landry et al. 1989, Breuer 1990, Blair et al. 1990, Bear et al. 1992, Reenstra et al. 1992). Furthermore, Ran and Benos (1992) recently discussed a 38 kDa putative epithelial chloride channel molecule from bovine trachea to be a homologue of "Porin 31HL".

3) Conductance in artificial membranes (Porin 31HL: Thinnes et al. 1989, Benz et al. 1992 - Chloride channels: Landry et al. 1989)

4) Ability to be modified by DIDS (Porin 31HL: Thinnes et al. 1990 - Chloride channels: Breuer 1990, Kanno et al. 1990, Ward et al. 1991, Pahapill et al. 1992, Egan et al. 1992)

5) Subject to regulation by modulators (Porin 31HL: Benz et al. 1988, Holden et al. 1988 - Chloride channels: Kunzelmann et al. 1989, Welsh et al. 1989, Welsh 1990).

On this basis, we formulated (Thinnes et al. 1990, 1991) and since refined a two-component model of a postulated VDAC-containing epithelial chloride channel complex, misregulated in CF (Thinnes 1992, Fig.1): in resting cells plasmalemma-associated VDAC is kept closed by interation with the CFTR molecule, here assumed to be expressed on the cytosolic side of the plasmalemma. In normal cells, stimulation induced changes of the conformation of the CFTR allow full opening of VDAC to the "open" anion (Cl^-) selective conductance state. In CF cells, the CFTR is mutated, this form only allowing a "closed" cation (Na^+) selective sub-conductance state of VDAC. The model thus explaines, at least formally, suppressed Cl^- efflux but increased Na^+ influx, as observed in CF epithelial cells. In immunocompetent B lymphocytes porin channels in the plasmalemma are likely to be regulated by distinct modulators (Thinnes et al. 1991).

Very recently, Egan et al. (1992) induced the appearance of small ohmic chloride channels and furthermore restored the activation of ORDIC channels by the expression of recombinant CFTR in CF bronchial epithelial cells, data suggesting that CFTR regulates ORDIC channels.

CONCLUSIONS AND PERSPECTIVES

Our model of a postulated VDAC-containing epithelial chloride channel complex, putatively affected in Cystic Fibrosis, needs modification or even may prove invalid. First of all, it was meanwhile verified that the CFTR molecule is a trans-plasmalemmal regulated low-conductance ohmic chloride channel itself (Anderson et al. 1991, Tilly et al. 1992, Bear et al. 1992). Secondly, our own experiments so far did not give clear indications of the involvement of porin channels in the CF syndrome. However, according to our model normal VDAC may be part of a functionally defective channel complex in which altered CFTR causes epithelial ionflux disturbances in CF. Mutations in the porin molecules would even be in contradiction with the autosomal recessive inheritance of the

CF gene, except an absolute linkage of VDAC and CFTR loci was assumed.

On the other hand, in CF chloride channels sharing traits with "Porin 31HL" are affected. Furthermore, our model on the interaction of VDAC with the CFTR might be a second example of the involvement of VDAC in channel complexes. There is recent evidence on a tight in situ association of VDAC with cytoplasmic membrane localized $GABA_A$-receptors in mammalian brain, ligand-gated channels themselves (Bureau et al. 1992).

Our model on the interaction of VDAC and CFTR, with the modification of CFTR now penetrating the plasmalemma of epithelial cells, may thus reconcile data on the involvement of outwardly rectifying chloride channels in CF chloride flux disturbances with those on ohmic low-conductance channels, more recently observed in CFTR expression studies. The model furthermore may account for the abnormal sodium transport in CF epithelial cells.

Recent data, indicating that extracellularly applied ATP increases the open probability of ORDIC in epithelia by direct stimulation (Stutts et al. 1992) may support our proposal of VDAC as part of a chloride channel complex. "Porin 31HL" shows a putative nucleotide-binding side in its N-terminal part (see Appendix), wich is expressed on the outer surface of cells (see above). We furthermore have first data indicating that channel active "Porin 31HL" reversibly binds ATP, as revealed by the fixation of the VDAC to ATP-agarose and its competitive release by ATP disodium salt (Thinnes, Flörke und Heiden, in preparation). The effect of ATP on the Cl^- flux, which is preserved in CF epithelial cells, points to an even more complicated regulation of VDAC in the plasmalemma than assumed in our model. The pharmacological manipulation of porin channels in cytoplasmic membranes might positively influence defective Cl^- fluxes in Cystic Fibrosis cells.

Literature cited:

Anderson, M.P., Gregory, R.J., Thompson, S., Souza, D.W., Paul, S., Mulligan, R.C., Smith, A.E. and Welsh, M.J. Demonstration That CFTR Is a Chloride Channel by Alteration of Its Anion Selectivity. Science 253:202, 1991.

Babel, D., Walter, G., Götz, H., Thinnes, F.P., Jürgens, L., König, U. and Hilschmann, N. Studies on Human Porin VI. Production and Characterization of Eight Monoclonal Mouse Antibodies against the Human VDAC "Porin 31HL" and Their Application for Histotopological Studies in Human Skeletal Muscle. Biol. Chem. Hoppe-Seyler 372:1027, 1991.

Bear, C.E., Li, C., Kartner, N., Bridges, R.J., Jensen, T.J., Ramjeesingh, M. and Riordan, J.R. Purification and Functional Reconstitution of the Cystic Fibrosis Transmembrane Conductance Regulator (CFTR). Cell 68:809, 1992.

Benz, R., Wojtczak, L, Bosch, W. and Brdiczka, D. Inhibition of adenine nucleotide transport through the mitochondrial porin by a synthetic polyanion. FEBS Letters 231:75, 1988.

Benz, R., Maier, E., Thinnes, F.P., Götz, H. and Hilschmann, N. Studies on Human Porin VII. The Channel Properties of the Human B-Lymphocyte Membrane-Derived "Porin 31HL" are Similar to those of Mitochondrial Porins. Biol. Chem. Hoppe-Seyler 373:295, 1992.

Blachly-Dyson, E. and Forte, M. Cloning of a human VDAC cDNA. Biophys. J. 59:216a, 1991.

Blachly-Dyson, E., Zambronicz, E.B., Yu, W.H., Adelman, J., Colombini, M. and Forte, M. Cloning and functional expression in yeast of two human isoforms of the outer mitochondrial membrane channel, VDAC. J. Biol. Chem., in press, 1992.

Blair, H.C. and Schlesinger, P.H. Purification of a stilbene sensitive chloride channel and reconstitution of chloride conductivity into phospholipid vesicles. Biochem. Biophys. Res. Comm. 171:920, 1990.

Blatz, A.L. and Magleby, K.L. Single voltage-dependent chloride-selective channels of large conductance in cultured rat muscle. Biophys. J. 43:237, 1983.

Bosma, M.M. Anion channels with multiple conductance levels in a mouse B lmpyhocyte cell line. J. Physiol. 410:67, 1989.

Boucher, R.C., Cotton, C.U., Gatzy, J.T., Knowles, M.R. and Yankaskas, J.R. Evidence for reduced CI^- and increased Na+ permeability in cystic fibrosis human primary cell cultures. J. Physiol. 405:77, 1988.

Breuer, W. Reconstitution of a kidney chloride channel and its identification by covalent labeling. Biochim. Biophys. Acta 1022:229, 1990.

Bureau, M.H., Khrestchatisky, M., Heeren, M.A., Zambrowicz, E.B., Kim, H., Grisar, T.M., Colombini, M., Tobin, A.J. and Olsen, R.W. Isolation and Cloning of a Voltage-dependent Anion Channel-like Mr 36,000 Polypeptide from Mammalian Brain. J. Biol. Chem. 267:8679, 1992.

Cabantchik, Z.I. and Greger R. Chemical probes for anion transporters of mammalian cell membranes. Am. J. Physiol. 262: C803, 1992.

Chen, J.H., Schulman, H. and Gardner, P. A cAMP-Regulated Chloride Channel in Lymphocytes That Is Affected in Cystic Fibrosis. Science 243:657, 1989.

Cole, T., Awni, L.A., Nyakatura, E. Götz, H., Walter, G., Thinnes, F.P. and Hilschmann, N. Studies on Human Porin VIII. Expression of "Porin 31HL" Channels in the Plasmalemma of the Acute-Lymphoblastic-Leukemia Cell Line KM3 as Revealed by Light- and Electron-Microscopy. Biol.Chem.Hoppe-Seyler 373: 891, 1992.

De Pinto, V., Prezioso, G., Thinnes, F., Link, T.A. and Palmieri, F. Peptide-Specific Antibodies and Proteases as Probes of the Transmembrane Topology of the Bovine Heart Mitochondrial Porin. Biochemistry 30:10191, 1991.

Egan, M., Flotte, T., Afione, S., Solow, R., Zeitlin, P.L., Carter, B.J. and Guggino, W.B. Defective regulation of outwardly rectifying Cl^- channels by protein kinase A corrected by insertion of CFTR. Nature 358:581, 1992.

Gallin, E.K. Ion Channels in Leukocytes. Physiol. Reviews 71:775 1991.

Garber, S.S. Outwardly Rectifying Chloride Channels in Lymphocytes. J. Membr. Biol. 127:49, 1992.

Gögelein, H. Chloride channels in epithelia. Biochim. Biophys. Acta 947:521, 1988.

Hein, A. Primärstrukturanalyse des plasmalemmständigen Porins aus EBV-transformierten B-Zellen eines Patienten mit Cystischer Fibrose. Diplom-Arbeit, Universität Göttingen, 1992.

Holden, M.J. and Colombini M. The mitochondrial outer membrane channel, VDAC, is modulated by a soluble protein. FEBS Letters 241:105, 1988.

Hyde, S.C., Emsley, P., Hartshorn, M.J., Mimmack, M.M., Gileadi, U., Pearce, S.R., Gallagher, M.P., Gill, D.R., Hubbard, R.E. and Higgins, C.F. Structural model of ATP-binding proteins associated with cystic fibrosis, multidrug resistance and bacterial transport. Nature 346:362, 1990.

Jessen, F., Sjøholm, C. and Hoffmann, E.K. Identification of the Anion Exchange Protein of Ehrlich Cells: A Kinetic Analysis of the Inhibitory Effects of 4,4'-Diisothiocyano-2,2'-Stilbene-Disulfonic Acid (DIDS) and Labeling of Membrane Proteins with ^{3}H-DIDS. J. Membr. Biol. 92:195, 1986.

Jorissen, M., Vereecke, J., Carmeliet, E., Van den Berghe, H. and Cassiman, J.-J. Non-selective cation and dysfunctional chloride channels in the apical membrane of nasal epithelial cells cultured from cystic fibrosis patients. Biochim. Biophys. Acta 1096:52, 1991.

Jürgens, L., Ilsemann, P., Kratzin, H.D., Hesse, D., Eckart, K., Thinnes, F.P. and Hilschmann, N. Studies on Human Porin IV. The Primary Structures of "Porin 31HM" Purified from Human Skeletal Muscle Membranes and of "Porin 31HL" Derived from Human B Lymphocyte Membranes are Identical. Biol. Chem. Hoppe-Seyler 372:455, 1991.

Kanno, T. and Takishima T. Chloride and Potassium Channels in U937 Human Monocytes. J. Membr. Biol. 116:149, 1990.

Kayser, H., Kratzin, H.D., Thinnes, F.P., Götz, H., Schmidt, W.E., Eckart, K. und Hilschmann, N. Zur Kenntnis der Porine des Menschen II. Charakterisierung und Primärstruktur eines 31-kDa-Porins aus menschlichen B-Lmyphozyten (Porin 31HL). Biol. Chem. Hoppe-Seyler 370:1265, 1989.

König, U., Götz, H., Walter, G., Babel, D., Hohmeier, H.-E., Thinnes, F.P. und Hilschmann, N. Zur Kenntnis der Porine des Menschen V. Die Plasmalemmständigkeit von "Porin 31HL" ist keine Folge einer Zell-Transformation. Biol. Chem. Hoppe-Seyler 372:565, 1991.

Kolb, H.A., Brown, C.D.A. and Murer, H. Identification of a voltage-dependent anion channel in the apical membrane of a Cl^--secretory epithelium (MDCK). Pflügers Arch. 403:262, 1985.

Kunzelmann, K., Pavenstädt, H. and Greger, R. Properties and regulation of chloride channels in cystic fibrosis and normal airway cells. Pflügers Arch. 415:172, 1989.

Landry, D.W., Akabas, M.H., Redhead, C., Edelman, A., Cragoe, E.J., Jr. and Al-Awqati, Q. Purification and Reconstitution of Chloride Channels from Kidney and Trachea. Science 244:1469, 1989.

Pahapill, P.A. and Schlichter L.C. Cl^- Channels in Intact Human T Lymphocytes. J. Membr. Biol. 125:171, 1992.

Quinton, P.M. Cystic fibrosis: a disease in electrolyte transport. FASEB J. 4:2709, 1990.

Ran S. and Benos, D.J. Immunopurification and Structural Analysis of a Putative Epithelial Cl^- Channel Protein Isolated from Bovine Trachea. J. Biol. Chem. 267:3618, 1992.

Reenstra, W.W., Sabolic, I., Bae, H.-R. and Verkman, A.S. Protein Kinase A Dependent Membrane Protein Phosphorylation and Chloride Conductance in Endosomal Vesicles from Kidney Cortex. Biochemistry 31:175, 1992.

Riordan, J.R., Rommens, J.M., Kerem, B.-S., Alon, N., Rozmahel, R., Grzelczak, Z., Zielenski, J., Lok, S., Plavsic, N., Chou, J.-L., Drumm, M.L, Iannuzzi, M.C., Collins, F.S., Tsui, L.-Ch. Identification of the Cystic Fibrosis Gene: Cloning and Characterization of Complementary DNA. Science 245:1066, 1989.

Stutts, M.J., Chinet, T.C., Mason, S.J., Fullton, J.M., Clarke, L.L. and Boucher, R.C. Regulation of Cl^- channels in normal and cystic fibrosis airway epithelial cells by extracellular ATP. Proc. Natl. Acad. Sci. USA 89:1621, 1992.

Thinnes, F.P. Evidence for Extra-Mitochondrial Localization of the VDAC/Porin Channel in Eucaryotic Cells. J. Bioenerg. Biomembr. 24:71, 1992.

Thinnes, F.P., Götz, H., Kayser, H., Benz, R., Schmidt, W.E., Kratzin, H.D. und Hilschmann, N. Zur Kenntnis der Porine des Menschen I. Reinigung eines Porins aus menschlichen B-Lymphozyten (Porin 31HL) und sein topochemischer Nachweis auf dem Plamsmalemm der Herkunftszelle. Biol. Chem. Hoppe-Seyler 370:1253, 1989.

Thinnes, F.P., Schmid, A., Benz, R. and Hilschmann, N. Studies on Human Porin III. Does the Voltage-Dependent Anion Channel "Porin 31HL" Form Part of the Chloride Channel Complex, which is Observed in Different Cells and Thought to be Affected in Cystic Fibrosis? Biol. Chem. Hoppe-Seyler 371:1047, 1990.

Thinnes, F.P., Babel, D., Hein, A., Jürgens, L., König, U., Schmid, A. und Hilschmann, N., Ein Flip-Flop-Modell des Chlorid-Kanal-Komplexes erklärt die Fehlregulation des Chloridflusses am Plasmalemm von Zellen bei der cystischen Fibrose. Klin. Wochenschr. 69:283, 1991.

Tilly, B.C., Winter, M.C., Ostedgaard, L.S., O'Riordan, C., Smith, A.E. and Welsh, M.J. Cyclic AMP-dependent Protein Kinase Activation of Cystic Fibrosis Transmembrane Conductance Regulator Chloride Channels in Planar Lipid Bilayers. J. Biol. Chem. 267:9470, 1992.

Ward, C.L., Krouse, M.E., Gruenert, D.C., Kopito, R.R. and Wine, J.J. Cystic fibrosis gene expression is not correlated with rectifying Cl^- channels. Proc. Natl. Acad. Sci. USA 88:5277, 1991.

Welsh, M.J., Li, M. and McCann, J.D. Activation of Normal and Cystic Fibrosis Cl^- Channels by Voltage, Temperature, and Trypsin. J. Clin. Invest. 84:2002, 1989.

Welsh, M.J. Abnormal regulation of ion channels in cystic fibrosis epithelia. FASEB J. 4:2718, 1990.

Wine, J.J. Basic Aspects of Cystic Fibrosis. Clin. Rev. Allergy 9:1, 1991.

LEGEND TO APPENDIX:

HVDAC1 : VDAC1 sequenced on cDNA level from a human pituary gland library (Blachly-Dyson et al. 1992)

HVDAC2 : VDAC2 sequenced on cDNA level from a human liver library (Blachly-Dyson et al. 1992)

RVDAC : Rat VDAC sequenced on cDNA level from a hippocampal library (Bureau et al. 1992)

Porin 31HL: VDAC purified from total membrane fractions of human B lymphocytes, sequenced on protein level (Kayser et al. 1989)

Porin 31HM: VDAC purified from total membrane fractions of human skeletal muscle, sequenced on protein level (Jürgens et al. 1991)

RaKmP : VDAC purified from mitochondria of rat liver, partially sequenced on protein level (De Pinto & Palmieri in Kayser et al. 1989)

BoHmP : VDAC purified from mitochondria of bovine heart muscle, partially sequenced on protein level (De Pinto & Palmieri in Kayser et al. 1989)

Bu-Ra-Bo : VDAC from rat and bovine nervous tissues partially sequenced on protein level (Bureau et al. 1992)

Dashes : identical amino acids in corresponding positions

Single line boxing: substitutions, uncharged to charged amino acids and vice versa

Double line boxing: substitutions, positively to negatively charged amino acids

°°°°°° underlining: parts of "Porin 31 HL" putatively outside the outer mitochondrial membrane bilayer (De Pinto et al. 1991)

Ac : acetylation of the N-terminus

n : consensus nucleotide-binding site

* : consensus sequences for glycosylation

Appendix: PRIMARY STRUCTURE COMPARISON OF KNOWN MAMMALIAN PORIN MOLECULES

```
                                                         -10                 -1
HVDAC1      :                                                                 M
HVDAC2      :                                              M A T H G Q T C A R P M
RVDAC       :                                            M A E C C V P V C Q R A I

                                  10                   20                  30
Porin 31HL:AcA V P P T Y A D L G K S A R D V F T K GnY GnF GnL I K L D L
Porin 31HM:Ac- - - - - - - - - - - - - - - - - - - - - - - - - - - - - -
HVDAC1     :  - - - - - - - - - - - - - - - - - - - - - - - - - - - - - -
RaKmP      :                              - - - - - - - - - - - - - - - -
HVDAC2     :  C I - - S - - - - - - V - - - I - N - - F - - - - V - - - V
RVDAC      :  C I - - P - - - - - - A - - - I - N - - F - - - - V - - - V
              ooooooooooooooooooooooooooooooooooooooooooo

                                  40                  50                  60
Porin 31HL:   K T K S|E|N G L E F T S S G S A N T E T T K V T G S L E T K
Porin 31HM:   - - - -|-|- - - - - - - - - - - - - - - - - - - - - - - - -
HVDAC1     :  - - - -|-|- - - - - - - - - - - - - - - - - - - - - - - - -
RaKmP      :  - - - -|-|E
HVDAC2     :  - - - -|C|S - V - - S T - - - S - - D - G - - - - T - - - -
RVDAC      :  - - - -|C|S - V - - S T - - - S - - D - G - - S - T - - - -
Bu-Ra-Bo                                             - - - - N - - - - - -
                  ooooooooooo

                                  70                  80                  90
Porin 31HL:   Y R W T E Y G L T F T E K W N T D N T L G T E I T V E D Q L
Porin 31HM:   - - - - - - - - - - - - - - - - - - - - - - - - - - - - - -
HVDAC1     :  - - - - - - - - - - - - - - - - - - - - - - - - - - - - - -
HVDAC2     :  - K - C - - - - - - - - - - - - - - - - - - - - A I - - - I
RVDAC      :  - K - C - - - - - - - - - - - - - - - - - - - - A I - - - I
Bu-Ra-Bo      - - I - - - - - - - - - - K - - - - A - T
                                        ooooooo

                                 100                 110                 120
Porin 31HL:   A|R|G L K L T F D S S F S P N T G K K N A K I K T G Y K R E
Porin 31HM:   -|-|- - - - - - - - - - - - - - - - - - - - - - - - - - - -
HVDAC1     :  -|-|- - - - - - - - - - - - - - - - - - - - - - - - - - - -
RaKmP      :     - - - - - - - - - - - - - - - - - - - - - - - - - - - -
HVDAC2     :  C|Q|- - - - - - - T T - - - - - - - - S G - - - S S - - - -
RVDAC      :  C|Q|- - - - - - - T T - - - - - - - - S G - - - S A - - - -
                                    ooooooooooooooooooooooooooooooooooooooo

                                 130                 140                 150
Porin 31HL:  |H|I N L G C D M D F D I A G P S I R G A L V L G Y E G W L A
Porin 31HM:  |-|- - - - - - - - - - - - - - - - - - - - - - - - - - - - -
HVDAC1     : |-|- - - - - - - - - - - - - - - - - - - - - - - - - - - - -
RaKmP      : |-|- - - - - - V - - - - - - - - - - - - - - - - - - - - - -
HVDAC2     : |C|- - - - - - V - - - F - - - A - H - S A - F - - - - - - -
RVDAC      : |C|- - - - - - V - - - F - - - A - H - S A - F - - - - - - -

              ooooooo                                              ooooooo
```

```
                                  160                         170                         180
Porin 31HL:  G Y Q M N F E T A K S R V T Q S N F A V G Y K T D E F Q L H
Porin 31HM:  - - - - - - - - - - - - - - - - - - - - - - - - - - - - - -
HVDAC1    :  - - - - - - - - - - - - - - - - - - - - - - - - - - - - - -
RaKmP     :  - - E - - K - - S ? -   - - - - - - - - - - - - - -
BoHmP              - - - - - - - - H - - - Q - - - - - - - - N - - - - -
HVDAC2    :  - - - - T - D S - - - K L - R N - - - - - - - R - G D - - - -
RVDAC        - - - - T - D S - - - K L - R - - - - - - - - R - G D - - - -
Bu-Ra-Bo                                                       - - - -
             ooooooooooooooooooooooooooooooooooooooooooooooo

                                  190                         200                         210
Porin 31HL:  T N V N D G T E F G G S I Y Q K V N K K L E T A V N L A W T
Porin 31HM:  - - - - - - - - - - - - - - - - - - - - - - - - - - - - - -
HVADC1    :  - - - - - - - - - - - - - - - - - - - - - - - - - - - - - -
BoHmP     :  -
HVDAC2    :  - - - - - - - - - - - - - - - - - C E D - D - S - - - - - -
RVDAC     :  - - - - N*- - - - - - - - - - - - C E D F D - S - - - - - -
Bu-Ra-Bo     - - - - - - - - - - -(T)- - - - S
                                               ooooooooo

                                  220                         230                         240
Porin 31HL:  A G N S N T R F G I A A K Y Q I D P D A C F S A K V N*N*S S
Porin 31HM:  - - - - - - - - - - - - - - - - - - - - - - - - - - - - - -
HVDAC1    :  - - - - - - - - - - - - - - - - - - - - - - - - - - - - - -
RaKmP     :                - - - - - ? - - V - S S
HVDAC2    :  S - T N*C - - - - - - - - - - L - - T - S I - - - - - - - -
RVDAC     :  S - T N*C - - - - - - - - - - L - - T - S I - - - - - - - -
               ooooooooo                         ooo

                                  250                         260                         270
Porin 31HL:  L I G L G Y T Q T L K P G I K L T L S A L L D G K N V N A G
Porin 31HM:  - - - - - - - - - - - - - - - - - - - - - - - - - - - - - -
HVDAC1    :  - - - - - - - - - - - - - - - - - - - - - - - - - - - - - -
HVDAC2    :  - - - V - - - - - - R - - V - - - - - - - V - - - S I - - -
RVDAC     :  - - - V - - - - - - R - - V - - - - - - - V - - - S F - - -
                             oooooooooooooo

                                  280
Porin 31HL:  G H K L G L G L E F Q A
Porin 31HM:  - - - - - - - - - - - -
HVDAC1    :  - - - - - - - - - - - -
HVDAC2    :  - - - V - - A - - L E -
RVDAC     :  - - - - - - A - - L E -
```

Printing: Druckhaus Beltz, Hemsbach
Binding: Buchbinderei Schäffer, Grünstadt

NATO ASI Series H

Vol. 21: **Cellular and Molecular Basis of Synaptic Transmission.** Edited by H. Zimmermann. 547 pages. 1988.

Vol. 22: **Neural Development and Regeneration. Cellular and Molecular Aspects.** Edited by A. Gorio, J. R. Perez-Polo, J. de Vellis, and B. Haber. 711 pages. 1988.

Vol. 23: **The Semiotics of Cellular Communication in the Immune System.** Edited by E.E. Sercarz, F. Celada, N.A. Mitchison, and T. Tada. 326 pages. 1988.

Vol. 24: **Bacteria, Complement and the Phagocytic Cell.** Edited by F. C. Cabello und C. Pruzzo. 372 pages. 1988.

Vol. 25: **Nicotinic Acetylcholine Receptors in the Nervous System.** Edited by F. Clementi, C. Gotti, and E. Sher. 424 pages. 1988.

Vol. 26: **Cell to Cell Signals in Mammalian Development.** Edited by S.W. de Laat, J.G. Bluemink, and C.L. Mummery. 322 pages. 1989.

Vol. 27: **Phytotoxins and Plant Pathogenesis.** Edited by A. Graniti, R. D. Durbin, and A. Ballio. 508 pages. 1989.

Vol. 28: **Vascular Wilt Diseases of Plants. Basic Studies and Control.** Edited by E. C. Tjamos and C. H. Beckman. 590 pages. 1989.

Vol. 29: **Receptors, Membrane Transport and Signal Transduction.** Edited by A. E. Evangelopoulos, J. P. Changeux, L. Packer, T. G. Sotiroudis, and K.W.A. Wirtz. 387 pages. 1989.

Vol. 30: **Effects of Mineral Dusts on Cells.** Edited by B.T. Mossman and R.O. Begin. 470 pages. 1989.

Vol. 31: **Neurobiology of the Inner Retina.** Edited by R. Weiler and N.N. Osborne. 529 pages. 1989.

Vol. 32: **Molecular Biology of Neuroreceptors and Ion Channels.** Edited by A. Maelicke. 675 pages. 1989.

Vol. 33: **Regulatory Mechanisms of Neuron to Vessel Communication in Brain.** Edited by F. Battaini, S. Govoni, M.S. Magnoni, and M. Trabucchi. 416 pages. 1989.

Vol. 34: **Vectors asTools for the Study of Normal and Abnormal Growth and Differentiation.** Edited by H. Lother, R. Dernick, and W. Ostertag. 477 pages. 1989.

Vol. 35: **Cell Separation in Plants: Physiology, Biochemistry and Molecular Biology.** Edited by D. J. Osborne and M. B. Jackson. 449 pages. 1989.

Vol. 36: **Signal Molecules in Plants and Plant-Microbe Interactions.** Edited by B.J.J. Lugtenberg. 425 pages. 1989.

Vol. 37: **Tin-Based Antitumour Drugs.** Edited by M. Gielen. 226 pages. 1990.

Vol. 38: **The Molecular Biology of Autoimmune Disease.** Edited by A.G. Demaine, J-P. Banga, and A.M. McGregor. 404 pages. 1990.

NATO ASI Series H

Vol. 39: **Chemosensory Information Processing.**
Edited by D. Schild. 403 pages. 1990.

Vol. 40: **Dynamics and Biogenesis of Membranes.**
Edited by J. A. F. Op den Kamp. 367 pages. 1990.

Vol. 41: **Recognition and Response in Plant-Virus Interactions.**
Edited by R. S. S. Fraser. 467 pages. 1990.

Vol. 42: **Biomechanics of Active Movement and Deformation of Cells.**
Edited by N. Akkas. 524 pages. 1990.

Vol. 43: **Cellular and Molecular Biology of Myelination.**
Edited by G. Jeserich, H. H. Althaus, and T. V. Waehneldt. 565 pages. 1990.

Vol. 44: **Activation and Desensitization of Transducing Pathways.**
Edited by T. M. Konijn, M. D. Houslay, and P. J. M. Van Haastert. 336 pages. 1990.

Vol. 45: **Mechanism of Fertilization: Plants to Humans.**
Edited by B. Dale. 710 pages. 1990.

Vol .46: **Parallels in Cell to Cell Junctions in Plants and Animals.**
Edited by A. W Robards, W. J . Lucas, J . D. Pitts, H . J . Jongsma, and D. C. Spray. 296 pages. 1990.

Vol. 47: **Signal Perception and Transduction in Higher Plants.**
Edited by R. Ranjeva and A. M. Boudet. 357 pages. 1990.

Vol. 48: **Calcium Transport and Intracellular Calcium Homeostasis.**
Edited by D. Pansu and F. Bronner. 456 pages. 1990.

Vol. 49: **Post-Transcriptional Control of Gene Expression.**
Edited by J. E. G. McCarthy and M. F. Tuite. 671 pages. 1990.

Vol. 50: **Phytochrome Properties and Biological Action.**
Edited by B. Thomas and C. B. Johnson. 337 pages. 1991.

Vol. 51: **Cell to Cell Signals in Plants and Animals.**
Edited by V. Neuhoff and J. Friend. 404 pages. 1991.

Vol. 52: **Biological Signal Transduction.**
Edited by E. M . Ross and K . W. A. Wirtz. 560 pages. 1991.

Vol. 53: **Fungal Cell Wall and Immune Response.**
Edited by J. P. Latge and D. Boucias. 472 pages. 1991.

Vol. 54: **The Early Effects of Radiation on DNA.**
Edited by E. M. Fielden and P. O'Neill. 448 pages. 1991.

Vol. 55: **The Translational Apparatus of Photosynthetic Organelles.**
Edited by R. Mache, E. Stutz, and A. R. Subramanian. 260 pages. 1991.

Vol. 56: **Cellular Regulation by Protein Phosphorylation.**
Edited by L. M. G. Heilmeyer, Jr. 520 pages. 1991.

NATO ASI Series H

Vol. 57: Molecular Techniques in Taxonomy.
Edited by G . M . Hewitt, A. W. B. Johnston, and J. P. W. Young .
420 pages. 1991.

Vol. 58: Neurocytochemical Methods.
Edited by A. Calas and D. Eugene. 352 pages. 1991.

Vol. 59: Molecular Evolution of the Major Histocompatibility Complex.
Edited by J. Klein and D. Klein. 522 pages. 1991.

Vol. 60: Intracellular Regulation of Ion Channels.
Edited by M. Morad and Z. Agus. 261 pages. 1992.

Vol. 61: Prader-Willi Syndrome and Other Chromosome 15q Deletion Disorders.
Edited by S. B. Cassidy. 277 pages. 1992.

Vol. 62: Endocytosis. From Cell Biology to Health, Disease and Therapie.
Edited by P. J. Courtoy. 547 pages. 1992.

Vol. 63: Dynamics of Membrane Assembly.
Edited by J. A. F. Op den Kamp. 402 pages. 1992.

Vol. 64: Mechanics of Swelling. From Clays to Living Cells and Tissues.
Edited by T. K. Karalis. 802 pages. 1992.

Vol. 65: Bacteriocins, Microcins and Lantibiotics.
Edited by R. James, C. Lazdunski, and F. Pattus. 530 pages. 1992.

Vol. 66: Theoretical and Experimental Insights into Immunology.
Edited by A. S. Perelson and G. Weisbuch. 497 pages. 1992.

Vol. 67: Flow Cytometry. New Developments.
Edited by A. Jacquemin-Sablon. 1993.

Vol. 68: Biomarkers. Research and Application in the Assessment of Environmental Health. Edited by D. B. Peakall and L. R. Shugart.
138 pages. 1993.

Vol. 69: Molecular Biology and its Application to Medical Mycology.
Edited by B. Maresca, G. S. Kobayashi, and H. Yamaguchi. 271 pages. 1993.

Vol. 70: Phospholipids and Signal Transmission.
Edited by R. Massarelli, L. A. Horrocks, J. N. Kanfer, and K. Löffelholz.
448 pages. 1993.

Vol. 71: Protein Synthesis and Targeting in Yeast.
Edited by A. J. P. Brown, M. F. Tuite, and J. E. G. McCarthy. 425 pages. 1993.

Vol. 72: Chromosome Segregation and Aneuploidy.
Edited by B. K. Vig. 425 pages. 1993.

Vol. 73: Human Apolipoprotein Mutants III. In Diagnosis and Treatment.
Edited by C. R. Sirtori, G. Franceschini, B. H. Brewer Jr. 302 pages. 1993.

NATO ASI Series H

Vol. 74: **Molecular Mechanisms of Membrane Traffic.**
Edited by D. J. Morré, K. E. Howell, and J. J. M. Bergeron. 429 pages. 1993.

Vol. 75: **Cancer Therapy. Differentiation, Immunomodulation and Angiogenesis.**
Edited by N. D'Alessandro, E. Mihich, L. Rausa, H. Tapiero, and T. R. Tritton. 299 pages. 1993.

Vol. 76: **Tyrosine Phosphorylation/Dephosphorylation and Downstream Signalling.**
Edited by L. M. G. Heilmeyer Jr. 388 pages. 1993.

Vol. 77: **Ataxia-Telangiectasia.** Edited by R. A. Gatti, R. B. Painter. 306 pages. 1993.

Vol. 78: **Toxoplasmosis.** Edited by J. E. Smith. 272 pages. 1993.

Vol. 79: **Cellular Mechanisms of Sensory Processing. The Somatosensory System.**
Edited by L. Urban. 514 pages. 1994.

Vol. 80: **Autoimmunity: Experimental Aspects.**
Edited by M. Zouali. 318 pages. 1994.

Vol. 81: **Plant Molecular Biology. Molecular Genetic Analysis of Plant Development and Metabolism.**
Edited by G. Coruzzi, P. Puigdomènech. 579 pages. 1994.

Vol. 82: **Biological Membranes: Structure, Biogenesis and Dynamics.**
Edited by Jos A. F. Op den Kamp. 367 pages. 1994.

Vol. 83: **Molecular Biology of Mitochondrial Transport Systems.**
Edited by M. Forte, M. Colombini. 420 pages. 1994.

Vol. 84: **Biomechanics of Active Movement and Division of Cells.**
Edited by N. Akkaş. 587 pages. 1994.